Energy Storage and Electric Power Systems: Theory, Methods, and Applications

Energy Storage and Electric Power Systems: Theory, Methods, and Applications

Guest Editors

George Caralis
Konstantinos Braimakis

MDPI

Basel • Beijing • Wuhan • Barcelona • Belgrade • Novi Sad • Cluj • Manchester

Guest Editors
George Caralis
School of Mechanical Engineering
National Technical University of Athens
Athens
Greece

Konstantinos Braimakis
School of Mechanical Engineering
National Technical University of Athens
Athens
Greece

Editorial Office
MDPI AG
Grosspeteranlage 5
4052 Basel, Switzerland

This is a reprint of the Special Issue, published open access by the journal *Applied Sciences* (ISSN 2076-3417), freely accessible at: https://www.mdpi.com/journal/applsci/special_issues/E101H6JL25.

For citation purposes, cite each article independently as indicated on the article page online and as indicated below:

Lastname, A.A.; Lastname, B.B. Article Title. *Journal Name* **Year**, *Volume Number*, Page Range.

ISBN 978-3-7258-7420-0 (Hbk)
ISBN 978-3-7258-7421-7 (PDF)
https://doi.org/10.3390/books978-3-7258-7421-7

Contents

About the Editors

George Caralis

George Caralis is a Mechanical Engineer with a PhD and Postdoctoral research experience at the National Technical University of Athens (NTUA), where he serves as Laboratory Teaching and Research Staff in the School of Mechanical Engineering. With more than 25 years of experience in wind energy and renewable energy systems, his research focuses on renewable energy sources, particularly wind energy, hydro-pumped storage, hybrid systems, and grid integration, as well as the techno-economic and socio-economic aspects of energy deployment. His doctoral work addressed wind energy integration with hydro-pumped storage in autonomous island systems, while his postdoctoral research at the Institute of Atmospheric Physics of the Chinese Academy of Sciences (IAP-CAS) focused on wind energy planning using mesoscale atmospheric modeling. His extensive academic contributions include more than 60 publications in scientific journals and international conferences. He is a frequent lecturer for prestigious academic programs, including the EUREC European Master in Renewable Energy since 2002 and the China-EU Institute for Clean and Renewable Energy (ICARE) since 2011. In addition to his academic roles, he has served as a consultant for the Greek Ministry of Environment, Energy and Climate Change, contributing to national renewable energy action plans and legislative frameworks. He has also been a lead technical researcher on numerous European Commission projects involving offshore grids, energy storage, and 100% RES supply for islands.

Konstantinos Braimakis

Konstantinos Braimakis is an Assistant Professor at the Laboratory of Refrigeration, Air Conditioning and Solar Energy of the School of Mechanical Engineering at the National Technical University of Athens (NTUA). He is also a Visiting Professor at the China-EU Institute for Clean and Renewable Energy (ICARE) at Huazhong University of Science and Technology, China.

He received his Diploma in Mechanical Engineering in 2013 and his PhD in 2018 from NTUA, with a dissertation focused on the thermodynamic and economic optimization of advanced Organic Rankine Cycle configurations. In 2014, he was awarded a KCORC Fellowship to conduct part of his doctoral research at the University of Bayreuth, in collaboration with the Technical University of Munich. His work has been recognized with several distinctions, including the Thomais Prize for peer-reviewed scientific publications (2014–2018) and the Dimitris Horafas Foundation Prize in Physics (2018) for his doctoral research.

He is currently Principal Investigator in two ongoing European projects and has previously coordinated four completed projects, while also participating in more than ten European and national research initiatives. He has authored over 40 articles in peer-reviewed international journals and more than 30 publications in international conference proceedings. His research interests include techno-economic assessment of energy storage systems, decentralized energy conversion systems, solar thermal energy utilization, heat pumps and advanced heating and cooling technologies, industrial refrigeration, waste heat recovery, multi-generation systems, and zero-energy buildings.

Preface

The transition toward a decarbonized energy future is one of the most significant challenges of our time. As nations strive to meet ambitious carbon neutrality targets, the rapid expansion of variable renewable energy has necessitated a fundamental shift in how we design and operate electric power systems. This Reprint, "Energy Storage and Electric Power Systems: Theory, Methods, and Applications", was conceived to address the critical need for flexibility and reliability in modern grids through the lens of advanced energy storage technologies.

Our primary motivation in compiling this work was to bridge the gap between emerging storage research—such as hydro-pumped storage, hydrogen fuel cells, and next-generation batteries - and their effective integration into the power system architecture. The scope of this collection encompasses a broad range of scientific inquiries, from fundamental theory to the practical implementation of hybrid solutions in both isolated and grid-connected environments.

As Guest Editors, we aimed to provide a holistic perspective on how diverse storage applications can mitigate the intermittency of renewables while ensuring power quality. This Reprint is addressed to researchers, energy engineers, and graduate students seeking to understand the symbiotic relationship between renewable generation and storage. We hope that this collection serves as a valuable foundation for those working to build a more resilient and sustainable global energy landscape.

George Caralis and Konstantinos Braimakis
Guest Editors

applied sciences

MDPI

Editorial

Energy Storage and Electric Power Systems: Theory, Methods, and Applications

George Caralis [1,*] and Konstantinos Braimakis [2]

1 Fluids Section, School of Mechanical Engineering, National Technical University of Athens, 9 Heroon Polytechniou Street, 15780 Zografou, Athens, Greece
2 Laboratory of Refrigeration, Air-Conditioning and Solar Energy, School of Mechanical Engineering, National Technical University of Athens, 9 Heroon Polytechniou Street, 15780 Zografou, Athens, Greece; mpraim@central.ntua.gr
* Correspondence: gcaralis@mail.ntua.gr

1. Introduction

The global energy landscape is undergoing a profound transformation driven by the urgent need to mitigate climate change and the ambitious targets set by nations worldwide to achieve carbon neutrality. Europe, in particular, is spearheading efforts to decarbonize its economy, aiming for a future dominated by renewable energy sources (RESs). The rapid proliferation of variable renewable energy technologies, especially wind and solar photovoltaics (PVs), has brought forth new design and operational challenges for electric power systems. As the use of these intermittent energy sources continues to grow, there is increasing concern regarding the stability, reliability, and flexibility of the grid.

In this context, energy storage systems (ESSs) have emerged as a cornerstone of the energy transition. They offer the necessary flexibility to balance supply and demand, manage congestion, and ensure power quality. From large-scale solutions like pumped hydro and compressed air energy storage to distributed technologies such as batteries and hydrogen fuel cells, the role of storage is expanding beyond merely being a back-up: it is becoming an integral component of modern power systems.

This Special Issue, "Energy Storage and Electric Power Systems: Theory, Methods, and Applications", was created to address these challenges. It aims to gather high-quality research exploring the symbiotic relationship between renewable energy generation and storage technologies. The articles collected herein cover a broad range of topics, including the optimization of hybrid systems, techno-economic assessment of novel storage solutions, and integration of storage in off-grid and grid-connected applications.

2. Overview of Published Articles

The contributions to this Special Issue provide valuable insights into the theoretical and practical aspects of energy storage integration. The articles published reflect the diverse methodologies and applications currently being explored in the field of energy engineering. They can be broadly categorized into four main areas reflecting the diversity of current research efforts: off-grid and hybrid system optimization, grid stability and protection, transportation and components, and energy harvesting and cogeneration.

2.1. Optimization of Hybrid and Off-Grid Systems

A significant portion of the research focuses on the optimization of hybrid systems for isolated or off-grid applications. Robert et al. [1] investigated the feasibility of integrating a redox flow battery (RFB) storage system to optimize wind energy utilization at a remote mining site. Their study developed a simulation model to evaluate diesel savings and CO_2 reduction, highlighting the potential of RFBs to enhance energy security in harsh environments. Similarly, Karapidakis et al. [2] examined the ambitious goal of achieving zero carbon emissions in interconnected islands, using the case of Crete as an example. Their work evaluated the potential of ultra-high RES penetration combined with hydrogen storage to achieve a carbon-neutral electricity system. Furthermore, Ma et al. [3] proposed a bi-objective optimization strategy for distribution networks incorporating distributed energy systems, utilizing a stepper search algorithm to balance economic and operational goals.

2.2. Grid Stability, Protection, and Power Quality

Ensuring the reliability of power systems with high RES penetration is a recurring theme. Blanco et al. [4] presented a power smoothing strategy for wave energy converters using energy storage systems. By employing forecasting-enhanced filtering algorithms, they demonstrated how storage requirements can be reduced while maintaining grid stability. Regarding protection, Grcić and Pandžić [5] addressed the challenge of detecting high-impedance faults in DC microgrids. They introduced a novel method based on open-set recognition that significantly improves fault detection accuracy in systems where conventional protection schemes often fail.

2.3. Transportation and Advanced Components

The electrification of the transportation sector and the development of advanced storage materials are also critical. Wang et al. [6] focused on the durability of fuel cells in commercial vehicles, proposing an improved multiple gray prediction method to estimate power demand and extend the service life of proton exchange membrane fuel cells (PEMFCs). Complementing the research on hydrogen technologies, Luo et al. [7] evaluated the accuracy of various gas equations of state (EOSs) for modeling thermodynamic conditions during hydrogen tank refueling. Their comparative analysis demonstrated that a polynomial EOS significantly outperforms traditional models like van der Waals, providing a critical theoretical basis for ensuring safety and precision in refueling protocols. In the field of materials, Del Bosque et al. [8] provided a comprehensive review of carbon fiber-based batteries. Their work discusses the trends and future perspectives of integrating energy storage functionalities into structural materials, a key innovation for lightweight electric mobility.

2.4. Energy Harvesting and Cogeneration

Finally, the Special Issue encompasses broader efficiency and harvesting technologies. Golonis et al. [9] performed a Life Cycle Cost Analysis (LCCA) of a biomass-driven Organic Rankine Cycle (ORC) cogeneration system, underscoring the importance of economic feasibility in sustainable agricultural applications. Additionally, Litak et al. [10] explored energy harvesting at a smaller scale, analyzing a rotational pendulum device with support from Finite Element Method (FEM) simulations to optimize power generation from ambient vibrations.

3. Conclusions

The research presented in this Special Issue confirms that energy storage is not a monolithic solution but a diverse field of technologies and applications that must be tailored to the needs of specific systems. The transition to a future of 100% renewable energy requires a holistic approach that brings together theory, advanced methods, and practical applications. From advanced prediction algorithms for fuel cells to the techno-economic optimization of large-scale storage in remote grids, the findings in this collection

https://doi.org/10.3390/app16031595

highlight the fact that there is no "one-size-fits-all" solution. Instead, a combination of appropriate storage technologies, optimized control strategies, and robust economic planning is essential for the future of electric power systems.

The Guest Editors would like to thank all the authors for their innovative contributions and the reviewers for the time dedicated to this collection and their constructive feedback, which ensured the high quality of this Special Issue. We hope that this collection will serve as a useful reference for researchers, engineers, and policymakers working towards a sustainable and resilient energy future.

Conflicts of Interest: The authors declare no conflicts of interest.

References

1. Robert, A.; Mungyeko Bisulandu, B.-J.R.; Ilinca, A.; Rousse, D.R. Hybrid Wind–Redox Flow Battery System for Decarbonizing Off-Grid Mining Operations. *Appl. Sci.* **2025**, *15*, 7147. [CrossRef]
2. Karapidakis, E.; Mozakis, I.; Nikologiannis, M.; Tsikalakis, A. Zero Carbon Emissions Due to Ultra-High RES Penetration in Interconnected Island. *Appl. Sci.* **2024**, *14*, 4668. [CrossRef]
3. Ma, S.; Meng, Z.; Cui, Y. A Bi-Objective Optimization Strategy of a Distribution Network Including a Distributed Energy System Using Stepper Search. *Appl. Sci.* **2024**, *14*, 9480. [CrossRef]
4. Blanco, M.; Mazorra, L.; Villalba, I.; Navarro, G.; Nájera, J.; Lafoz, M. Power Smoothing in a Wave Energy Conversion Using Energy Storage Systems: Benefits of Forecasting-Enhanced Filtering for Reduction in Energy Storage Requirements. *Appl. Sci.* **2025**, *15*, 11106. [CrossRef]
5. Grcić, I.; Pandžić, H. High-Impedance Fault Detection in DC Microgrid Lines Using Open-Set Recognition. *Appl. Sci.* **2025**, *15*, 193. [CrossRef]
6. Wang, Y.; Li, Y.; Lu, J.; Zhou, H. Estimation of Fuel Cell Power Demand on Commercial Vehicles Based on Improved Multiple Grey Prediction Method Considering Dynamic Time Window. *Appl. Sci.* **2025**, *15*, 1213. [CrossRef]
7. Luo, H.; Xin, Q.; Yao, C.; Li, C.; Yang, T.; Wu, X.; Chahine, R.; Xiao, J. Effect of Real Gas Equations on Calculation Accuracy of Thermodynamic State in Hydrogen Storage Tank. *Appl. Sci.* **2025**, *15*, 11151. [CrossRef]
8. Del Bosque, A.; Vergara, D.; Fernández-Arias, P. Energy Storage in Carbon Fiber-Based Batteries: Trends and Future Perspectives. *Appl. Sci.* **2024**, *14*, 10034. [CrossRef]
9. Golonis, C.; Tyris, D.; Skiadopoulos, A.; Bilalis, D.; Manolakos, D. Life Cycle Cost Analysis of a Biomass-Driven ORC Cogeneration System for Medical Cannabis Greenhouse Cultivation. *Appl. Sci.* **2025**, *15*, 12085. [CrossRef]
10. Litak, G.; Kondratiuk, M.; Wolszczak, P.; Ambrożkiewicz, B.; Giri, A.M. Energy Harvester Based on a Rotational Pendulum Supported with FEM. *Appl. Sci.* **2024**, *14*, 3265. [CrossRef]

https://doi.org/10.3390/app16031595

applied sciences

MDPI

Article

Life Cycle Cost Analysis of a Biomass-Driven ORC Cogeneration System for Medical Cannabis Greenhouse Cultivation

Chrysanthos Golonis [1], Dimitrios Tyris [1], Anastasios Skiadopoulos [1], Dimitrios Bilalis [2] and Dimitris Manolakos [1,*]

[1] Department of Natural Resources Development and Agricultural Engineering, Agricultural University of Athens, Iera Odos, 75-11855 Athens, Greece; chrysanthosgolonis@aua.gr (C.G.); dtyris@aua.gr (D.T.); tskiado@aua.gr (A.S.)

[2] Department of Crop Science, Agricultural University of Athens, Iera Odos, 75-11855 Athens, Greece; bilalis@aua.gr

* Correspondence: dman@aua.gr

Abstract

Medical cannabis cultivation requires substantial energy for heating, lighting, and climate control. This study evaluates the economic feasibility of an innovative biomass-fired micro-CHP system in a greenhouse facility for medicinal cannabis cultivation. The system comprises an 80 kW_{th} boiler retrofitted for biomass and a 7 kW_{el} ORC engine and is assessed against a diesel-boiler Business-As-Usual (BAU) benchmark. Thermal load simulations for two growing periods (1 March–30 June and 1 September–30 December) estimate an annual heating demand of 91,065.20 kWh_{th}. The micro-CHP system delivers 8195.87 kWh_{el} per year, exceeding the greenhouse's 7839.90 kWh_{el} consumption. Over a 30-year lifespan at a 7% discount rate, Life Cycle Costing yields EUR 196,421.33 for micro-CHP versus EUR 229,468.46 for BAU, a 14.4% reduction. Under all-equity financing, the project achieves an NPV of EUR 59,591.88, IRR of 27.32%, and a DPBP of 12.1 years; with 70% debt financing, NPV rises to EUR 61,211.39 and DPBP shortens to 10.5 years. Levelized Cost of Energy (LCOE) and Heat (LCOH) are EUR 0.122 per kWh_{el} and EUR 0.062 per kWh_{th}, respectively. While the LCOE is below the Greek and EU non-household averages (EUR 0.1578 and EUR 0.1515 per kWh_{el}), the LCOH exceeds the corresponding heat price benchmarks (EUR 0.0401 and EUR 0.0535 per kWh_{th}). These results indicate that, in the modeled context, biomass-ORC cogeneration can be a financially attractive and lower-carbon option for medicinal cannabis greenhouse operations.

Keywords: organic rankine cycle (ORC); micro-combined heat and power (CHP); medical cannabis cultivation; greenhouse heating; sustainable agriculture

1. Introduction

1.1. Medical Cannabis Cultivation

Cannabis (*Cannabis sativa* L.) is a plant primarily known for its psychoactive properties, which are caused by tetrahydrocannabinol (THC). THC is the primary psychoactive compound and the most recognizable, but beyond this well-known facet, the therapeutic potential of cannabis has emerged as a global focus, placing the species among the most discussed in botanical and medical sciences [1]. In recent years, cannabis has raised public awareness in many countries worldwide about its medical uses, as an expanding body of research and regulatory shifts has underscored the plant's value for therapeutic purposes. This has subsequently led to a number of jurisdictions either decriminalizing

https://doi.org/10.3390/app152212085

cannabis or legalizing it for medical use, with some moving toward full legalization [2,3]. As societal and legislative landscapes continue to evolve and public acceptance grows, a dedicated cannabis industry has emerged to meet consumer demand and preferences, such as demand for cannabis-based medicines [4].

In March of 2018, Greece introduced Law 4523/2018, which regulates the production and processing of medical cannabis. This law was applied due to the desire to revitalize the crop industry and to trade in cannabis products [5]. For the granting of approval for installation and authorization, the applicant must submit documents for all stages (from cultivation to processing and production of final medical cannabis products). On the basis of legislation, the cultivation will take place under greenhouse or indoor conditions [6]. This legal requirement aligns with a broader modernization trend in agriculture toward controlled-environment agriculture (CEA). In practice, high-value cannabis is primarily grown in controlled settings using year-round production schedules, soilless systems, and data-driven climate control [7]. Recent analyses identify greenhouses as a means for continuous production, precise environmental management, and improved resource efficiency [8,9]. At the same time, greenhouse and high-tunnel studies on cannabis/hemp demonstrate that protected cultivation can improve crop uniformity and influence cannabinoid and aromatic profiles compared with open-field systems, which helps explain why medicinal cannabis projects gravitate to greenhouses even when costs are higher [10,11]. In this sense, the legal framework does not create an isolated practice but rather situates medicinal cannabis production within the ongoing move toward greenhouse-based CEA.

1.2. Energy Use in Medicinal Cannabis Greenhouses

Greenhouses, particularly in Mediterranean climates, are widely used for crop cultivation due to their ability to regulate environmental conditions and enhance productivity [12]. They are, however, energy-intensive systems, especially when used for high-demand crops like cannabis. Heating in conventional greenhouses is typically achieved through hot air or hot water systems, and energy use has become a growing concern due to rising costs and environmental impacts. This has driven increased interest in renewable energy sources (RES) integration to reduce reliance on non-renewable sources [13].

Legalized cannabis cultivation in various regions is a particularly energy-intensive greenhouse application. Basic energy consumption includes lighting, dehumidification, heating, and cooling; yet recent reviews and program reports highlight a persistent lack of standardized, publicly available energy-use data for cannabis cultivation, particularly outside North America, underscoring the need for benchmarking and comparable metrics to support robust techno-economic and environmental analyses [14–17].

For illustration, cannabis cultivation accounted for 45.5% of Denver's 1.2% year-over-year increase in electricity demand in 2014 [18], and in California, indoor cannabis uses about 3% of the state's and 1% of the nation's electricity [19]. Reported electricity intensity spans ~21.7 $kWh_{el}\cdot m^{-2}$ for outdoor production to ~2820 $kWh_{el}\cdot m^{-2}$ for indoor facilities, whereas greenhouse vegetables and flowers typically consume ~167–785 $kWh_{th}\cdot m^{-2}\cdot yr^{-1}$ [20]. In indoor cannabis operations, HVAC and lighting commonly account for ~51% and ~38% of electricity use, respectively [19], while electricity represents only ~10–15% of total energy in conventional greenhouses [21]. Complementing these benchmarks, modeled U.S. indoor scenarios estimate total site energy at approximately 2.3–6.0 $MWh\cdot kg^{-1}$ (1.8–4.6 $MWh\cdot kg^{-1}$ of electricity, plus space-heating fuels of 0.47–1.47 $MWh\cdot kg^{-1}$) [22]. Greenhouse compilations [23] report electricity productivity of approximately 0.68 $g\cdot kWh^{-1}$ (i.e., 1.47 MWh of electricity per kg of dried flower) and heating fuel use around 552 $kWh\cdot m^{-2}\cdot yr^{-1}$, noting that system boundaries vary across studies. European figures, though sparse, align with high energy demand under protected cultivation. A Portuguese medical cannabis greenhouse (15,000 m^2) was estimated at ~2209 $MWh \cdot yr^{-1}$ (~147 $kWh \cdot m^{-2} \cdot yr^{-1}$), with HVAC and pumping totaling to ~87.5 kWh $m^{-2} \cdot yr^{-1}$ [24]. Moreover, a Dutch desk study calculates a direct energy requirement of ~2.12 $kWh \cdot g^{-1}$ (~6366 $kWh \cdot m^{-2} \cdot yr^{-1}$ under a stylized high-intensity regime), highlighting both the potential magnitude of loads and the current reliance on modeled estimates in Europe [25].

Cannabis cultivation progresses through three main stages—propagation (or seedling), vegetative, and flowering—each stage has distinct requirements for light, temperature, and humidity, which result in varying energy demands (Table 1). The propagation stage typically lasts 1 to 2 weeks and involves low-intensity lighting (54–431 $W \cdot m^{-2}$) for 18 to 24 h per day. This stage accounts for less than 5% of total electricity consumption [26]. Optimal environmental conditions require a temperature range between 24 and 26 °C and relative humidity (RH) of 70–75%, which can be maintained using split air conditioning units [27]. The vegetative stage lasts 2–3 weeks in indoor setups (or 4 to 6 weeks in greenhouses), using medium-intensity lighting (161–753 $W \cdot m^{-2}$) for 18 h daily. This phase consumes approximately 30–40% of total electricity [26]. While metal halide (MH) and high-pressure sodium (HPS) lamps have been traditionally used, modern operations increasingly adopt LED or ceramic metal halide (CMH) lighting. Climate conditions during this stage are typically regulated with rooftop HVAC units. The flowering stage is the most energy-intensive, accounting for 50–65% of total electricity use. It lasts 7–11 weeks and requires 12 h of lighting per day, at an intensity of 40–70 W m^{-2}. Environmental control is maintained using rooftop HVAC systems and dehumidifiers to ensure proper temperature and humidity levels [26].

Table 1. Requirements for different stages of cannabis cultivation.

Cultivation Stage	Duration	Cultivation Stages' Needs		
		Lighting	Dehumidification	Cooling
Propagation	1–2 weeks	54–431 $W \cdot m^{-2}$, 18–24 h	Low	Low
Vegetative	2–6 weeks	161–753 $W \cdot m^{-2}$, 18 h	Medium	High
Flowering	6–11 weeks	40–70 $W \cdot m^{-2}$, 12 h	High	Very high

Facility type, of course, strongly influences energy use. As of 2017, 60% of U.S. legal cannabis electricity demand was from indoor operations and 37% from greenhouses. Energy costs in indoor cultivation can reach 20–50% of operating expenses, equating to roughly 1614 $kWh \cdot m^{-2}$ per year. Despite the lower energy intensity of outdoor cultivation, indoor and greenhouse methods are favored for yield control and consistency. Consequently, utilities and planners must understand this load's profile [28].

Incentivizing best practices, such as LED adoption, off-peak scheduling, and cogeneration solutions such as biomass-fired ORC systems, can reduce the energy footprint of cannabis greenhouses. Emerging regulations that mandate energy plans, monitoring, and reporting, alongside local policies, can foster efficiency, build consumption databases, and help set realistic energy use targets.

1.3. Micro-CHP (Combined Heat and Power) in Greenhouse Applications

Combined Heat and Power (CHP), or cogeneration, holds significant potential across a range of industries. Among them are greenhouses, one of the most energy-intensive segments of modern agriculture. CHP systems are already in use across several greenhouse complexes, particularly in the Netherlands, where they have shown high efficiency and flexibility. As Canova et al. [29] point out, cogeneration is widely recognized for its superior fuel efficiency compared to the separate production of heat and electricity.

Greenhouses, as a form of CEA, offer a way to optimize growing conditions year-round, providing high productivity while improving resource efficiency. They can achieve high yields while substantially reducing water consumption, land use, and chemical inputs. However, these benefits come with significant energy demands, particularly for heating, lighting, and ventilation, making energy efficiency a critical concern for long-term sustainability. In this context, CHP systems can generate electricity on-site while utilizing the resulting heat for immediate use, efficiently meeting the diverse energy needs of greenhouse operations. Compared to traditional setups that rely on grid electricity and separate boilers, CHP systems reduce energy losses and environmental impact [30]. They can also be paired with complementary technologies like absorption cooling or carbon capture and utilization (CCU), expanding their benefits even further.

In practice, CHP is already being used successfully in greenhouse cultivation, particularly in Northern European countries. These systems often supply electricity to the grid while using the heat and CO_2 byproducts to enhance the indoor environment [31]. When the on-site electricity demand is lower than the system's output, excess power can be sold back to the grid, providing an additional income stream.

Beyond their operational benefits, CHP systems are increasingly seen as a strategic tool for reducing the carbon footprint of energy-intensive agricultural practices [32]. Their ability to improve efficiency, lower emissions, and integrate with existing greenhouse infrastructure makes them especially well-suited for sustainable greenhouse cultivation. A key challenge, however, lies in developing advanced CHP systems that can effectively integrate renewable energy sources and adapt to variable weather patterns, while maintaining a reliable balance between energy supply and demand [33].

The integration potential of micro-CHP systems with other RES-assisted and energy-efficient solutions has been shown to further enhance energy autonomy and sustainability. Moreover, continuous progress in fuel cell technology has motivated researchers and industry to consider adopting fuel cell-based micro-CHP systems as promising candidates for future sustainable energy solutions [34]. Hybrid micro-CHP systems that combine photovoltaic (PV) systems, battery storage, and predictive control algorithms can dynamically balance supply and demand, achieving near-zero grid dependency [35]. The coupling of photovoltaic–thermal (PVT) collectors with Stirling- and ORC-based micro-CHP units has also emerged as a promising configuration for residential and agricultural buildings, stably delivering power and heat even under variable solar conditions [36]. These configurations can anticipate energy fluctuations and ensure that the micro-CHP unit operates only when required, minimizing both operational costs and carbon emissions.

From a broader perspective, international experiences reflect the growing attention that micro-CHP technologies receive due to policy incentives and technological advances, which provide higher electrical efficiencies and lower CO_2 emissions compared to conventional boilers [37]. Studies on distributed renewable-based CHP systems have further shown that hybridizing with solar and wind power can meet up to 79% of community-level energy demand while reducing over 4000 tCO_2 annually [38]. The inclusion of biofuels such as biogas or biodiesel in CHP operation has also proven effective for reducing fossil fuel dependence and enhancing combustion performance [39].

The Organic Rankine Cycle (ORC) Technology

The organic Rankine cycle (ORC) is a closed thermodynamic process used for power generation. Although it follows the same basic principles as the Clausius–Rankine cycle, it uses organic working fluids instead of water or steam. The ORC exploits heat from a high-temperature source into mechanical work, while releasing excess heat to a cold sink, typically the environment. It is commonly applied for recovering energy from low-grade heat sources such as waste heat, solar power, geothermal energy, biomass, and hybrid systems [40]. The main components of a basic ORC system, namely the feed pump, evaporator, expander (or turbine), and condenser, are illustrated in Figure 1 [41]. The corresponding thermodynamic cycle comprises liquid compression (1–2), phase change in the evaporator (2–3), expansion in the expander or turbine (3–4), and condensation (4–1). The heat source and heat sink are finite thermal reservoirs, indicated by lines (5–6) and (7–8), respectively.

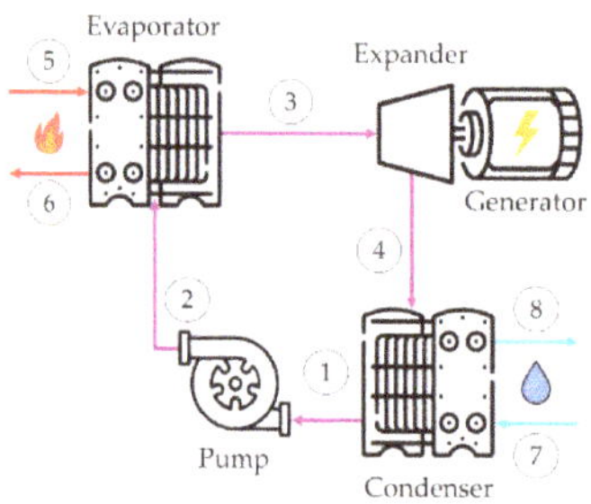

Figure 1. Schematic representation of a simple ORC layout. (1–2): Compression, (2–3): Evaporation, (3–4): Expansion, (4–1): Condensation, (5–6): Heat source, (7–8): Heat sink.

In commercial applications, for small-scale (up to a few tens of kW) and relatively small-scale (up to a few tens of kW) at low temperature (below 100–120 °C), the basic ORC setup (Figure 1) is almost exclusively used. With proper selection of working fluids for each application, the ORC can outperform the conventional water/steam Rankine cycle for waste heat recovery at temperatures up to 300 °C. This is mainly due to the lower critical temperatures of organic fluids, which allow for more efficient heat transfer and optimized system design [42]. The sustainability of the ORC technology is clearly demonstrated by the growing number of installations. Over the past decade, 759 installations have been completed, with an additional 63 currently under construction [43]. Nearly one-third of these installed systems (31.5%) are biomass-based.

1.4. Research Objectives and Scope

This study will investigate the economic feasibility of an innovative micro-CHP system, incorporating a biomass boiler and an ORC engine for a greenhouse facility intended for cultivating medicinal cannabis. Techno-economic metrics, such as Life Cycle Cost (LCC), Net Present Value (NPV), Discounted Payback Period (DPBP), and Internal Rate of Return (IRR), will be evaluated to determine the economic viability for a medicinal cannabis cultivator to invest in replacing the conventional system with the innovative micro-CHP configuration.

To this end, two main scenarios are analyzed. The first is the Business-As-Usual (BAU) scenario, in which the greenhouse relies exclusively on a diesel-fired boiler for heating and draws electricity from the grid. The second is the innovative micro-CHP (m-CHP) scenario, which features a boiler retrofitted for biomass combustion, integrated with an ORC engine for on-site power generation. In this setup, the ORC engine operates in parallel with the boiler, producing electricity whenever the boiler is active. In both scenarios, electricity consumption is supplied by the grid; however, in the m-CHP scenario, the electricity generated on-site is fed into the grid under a net metering arrangement. Following the simulation of the greenhouse's thermal load profile, the corresponding thermal and electrical energy outputs of the m-CHP system are estimated.

2. Materials and Methods

2.1. Description of the Micro-CHP and Greenhouse Setup

The study is based on the performance of an existing micro-CHP system installed at the Agricultural University of Athens (AUA). The system comprises (i) a fire-tube biomass boiler with modular, water-jacketed sections and (ii) an ORC unit for power generation, sized to meet the thermal requirements of a small greenhouse. As illustrated in Figure 2, the biomass boiler heats a glycol-based working fluid to 130 °C, providing up to 200 kW_{th} of thermal capacity. The heated working fluid then flows into the hot side of the ORC evaporator, where heat is transferred to the R1233zd(E) refrigerant. A single cooling loop condenses the refrigerant in the ORC condenser, and the rejected heat from this condenser is used to warm the greenhouse air.

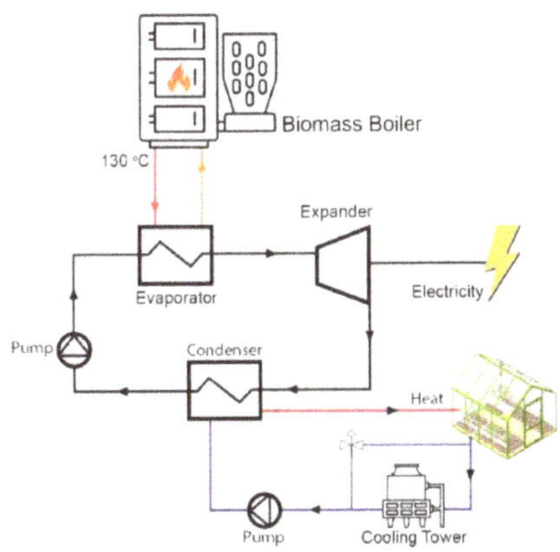

Figure 2. Flow diagram of the micro-CHP experimental setup.

Figure 3 presents photographs of the biomass system components, including the storage hoppers, comprising the main silo and the smaller open tank, boiler, and the control panel, installed in a dedicated enclosure adjacent to the greenhouse. The biomass feedstock used consists of wood pellets that are commercially available in Greece, typically produced from debarked coniferous residues sourced domestically or from neighboring Balkan regions. These pellets are delivered in bulk over transport distances of approximately 100–150 km. From the open tank, pellets are moved to the feeder and then to the boiler's combustion chamber by screw conveyors with stainless-steel blades, controlled by level-sensor readings. As a standardized market product, no on-site drying or special storage procedures are applied, apart from standard covered storage at the boiler facility to protect the pellets from moisture and weather exposure. The main technical specifications of the biomass boiler and fuel storage system are summarized in Table 2.

Table 2. Main technical specifications of the biomass boiler and fuel storage system.

Specification	Value
Boiler efficiency	85%
Hopper volume (silo and open tank) [L]	5210.00
Boiler operating autonomy [h] [a]	~34

[a] Considering biomass LHV and density of 5.2 kWh·kg^{-1} and 0.6 kg·L^{-1}, respectively.

(a) (b) (c) (d)

Figure 3. The biomass boiler system component. (**a**) Open tank; (**b**) main silo (feeder); (**c**) boiler; (**d**) control panel.

The ORC engine operates at fixed frequencies for both the pump ($Freq_{pu}$) and the expander ($Freq_{exp}$). The system achieves a maximum net electrical output ($P_{net,ORC}$) of 20.39 kW when the pump operates at 42.70 Hz and the expander at 60.00 Hz. Additional test data across various frequency settings for both components are presented in Table A1 (Appendix A). The recorded parameters include the condenser inlet temperature ($T_{con,in}$ in °C), the expander's electrical power output ($P_{el,exp}$ in kW), and the pump's electrical power consumption ($P_{el,pu}$ in kW). Figure 4 illustrates 3D models of the ORC unit designed to produce 20 kW_{el}, showing its primary cycle components—namely the condenser, compressor, evaporator, and expander.

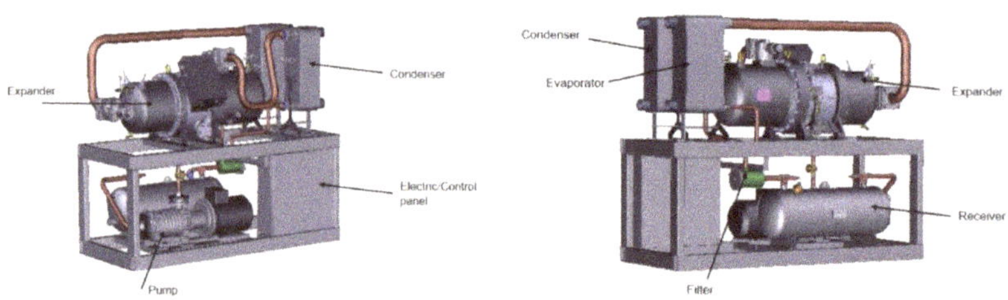

(a) (b)

Figure 4. 3D models of the ORC engine (key components labeled). (**a**) Front view; (**b**) back view.

An experimental greenhouse facility located adjacent to the micro-CHP system served as the reference case for the simulations conducted in this study. The structure is an even-span greenhouse constructed with polycarbonate panels, featuring a length of 30 m (L), a width of 13.5 m (W), an eave height of 3.15 m (H_1), and a ridge height of 5.00 m (H_2), as illustrated in Figure 5a. The total surface area is 405 m^2, while its volume totals 1648.50 m^3. Climate control within the greenhouse is achieved through forced ventilation via top windows oriented to the north and south, and space heating provided by fan coil-type air heaters (Figure 5b).

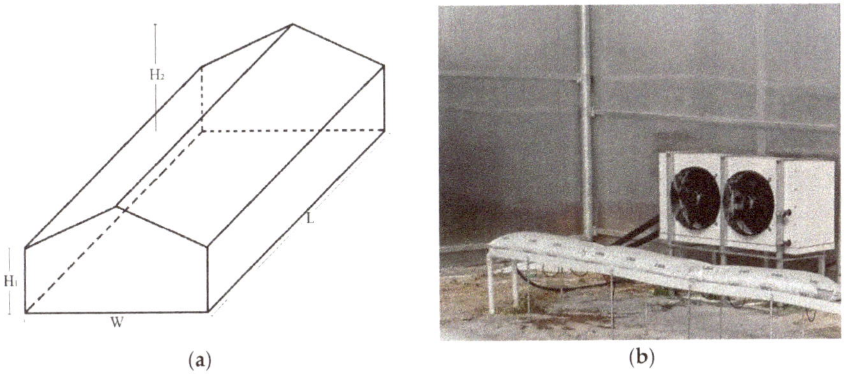

Figure 5. (**a**) Dimension of the AUA greenhouse; (**b**) inside view of the greenhouse.

2.2. Thermal and Electrical Energy Requirements Modeling

Calculating the greenhouse energy balance requires quantifying thermal interactions between exterior and interior environments, as well as moisture exchanges. Figure 6 illustrates the primary heat transfer mechanisms. Because these heat and mass fluxes vary continuously, dynamic calculations are needed to determine both thermal behavior and heating or cooling requirements.

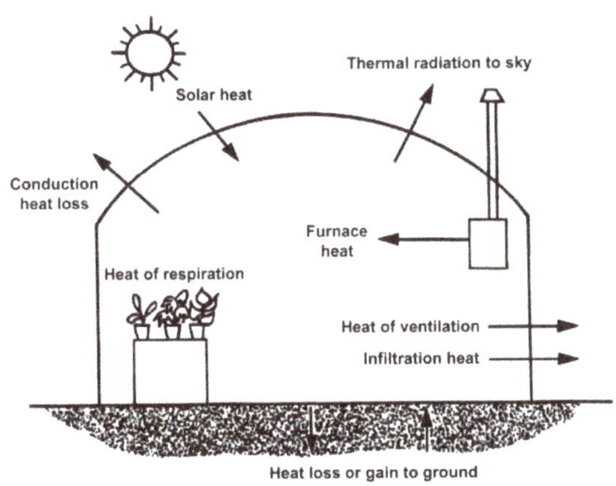

Figure 6. Heat transfer mechanisms in a greenhouse. Reproduced from [44].

Solar radiation reaching a greenhouse is partially absorbed or scattered by the atmosphere before striking the glazing. A portion is reflected, a small fraction absorbed by the glazing, and the remainder transmitted indoors. Heat exchange occurs via convection between the glazing and air, conduction through framing and insulation, and radiation between the glazing and internal surfaces. Transmitted solar energy is absorbed by plant canopies, soil, and internal structures, with retention aided by internal reflections and the greenhouse effect. Inside, plants exchange sensible heat with air and surfaces via convection and radiation, while evapotranspiration cools the leaf boundary through latent heat consumption. Similarly, the soil transfers heat by conduction, convection, and radiation, and loses latent heat through evaporation. Supplemental heating adds sensible heat to the air, while natural or forced ventilation removes both sensible and latent heat. Irrigation and fertigation influence soil moisture and humidity, affecting the greenhouse's energy balance.

This work focuses on the economic feasibility of a CHP system, and for this reason, only the sensible heat loads are estimated. Accordingly, a simplified energy balance approach is adopted (as described in [45]). Outdoor temperature and solar radiation data from a typical meteorological year were used to calculate the heating load per unit of greenhouse surface area, as shown in Equation (1). The total required heating load ($\dot{Q}_h$) is then calculated based on Equation (2).

$$H = \left[U \cdot \frac{A_c}{A_f} \cdot (T_{i,s} - T_o) \right] - (1 - f) \cdot \tau \cdot (1 - \rho) \cdot G_h \tag{1}$$

$$\dot{Q}_h = H \cdot A_f \tag{2}$$

where

- H is the greenhouse heating load per sq. meter ($W \cdot m^{-2}$);
- U is the overall thermal transmittance ($W \cdot m^{-2} \cdot {}^\circ C^{-1}$);
- A_c is the cover and wall area (m^2);
- A_f is floor area (m^2);
- $T_{i,s}$ is the indoor air temperature (setpoint) (°C);
- T_o is the outdoor air temperature (°C);
- f is the radiation to latent heat conversion factor inside the greenhouse (between 0.5 and 0.7);
- τ is the permeability to solar radiation factor (indicative value: 0.5);
- ρ is the reflectance coefficient of the interior (indicative value: 0.5);
- G_h is the solar radiation per sq. meter ($W \cdot m^{-2}$).

The parameters used in this simplified model are listed in Table 3.

Table 3. The constant parameters in the model.

Coefficients	Symbol	Value	Unit
Overall thermal transmittance	U	4.20	$W \cdot m^{-2} \cdot {}^\circ C^{-1}$
Cover area	A_c	783.13	m^2
Floor area	A_f	405.00	m^2
Covers-to-Floor ratio	A_c/A_f	1.93	[-]
Indoor air temperature (setpoint)	$T_{i,s}$	23.00	[°C]
Radiation to latent heat conversion factor inside the greenhouse	f	0.60	[-]
Permeability to solar radiation factor	τ	0.50	[-]
Reflectance coefficient of the interior	ρ	0.20	[-]

Electricity consumption has to be included in the calculations as well. This was performed by considering each electrically powered subsystem's nominal power rating (e.g., lighting, irrigation pumps, shading system, dehumidifiers) and its daily operating hours throughout each cultivation stage. Section 3.2 elaborates on the electrical energy needs calculations for each growing stage.

2.3. Life Cycle Cost Analysis and Economic Indicators

Life Cycle Costing (LCC) is a method for estimating the total economic cost of a project by accounting for all resources consumed over its lifetime [46]. The cost categories and parameters required for the LCC calculation are [47,48] as follows:

- Initial capital cost (i.e., equipment purchase and installation);
- Operating cost (i.e., energy and fuel consumption);
- Maintenance and repair (OM&R) costs, including both annual and non-annual expenses;
- End-of-life cost (i.e., disposal or decommissioning).

In addition to these cost items, the analysis includes the equipment's service life, the inflation rate, which is used to project future nominal costs, and the discount rate, which converts future expenditures to present value. The *LCC* is calculated using Equation (3), where C_{INV} denotes the initial investment costs, t the year in which a cost occurs, T is the system's lifetime in years, C_{tot} is the cost, i.e., cash outflow, incurred in year t, and r is the discount rate [49].

$$LCC = C_{INV} + \sum_{t=0}^{T} \frac{C_{out,tot}}{(1+r)^t} \tag{3}$$

The total annual costs are the sum of the operating costs (C_O), maintenance (C_M), and end-of-life costs (C_E)—Equation (4). An inflation rate is applied to all cost parameters to project their future nominal values over the equipment's lifetime.

$$C_{out,tot} = C_O(t) + C_M(t) + C_E(t) \tag{4}$$

Additionally, to provide a comprehensive assessment of economic viability, future costs are evaluated using three primary indicators, Net Present Value (NPV), Internal Rate of Return (IRR), and Discounted Payback Period (DPBP) [50], while applying a discount rate to convert future expenditures into present value.

In general, *NPV* is calculated by discounting all expected cash inflows (C_{in}) and outflows (C_{out}) over the project's lifetime to their present values and then summing them. A positive *NPV* indicates that the present value of benefits exceeds that of costs, signaling a financially viable investment at the given discount rate. In the present study, *NPV* is applied to quantify the savings between the *m-CHP* and *BAU* scenarios (Equation (5), adapted from [51]), calculated by subtracting the sum of net cash flows (NCF) of the BAU scenario NCF_{BAU} from that of the *m-CHP* scenario $NCF_{m\text{-}CHP}$.

$$NPV = \sum_{t=0}^{T} \frac{(NCF_{m-CHP,t} - NCF_{BAU,t})}{(1+r)^t} \tag{5}$$

The Internal Rate of Return (IRR) is defined as the discount rate that makes the *NPV* of all project cash flows equal to zero, or, in other words, the IRR is the break-even discount rate where discounted inflows exactly offset discounted outflows. If the IRR exceeds the project's required rate of return, the investment is considered acceptable; if it falls below, the project is not justified.

The Discounted Payback Period (DPBP) measures the length of time required for the discounted cumulative cash flows to recover the initial investment Equation (6). Unlike the simple payback period, DPBP thus accounts for the time value of money.

$$\sum_{t=0}^{T} \frac{(NCF_{m-CHP,t} - NCF_{BAU,t})}{(1+r)^t} \geq 0 \tag{6}$$

Finally, the Levelized Cost of Energy (LCOE, Equation (7)) and Levelized Cost of Heat (LCOH, Equation (8)) are calculated. The equations are adapted from [52,53] and represent the lifetime cost per unit of energy output, defined as the system's total LCC divided by its lifetime electrical energy generation (E_{tot}) and heat (Q_{tot}), respectively. Thus, LCOE is the cost per kilowatt-hour of electricity (EUR per kWh_{el}), and LCOH is the cost per kilowatt-hour of heat (EUR per kWh_{th}). The symbol a_{el} represents the cost allocation factor for the electricity generated from the micro-CHP unit, while a_{he} stands for the cost allocation factor for the heat generated by the cogeneration unit.

$$LCOE = a_{el} \frac{LCC}{\sum_{t=0}^{T} E_{tot}} \tag{7}$$

$$LCOH = a_{he} \frac{LCC}{\sum_{t=0}^{T} Q_{CHP}} \quad (8)$$

The values of a_{el} and a_{he} are determined in this work by applying the alternative generation method [52,53]. By following this approach, a_{el} and a_{he} are given by Equations (9) and (10), where $n_{el,ref}$ (0.425) is the reference efficiency for electricity generation in Greece, and $n_{he,ref}$ (0.85) is the efficiency of a typical natural gas boiler for heat generation [54].

$$a_{el} = \frac{E_{tot}/n_{el,ref}}{E_{tot}/n_{el,ref} + Q_{CHP}/n_{he,ref}} \quad (9)$$

$$a_{he} = \frac{Q_{CHP}/n_{he,ref}}{E_{tot}/n_{el,ref} + Q_{CHP}/n_{he,ref}} \quad (10)$$

2.4. *Life Cycle Inventory*

2.4.1. Initial Investment Costs

Initial investment costs are incurred only in the m-CHP scenario and cover one-time expenses at project inception. These include the ORC unit purchase (EUR 23,179.70), transportation of materials and installation of the biomass boiler, which replaces the conventional unit from the BAU scenario (EUR 1300.00), and a biomass storage tank for the boiler (EUR 500.00). The existing greenhouse heating distribution network requires no replacement or expansion; only minor modifications, which are already included in the retrofit and installation costs, are needed to connect the biomass boiler. The total initial investment of EUR 24,979.70 is summarized in Table 4, with a detailed breakdown of ORC engine component costs provided in Table A3 of Appendix A.

Table 4. Initial investment cost breakdown.

Cost Type	Cost (EUR)
ORC engine purchase costs	23,179.70
Transportation of equipment and materials	200.00
Boiler retrofitting and installation costs	1100.00
Biomass tank purchase costs	500.00
Total	24,979.70

2.4.2. Operation and Maintenance Phase

During operation in both BAU and m-CHP scenarios, the unit costs for fuel (i.e., diesel and biomass) and electricity from the national grid are taken into account (Table 5), determining the ongoing expenses to meet the greenhouse's energy demand. Finally, annual maintenance is estimated at EUR 200.00 for the BAU scenario, while for the m-CHP scenario, it amounts to EUR 270.00, including boiler cleaning (EUR 50.00), nozzle replacement (EUR 70.00), chimney cleaning (EUR 100.00), and working fluid refilling (EUR 50.00), summarized in Table 5 as well.

Table 5. Operation and maintenance costs under the BAU and m-CHP scenarios.

Operating Costs	Unit Cost
Diesel fuel (BAU)	EUR $1.12 \cdot L^{-1}$
Biomass fuel (BAU)	EUR $0.40 \cdot kg^{-1}$
Grid electricity (BAU and m-CHP)	EUR $0.1826 \cdot kWh_{el}^{-1}$
Maintenance Costs	**Cost (EUR)**
BAU scenario	200.00
m-CHP scenario	270.00

2.4.3. End-of-Life Stage

After 30 years of operation, the micro-CHP system is decommissioned, dismantled, and recycled. Modernization costs are prohibitively high, and when combined with its performance, continued operation is deemed uneconomical. The dismantling process itself incurs no significant costs. To enhance sustainability and generate profit, easily collectible and classifiable waste materials, specifically aluminum alloy, steel, plastic, and glass components, are recovered. Recovery rates for these materials are sourced from the literature [55], and their recycling values are based on average EU market prices shown in Table 6. At its end-of-life, the system yields 3000.00 kg of steel and 900.00 kg of aluminum solid waste. Recycling this equipment generates approximately EUR 4038.00 in revenue.

Table 6. Materials recovered at the end-of-life of the micro-CHP system.

Material	Recovery Yields	Recovery Price (EUR·ton^{-1})
Plastic	90%	244.70
Aluminum	86%	2820.00
Steel	90%	500.00
Glass	9%	55.10

2.4.4. Economic Parameters

To estimate the annual cash flows and evaluate the economic performance of both scenarios over the system's lifetime, several assumptions were made regarding key financial parameters. Fuel and energy prices are projected to rise annually, reflecting market forecasts, rather than remaining constant over the 30-year lifetime of the system. An inflation rate was set at 2.5%, in line with the Bank of Greece's projection for the next years. Additionally, a 7% discount rate was applied to account for the time value of money, within the typical range of 6–8% for investments of this nature.

Although renewable energy projects in Greece are often eligible for EU and national subsidies, this study assumes project financing is provided either solely from own funds or through a combination of own funds and bank debt. In the debt-financed scenario, bank borrowing constitutes 70% of the initial investment. The interest rate for the loan was set at 5.5%, based on current interbank lending rates for a 15-year repayment period. The complete set of financial input data is summarized in Table 7.

Table 7. Financial analysis input data.

Description	Value
Inflation	2.5%
Discount rate (r)	7%
Lifespan (years)	30
Loan-to-Cost Ratio (LTC)	0% or 70%
Loan interest rate	5.5%
Repayment term (years)	15

3. Results and Discussion

3.1. Greenhouse Heating Requirements and ORC Power Generation

For the simulations, two standard growing periods (GPs) within the year, i.e., 1st March to 30th June (GP1), and 1st September to 30th December (GP2), were considered. It should be noted that the greenhouse remains non-operational during July and August. Optimal indoor air temperature for cannabis growth typically ranges between 22 °C and 30 °C, while relative humidity is maintained between 65% and 70% [27]. Additionally, the ideal soil temperature ranges from 12 °C to 14 °C and should not fall below 6–8 °C. For the purposes of this study, the heating system setpoint temperature was fixed at 23 °C.

Based on Equation (2), the estimated thermal load profiles required for greenhouse heating throughout the two GPs over a full year are shown in Figure 7. Heating is required for a total of 3733 h annually, during which 91,065.20 kWh_{th} (224.58 kWhth·m^{-2}) of thermal energy is supplied. The most energy-intensive months in GP1 are March (20,841.91 kWh_{th}) and April (13,914.90 kWh_{th}), while in GP2, the highest demand occurs in November (20,005.57 kWh_{th}) and December (22,628.34 kWh_{th}).

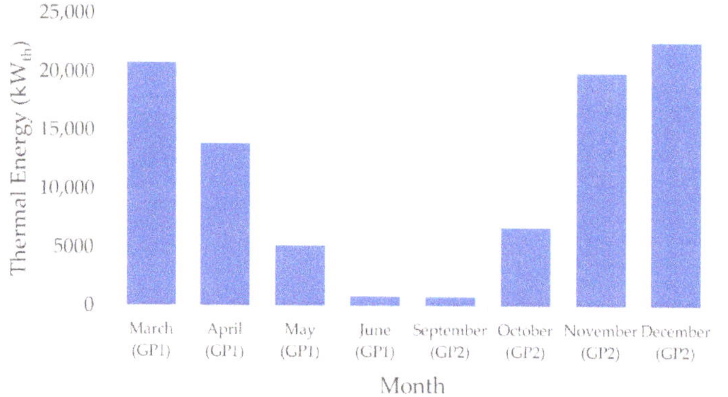

Figure 7. Monthly thermal energy demand of the greenhouse over the year.

After estimating the total required heating load, the boiler's fuel consumption was calculated (expressed as L·h^{-1} for diesel and kg·h^{-1} for biomass) using the assumed boiler efficiency, lower heating values (LHV), and densities listed in Table 8.

Table 8. Fuel consumption-related assumptions of micro-CHP and BAU scenarios.

Coefficients	Value	Unit
Boiler Efficiency (diesel)	90%	[-]
Diesel LHV	11.62	kWh·kg^{-1}
Diesel density	0.835	kg·L^{-1}
Boiler Efficiency (biomass)	85%	[-]
Biomass LHV [a]	5.20	kWh·kg^{-1}
Biomass pellet density	0.600	kg·L^{-1}

[a] Considering ENplus A1 class wood pellets, with a moisture content below 10% (wet basis, determined according to ISO 18134-1 [56] and compliant with ISO 17225-2 [57]).

In the BAU scenario, the boiler requires an annual diesel supply of 10,428.39 L, whereas in the m-CHP scenario, the biomass boiler requires 20,602.99 kg to deliver the same thermal energy. The estimated monthly fuel supply profiles for each scenario are illustrated in Figure 8.

Based on the calculations performed, an 80 kW_{th} boiler is deemed sufficient for the specific application. To model the micro-CHP system's hourly power generation and resulting electrical output, an average thermal efficiency of 9% was applied for a 7 kW_{el} ORC engine, considered reasonable based on available experimental data. Operating the proposed micro-CHP system with priority on covering the greenhouse heating is consistent with achieving high overall energy utilization while accepting the typical conversion efficiency at this scale. Specifically, lab-scale tests of a biomass-fired micro-ORC report net conversion efficiency up to 8.55% [58], whereas a hybrid solar–biomass micro-CHP achieved total energy efficiency up to 90.74% (exergy 15.05%) in a residential application [52]. The estimated monthly power generation output for the m-CHP scenario is illustrated in Figure 9, along with the corresponding monthly operating hours. On an annual basis, this corresponds to an estimated electrical energy production of 8195.87 kWh_{el}.

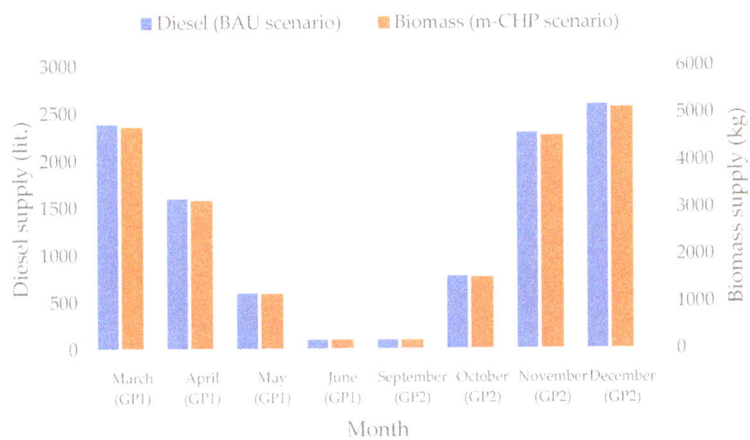

Figure 8. Monthly fuel supply to cover thermal energy demands.

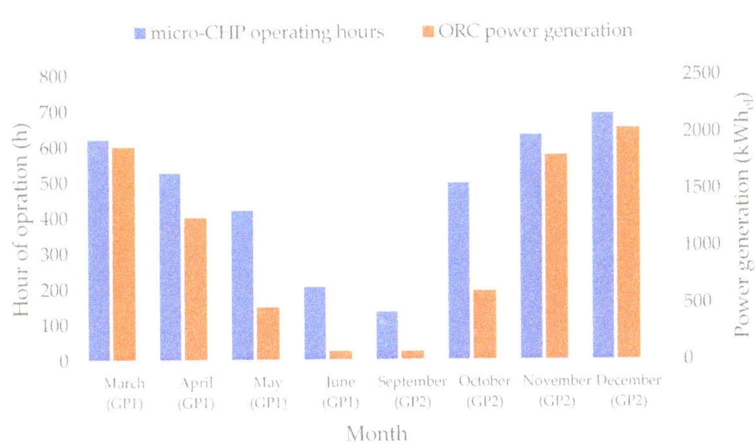

Figure 9. Operating hours and corresponding electricity production from March to December.

It should be noted that the system's operating hours are influenced by the climatic conditions of the region where the greenhouse and the micro-CHP system are located. In colder regions, the system is expected to operate for more hours, as heating demand increases. Since power generation is directly linked to the operation of the heating system, this could potentially result in higher electricity output.

3.2. Greenhouse Electrical Energy Needs

The electrical energy requirements for the vegetative and flowering stages are presented in detail in Tables 9–12. During the September to December cultivation period, electricity consumption was 1933.40 kWh_{el} for the vegetative stage (Table 9) and 2506.00 kWh_{el} for the flowering stage (Table 10). In the March to June period, the vegetative stage required 1253.00 kWh_{el} (Table 11), while the flowering stage consumed 1825.60 kWh_{el} (Table 12).

Table 9. Electrical energy needs for the vegetative stage from September to December (28 days).

System	Power (kW)	Operating Hours (h/day)	Daily Energy Consumption (kWh_{el})	Total Consumption (kWh_{el})
Lighting System	6.08	8.00	48.60	1360.80
Water pump	1.50	0.30	0.45	12.60
Shading system	5.00	2.00	10.00	280.00
Dehumidifiers	5.00	2.00	10.00	280.00
Total	-	-	-	1933.40

Table 10. Electrical energy needs for the flowering stage from September to December (56 days).

System	Power (kW)	Operating Hours (h/day)	Daily Energy Consumption (kWh_{el})	Total Consumption (kWh_{el})
Lighting System	6.08	4.00	24.30	1360.80
Water pump	1.50	0.30	0.45	25.20
Shading system	5.00	2.00	10.00	560.00
Dehumidifiers	5.00	2.00	10.00	560.00
Total	-	-	-	2506.00

Table 11. Electrical energy needs for the vegetative stage from March to June (28 days).

System	Power (kW)	Operating Hours (h/day)	Daily Energy Consumption (kWh_{el})	Total Consumption (kWh_{el})
Lighting System	6.08	4.00	24.30	680.40
Water pump	1.50	0.30	0.45	12.60
Shading system	5.00	2.00	10.00	280.00
Dehumidifiers	5.00	2.00	10.00	280.00
Total	-	-	-	1253.00

Table 12. Electrical energy needs for the flowering stage from March to June (56 days).

System	Power (kW)	Operating Hours (h/day)	Daily Energy Consumption (kWh_{el})	Total Consumption (kWh_{el})
Lighting System	6.075	2.00	12.15	680.40
Water pump	1.50	0.30	0.45	25.20
Shading system	5.00	2.00	10.00	560.00
Dehumidifiers	5.00	2.00	10.00	560.00
Total	-	-	-	1825.60

Although the propagation stage involves significantly lower energy demands, it is not entirely negligible. To ensure a more accurate estimate of total annual energy use, an additional 5% has been added to account for the electrical energy consumed during propagation [26], resulting in a total annual electricity demand for operating environmental control systems and irrigation equipment estimated at 7893.90 kWh_{el} (19.36 $kWh_{el}\cdot m^{-2}$). The annual thermal energy and fuel consumption figures for the micro-CHP and BAU scenarios are summarized in Table 13. For a more detailed presentation of monthly electrical energy needs in each GP, refer to Table A2 of Appendix A.

Table 13. Annual thermal energy and fuel consumption for the micro-CHP and BAU scenarios.

Parameter	Value	Unit
Thermal energy needs	91,065.20	kWh_{th}
Boiler capacity	80.00	kW_{th}
Diesel supply	10,428.39	L
Biomass supply	20,602.99	kg
Electrical energy needs	7893.90	kWh_{el}
ORC unit capacity	7.00	kW_{el}
ORC unit efficiency	9%	[-]
Power generation	8195.87	kWh_{el}

3.3. Operating Costs Under the BAU and m-CHP Scenarios

In the BAU scenario, meeting the greenhouse's annual heating demand of 91,065.20 kWh_{th} requires diesel consumption that results in an annual fuel cost of EUR 11,679.80, based on an average diesel price of EUR $1.12 \cdot L^{-1}$. Additionally, the total electricity consumption (7893.90 kWh_{el}), at an average price of EUR $0.1826 \cdot kWh_{el}^{-1}$ [59], leads to an additional cost of EUR 1441.43. This brings the total annual operational expense to EUR 12,921.22. These figures are summarized in Table 14.

Table 14. Annual operating costs under the BAU scenario.

Category	Quantity	Unit Cost	Total Cost (EUR)
Operating Costs			
Diesel fuel consumption	10,428.39 L	EUR $1.12 \cdot L^{-1}$	11,679.80
Grid electricity consumption	7893.90 kWh_{el}	EUR $0.1826 \cdot kWh_{el}^{-1}$	1441.43
Total annual operating cost	—	—	12,921.22

Under the m-CHP scenario (see Table 15), the greenhouse requires 20,602.99 kg of biomass annually to meet its heating demand. At a retail price of EUR 0.40 per kg for biomass pellets, this results in a fuel cost of EUR 8241.19. Electricity consumption from the grid incurs the same cost as in the BAU scenario (EUR 1441.43), yielding a total annual expense of EUR 10,920.66. However, the ORC engine generates 8195.87 kWh_{el} of electricity annually, which, when sold to the local grid at a rate of EUR 0.1840 per kWh_{el} [59], applicable to non-household consumers, yields an additional gross revenue of EUR 1508.04 (Table 15).

Table 15. Annual operating costs and revenues under the m-CHP scenario.

Category	Quantity	Unit Cost (EUR)	Total Cost or Revenue (EUR)
Operating Costs			
Biomass fuel consumption	20,602.99 kg	EUR $0.40 \cdot kg^{-1}$	8241.19
Grid electricity consumption	7893.90 kWh_{el}	EUR $0.1826 \cdot kWh_{el}^{-1}$	1441.43
Total annual operating cost	—	—	10,920.66
Generated Revenue			
Electricity fed into the grid	8195.87 kWh_{el}	EUR $0.1840 \cdot kWh_{el}^{-1}$	—
Total annual generated revenue	—	—	1508.04

3.4. Life Cycle Cost (LCC), Levelized Cost of Energy (LCOE), and Levelized Cost of Heat (LCOH)

Using Equations (3) and (4), the LCC for the m-CHP scenario is calculated at EUR 196,421.33 over the system's lifetime, representing a 14.4% reduction compared to the BAU scenario (EUR 229,468.46). Based on the undiscounted annual net cash flows, electricity revenues from the m-CHP system, and lower fuel costs yield cumulative savings of EUR 193,156.15. As shown in Figure 10, fuel and energy costs increase steadily on an annual basis due to the projected escalation of prices. In the BAU case, costs are largely driven by diesel purchases, whereas in the m-CHP case, the annual expenditure on fuel (biomass pellets) is ~29% lower. The two net savings curves reflect different financing structures: in the own-funds case, after the higher upfront investment in year 0, savings begin immediately and grow gradually, while in the loan-financed case, early-year savings are smaller due to debt service but converge after the loan is repaid (year 15). The steeper change in slope between year 0 and year 1 is driven by the difference in initial investment between the two cases, which affects annual net cash flows. Under 70% debt CAPEX financing (see Table 7), cumulative undiscounted savings decreased by EUR 8644.7, to EUR 184,511.45, reflecting the total principal and interest repayments.

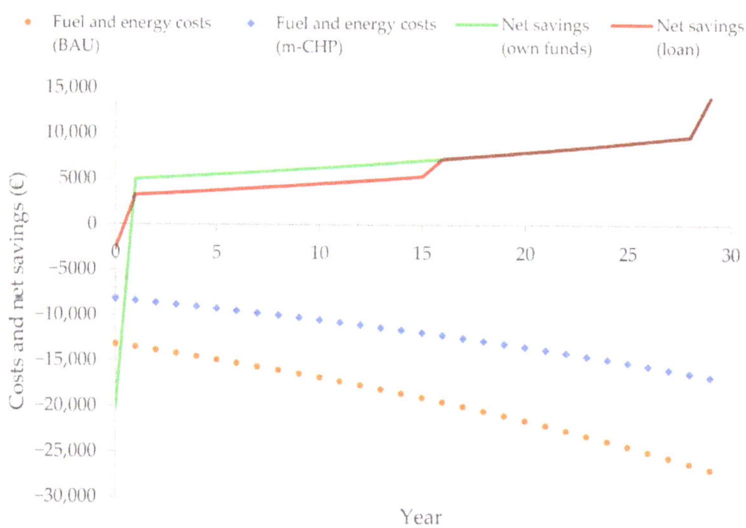

Figure 10. Fuel and energy costs under m-CHP and BAU scenarios and the resulting net savings.

The resulting LCOE and LCOH, EUR 0.122 per kWh_{el} and EUR 0.062 per kWh_{th}, respectively (Table 16), are highly competitive against current retail electricity and natural-gas prices for Greek non-household consumers (EUR 0.1578 per kWh_{el} and EUR 0.0401 per kWh_{th}), as well as against the EU averages (EUR 0.1515 per kWh_{el} and EUR 0.0535 per kWh_{th}) [60,61]. The cost allocation factors for electricity (a_{el}) and heat (a_{he}) are derived from Equations (9) and (10) based on the estimated total electricity (E_{tot}) and heat (Q_{CHP}) generated over the 30-year analysis period. If the loan on the initial investment is considered, both LCOE and LCOH are not significantly affected, showing only a reduction of 0.82% and 1.64%, respectively.

Table 16. Levelized Costs of Electricity (LCOE) and Heat (LCOH) in the micro-CHP scenario, alongside the key input parameters used for their calculation.

Parameter	Symbol	Value	Unit
Levelized Cost of Energy	LCOE	0.122	$EUR{\cdot}kWh_{el}{}^{-1}$
Levelized Cost of Heat	LCOH	0.062	$EUR{\cdot}kWh_{th}{}^{-1}$
Total electricity generated	E_{tot}	245,876.04	kWh_{el}
Total heat generated	Q_{CHP}	2,731,956.03	kWh_{th}
Reference efficiency for electricity generation	$n_{el,ref}$	0.425	-
Reference efficiency for heat generation	$n_{he,ref}$	0.850	-
Cost allocation factor for electricity	a_{el}	0.150	-
Cost allocation factor for heat	a_{he}	0.850	-

3.5. Net Present Value (NPV), Internal Rate of Return (IRR), and Discounted Payback Period (DPBP)

Under an all-equity financing structure, the m-CHP scenario delivers an NPV of EUR 59,591.88 and an un-levered IRR of 27.32%, indicating a highly attractive investment. When 70% of the capital expenditure is financed at 5.5% over 15 years (see Table 17), the NPV increases by EUR 1619.51 to EUR 61,211.39. Thus, although servicing the debt reduces gross cash flows, it amplifies the project's risk-adjusted returns. Finally, the DPBP is 12.1 years under the equity-financed scenario and 10.5 years under the debt-financed scenario, indicating a shorter capital repayment horizon under the leveraged structure. Table 15 presents a summary of the key financial performance indicators used to evaluate the m-CHP investment scenarios in comparison to the BAU case.

Table 17. Summary of key financial performance indicators for the m-CHP investment scenario compared to the BAU scenario.

Indicator	Scenario		
	BAU	m-CHP (Own-Funds)	m-CHP (Loan)
LCC	EUR 229,468.46	EUR 196,421.33	EUR 194,801.82
NPV	-	EUR 59,591.88	EUR 61,211.39
DPBP	-	12.1 years	10.5 years

3.6. Sensitivity Analysis

While baseline projections assumed prices rising with inflation, a sensitivity analysis investigating plausible variations in key cost drivers (i.e., diesel, grid electricity, biomass costs, and RES electricity revenue) was also deemed necessary to provide additional insight into their effect on the financial performance of the innovative micro-CHP system. Table 18 summarizes the sensitivity analysis scenarios and resulting key indicators (i.e., LCOE, LCOH, LCC, NPV, and DPBP). Using the m-CHP system (own-funds case) as the baseline, the main cost parameters were first varied individually within defined ranges to evaluate their impact on overall system performance. Subsequently, two scenarios were developed: one representing a condition more favorable to the m-CHP configuration and another favoring the BAU setup. In these, the most influential parameters were adjusted by approximately ±10% to reflect realistic shifts that would, respectively, benefit or challenge each configuration.

Table 18. Sensitivity analysis scenarios and resulting key indicators.

Sensitivity Analysis Scenario			LCOE ($EUR{\cdot}kWh_{el}{}^{-1}$)	LCOH ($EUR{\cdot}kWh_{he}{}^{-1}$)	LCC (EUR)	NPV (EUR)	DPBP (Years)
Baseline	Diesel cost ($EUR{\cdot}L^{-1}$) Grid electricity cost ($EUR{\cdot}kWh^{-1}$) Biomass cost ($EUR{\cdot}kg^{-1}$) RES electricity revenue ($EUR{\cdot}kWh^{-1}$)	1.12 0.1826 0.40 0.1840	0.122	0.062	196,421.33	59,591.88	12.13
Varying costs	Diesel cost ($EUR{\cdot}L^{-1}$)	0.90–1.15	0.122	0.062	196,421.33	20,071.70 to 127,854.00	>30 to 5.32
	Grid electricity cost ($EUR{\cdot}kWh^{-1}$)	0.15–0.25	0.119 to 0.128	0.060 to 0.64	191,988.43 to 205,586.29	59,591.88	12.13
	Biomass cost ($EUR{\cdot}kg^{-1}$)	0.30–0.50	0.100 to 0.144	0.050 to 0.072	160,931.09 to 231,911.58	95,082.12 to 24,101.63	7.41 to >30
	RES electricity revenue ($EUR{\cdot}kWh^{-1}$)	0.15–0.25	0.122	0.061	196,421.33	54,791.75 to 68,909.77	13.22 to 10.43
m-CHP favorable	Diesel cost ($EUR{\cdot}L^{-1}$) Grid electricity cost ($EUR{\cdot}kWh^{-1}$) Biomass cost ($EUR{\cdot}kg^{-1}$) RES electricity revenue ($EUR{\cdot}kWh^{-1}$)	1.232 0.1643 0.36 0.2024	0.112	0.056	179,736.83	96,505.05	7.06
BAU favorable	Diesel cost ($EUR{\cdot}L^{-1}$) Grid electricity cost ($EUR{\cdot}kWh^{-1}$) Biomass cost ($EUR{\cdot}kg^{-1}$) RES electricity revenue ($EUR{\cdot}kWh^{-1}$)	1.008 0.2009 0.44 0.1062	0.132	0.066	213,105.84	14,292.6	>>30

Although some indicators, such as LCOE, LCOH, and LCC, remained nearly constant in certain cases due to their definition and the system's configuration, the sensitivity analysis highlights that fuel-related parameters exert a stronger influence on the project's financial viability. Variations in diesel cost notably altered the DPBP from over 30 years to as low as 5.3 years, revealing how changes in fuel price can strongly affect investment attractiveness. Similarly, changes in biomass price significantly impacted both the levelized costs and the NPV, demonstrating that the system's profitability is highly dependent on feedstock economics. In contrast, variations in grid electricity price and RES electricity revenue had comparatively modest effects on overall performance, as their relative contribution to total costs and revenues was smaller due to the energy exchange arrangement with the grid.

In the m-CHP favorable scenario, the combination of a higher diesel price, slightly lower grid cost, lower biomass price, and higher RES electricity revenue represents conditions that favor the micro-CHP system. Under these assumptions, both the LCOE and LCOH decrease, while the NPV rises substantially and the DPBP shortens to just over 7 years. This demonstrates a strong improvement in system profitability and cost-effectiveness, confirming that when RES and electricity revenues align favorably, the micro-CHP configuration becomes highly competitive and financially robust. On the other hand, the scenario representing conditions more advantageous to the conventional system (BAU approach) results in poorer economic performance. This case yields higher LCOE and LCOH values, a higher LCC, a substantially lower NPV, and an exceedingly long payback period exceeding 30 years. These results clearly indicate that, under these assumptions, the conventional configuration is less economically attractive compared to the m-CHP alternative.

3.7. Environmental Aspects

Along with the generally positive assessment of the economic feasibility of the innovative biomass-fired micro-CHP system for medicinal cannabis cultivation, the environmental performance of the greenhouse facility also shows a significant improvement. The transition from diesel to sustainably sourced biomass substantially reduces GHG emissions. As presented in Table 19, replacing diesel with ENplus A1 wood pellets effectively eliminates net fossil CO_2 emissions (26.97 tCO_2), resulting in an estimated reduction of approximately 26.33 tCO_2-eq (97.3%) in reported GHG emissions, assuming the biomass is sustainably sourced.

Table 19. Fuel characteristics, emission factors, and calculated greenhouse gas emissions for diesel and wood pellet combustion.

Parameter	Unit	Diesel Fuel	Wood Pellet Biomass
Fuel consumption	kg	8707.71	20,602.99
Energy content	$kWh \cdot kg^{-1}$	11.62	5.2
	$TJ \cdot kg^{-1}$	4.18×10^{-5}	1.87×10^{-5}
Total energy used	TJ	0.36	0.39
CO_2 GWP [a]	$kg \cdot TJ^{-1}$	74,100.00	112,000.00
NH_4 GWP [a]	$kg \cdot TJ^{-1}$	3.00	30.00
N_2O GWP [a]	$kg \cdot TJ^{-1}$	0.60	4.00
Emitted CO_2	kg	26,972.40	0.00 [b]
Emitted CH_4	kg	1.09	11.58
Emitted N_2O	kg	0.22	1.54
$CO_2 \times$ GWP	kg CO_2-equivalent	26,972.40	0
$CH_4 \times$ GWP [c]	kg CO_2-equivalent	30.58	324.24
$N_2O \times$ GWP [c]	kg CO_2-equivalent	57.88	409.16
Total GHG emissions	kg CO_2-equivalent	27,060.85	733.40

[a] IPCC 2006 defaults for stationary combustion on a net calorific basis [62]. [b] The biogenic CO_2 emissions from pellet combustion ($\approx$43.2 tCO_2) are not included in the net fossil totals, consistent with IPCC reporting practice that considers CO_2 from sustainably sourced biomass as biogenic and carbon-neutral, while excluding upstream supply-chain emissions from the combustion inventory. [c] GWP values are from the IPCC AR5 (CH_4 = 28, N_2O = 265).

Additional CO_2 emission reductions can be attributed to RES electricity generation, which is treated as avoided grid electricity production. Specifically, assuming an average national grid carbon intensity of 269 $gCO_2 \cdot kWh^{-1}$ for 2024 [63], the avoided generation of 301.97 kWh_{el} (Table 12) corresponds to approximately 81.23 kg of avoided CO_2 emission, resulting in a total of 27.01 tCO_2.

3.8. Comparative Analysis of Alternative Energy Technologies for Greenhouse Applications

While not strictly comparable, the studied biomass-fired micro-CHP system appears to occupy a middle ground in techno-economic performance relative to both innovative and conventional technologies used in greenhouse settings. Although it offers a slower payback than biomass systems providing only heating, the on-site electricity generation could offset most purchased electricity and reduce exposure to grid price fluctuations. Compared with heat pump (HP) options, it is more resilient in contexts where power tariffs are high or drilling is costly. Unlike natural gas (NG) CHP, it avoids exposure to fossil fuel price volatility and potential carbon-pricing schemes, while remaining financially attractive in the modeled context.

More specifically, biomass heating-only options that use flue gases for soil or air heating in greenhouses in China have demonstrated payback periods (DPBPs) of less than three years when biomass is inexpensive [64]. Similar findings were reported for a 1000 m^2 greenhouse in Crete, Greece, where a solid biomass boiler achieved a payback of 2.95 years [65]. An earlier Canadian study [66] also showed that installing a biomass boiler to meet 40% of annual heating demand was more economical than using a natural gas boiler to supply the full load. However, these systems do not produce electricity.

The economics of gas-engine CHP systems are primarily governed by the spark ratio (i.e., the ratio of electricity price to natural gas price) and the potential to monetize CO_2 for enrichment. A pan-European analysis found that when the spark ratio exceeds 4, negative net heat costs can be achieved (i.e., the combined electricity and CO_2 value exceeds the fuel cost) [67]. Depending on operational strategy, daily dynamic optimization in Dutch greenhouses equipped with CHP, boilers, HPs, thermal buffers, and aquifer storage reduced energy costs by up to 29% compared to baseline grower dispatch [68].

For air-source heat pumps (ASHPs), measured Seasonal Coefficients of Performance (SCOPs) between 1.5 and 3.0 can yield relatively short payback periods (around 5.5 years). Ground-source heat pumps (GSHPs), while offering lower operating costs, typically have paybacks of around 18 years unless cheap electricity or favorable geothermal conditions are available [69,70]. Hybrid systems combining photovoltaic–thermal (PVT) panels with GSHPs have achieved COPs of 3.38 and 78% heating cost savings relative to kerosene, although their electricity generation offsets only 36.72% of HP electricity consumption, and performance strongly depends on hydrogeological conditions and local fuel benchmarks [71].

Finally, a recent feasibility study on greenhouse–fuel cell convergence systems reported benefit–cost ratios of 0.62–0.65 for farmer-led models and 1.19–1.86 for utility-led models [72]. In that study, the fuel-cell CHP was benchmarked against geothermal HP greenhouse systems, highlighting capital expenditure and policy support as decisive factors.

Beyond the reference case, biomass-ORC micro-CHP could be most advantageous in colder climates or in places with long, steady heating seasons, where higher run hours would increase annual electricity yield and shorten the payback period. In warmer climates, PV/PVT-assisted, high-SCOP heat-pump architectures may outperform, particularly where electricity tariffs are low. The micro-CHP configuration could also suit grid-constrained or outage-prone locations, utilizing on-site generation, electricity storage, and modest thermal buffering. A detailed, case-specific techno-economic assessment should be conducted beforehand to assess performance, as investment costs can be substantial.

If biomass-ORC micro-CHP were widely adopted, potential benefits could arise at multiple levels. By substituting biomass, substantial fossil CO_2 abatement could be achieved. Moreover, where biomass residues are abundant, local bioenergy value chains and regional circularity could be strengthened. However, using residual biomass rather than certified pellets introduces quality risks (e.g., lower energy content, ash) and necessitates the use of emission-certified boilers. This may also increase O&M needs (ash handling, fouling control, periodic cleaning) and entail local air-quality compliance obligations. Moreover, renewable-electricity production, paired with storage or a favorable interconnection/metering arrangement, could considerably reduce exposure to electricity-price volatility and improve the greenhouse sector's resilience.

4. Conclusions

The techno-economic assessment of a biomass-fired micro-CHP system integrated with an ORC unit for a medicinal cannabis greenhouse demonstrates clear financial advantages over the conventional diesel-boiler configuration. For the greenhouse studied, the estimated annual energy intensities are consistent with the recent literature on protected cannabis cultivation once differences in operational practices and local climate are considered, with an estimated annual heating demand of 224.58 $kWh_{th} \cdot m^{-2}$ (91,065.20 kWh_{th}), and electricity use of 19.36 $kWh_{el} \cdot m^{-2}$ (7839.90 kWh_{el}). Over a thirty-year lifetime, the micro-CHP scenario (80 kW_{th} boiler retrofitted for biomass and a 7 kW_{el} ORC engine) reduces the life-cycle cost from EUR 229,468.46 to EUR 196,421.33 when fully equity-financed, and to EUR 194,801.82 with 70% debt financing, representing savings up to approximately 14.4%. These savings translate into a positive net present value of EUR 59,591.88 (own-funds) and EUR 61,211.39 (leveraged), while the discounted payback period shortens from 12.1 years to 10.5 years under the debt-financed case. The internal rate of return of 27.32% under all-equity financing further confirms the project's attractiveness.

In addition to cost reductions, the micro-CHP configuration yields highly competitive levelized costs of electricity (EUR 0.122·kWh_{el}) and heat (EUR 0.062·kWh_{th}). These figures undercut typical retail electricity and natural-gas prices in Greece and compare favorably to EU averages. Though the on-site generation of 8195.87 kWh_{el} per year exceeds just by 4% the greenhouse's electricity requirements, it generates additional revenues for the producer. From an environmental standpoint, switching from diesel to biomass, ideally sourced from cannabis residue or local agricultural waste, reduces reliance on fossil fuels and lowers CO_2 emissions (27.01 tCO_2 reduction) and total GHG emissions. When combined with other best practices such as LED lighting, off-peak scheduling, and CO_2 enrichment via waste-heat recovery, the micro-CHP system advances the greenhouse toward a circular and sustainable operation.

The study evaluates an innovative micro-CHP system in a high-value greenhouse operation, providing insights into techno-economic performance, system optimization, and environmental impact. Socially, it highlights tangible benefits in operational cost reduction, energy autonomy, emission mitigation, and circular resource utilization, contributing to more sustainable, resilient, and economically viable practices in the agricultural sector. Biomass-ORC micro-CHP systems are well-suited to colder or grid-constrained sites with long heating seasons and reliable biomass supply. At scale, they cut GHG emissions and moderate power-price volatility, but require attention to feedstock quality, O&M practices, air-quality compliance, and storage/safety requirements. Future research should focus on dynamic control strategies and energy storage, both electrical and thermal, options that optimize ORC operation under real-time thermal loads, undertake detailed life-cycle carbon-footprint analysis, and evaluate at the techno-economic level the combined heat, power, and CO_2 utilization schemes in cannabis greenhouses.

Author Contributions: Methodology, C.G., D.B. and D.M.; software, C.G., D.T. and A.S.; validation, D.B. and D.M.; formal analysis, C.G. and D.T.; investigation, C.G. and A.S.; resources, D.M.; data curation, C.G. and A.S.; writing—original draft preparation, C.G. and D.T.; writing—review and editing, D.T., D.B. and D.M.; visualization, D.T.; supervision, D.M.; project administration, D.M. All authors have read and agreed to the published version of the manuscript.

Funding: This research received no external funding.

Institutional Review Board Statement: Not applicable.

Informed Consent Statement: Not applicable.

Data Availability Statement: The original contributions presented in this study are included in the article. Further inquiries can be directed to the corresponding authors.

Conflicts of Interest: The authors declare no conflicts of interest.

Abbreviations and Symbols

The following abbreviations and symbols are used in this manuscript:

Abbreviations	
ASHP	Air-Source Heat Pump
BAU	Business-As-Usual
CCU	Carbon Capture and Utilization
CEA	Controlled Environment Agriculture
CHP	Combined Heat and Power

CMH	Ceramic Metal Halide
COP	Coefficient of Performance
DPBP	Discounted Payback Period
EU	European Union
GHG	Greenhouse Gas
GSHP	Ground-Source Heat Pump
GWP	Global Warming Potential
HP	Heat Pump
HPS	High Pressure Sodium
HVAC	Heating, Ventilation, and Air Conditioning
IPCC (AR5)	United Nations Intergovernmental Panel on Climate Change (5th Assessment Report)
IRR	Internal Rate of Return
ISO	International Organization for Standardization
LCC	Life Cycle Cost
LCOE	Levelized Cost of Electricity
LCOH	Levelized Cost of Heat
LED	Light Emitting Diode
LHV	Lower Heating Value
m-CHP	Micro Combined Heat and Power
NCF	Net Cash Flow
NG	Natural Gas
MH	Metal Halide
NPV	Net Present Value
OM&R	Operation, Maintenance, and Repair
ORC	Organic Rankine Cycle
PLC	Programmable Logic Controller
PV	Photovoltaic
PVT	Photovoltaic-Thermal
RES	Renewable Energy Sources
RH	Relative Humidity
SCOP	Seasonal Coefficient of Performance
Symbols	
A_c	Cover and wall area
a_{el}	Cost allocation factor for electricity
A_f	Floor area
a_{he}	Cost allocation factor for heat
C_{INV}	Initial investment cost
$C_{out,tot}$	Total annual cost
C_O	Operating costs
C_M	Maintenance costs
C_E	End-of-life costs
E_{tot}	Total electricity generated over lifetime
$Freq_{pu}$	Pump operating frequency
$Freq_{exp}$	Expander operating frequency
f	Radiation to latent heat conversion factor
G_h	Solar radiation per square meter
H	Greenhouse heating load per square meter
NCF_{BAU}	Net cash flow for BAU scenario
$NCF_{m\text{-}CHP}$	Net cash flow for m-CHP scenario
$n_{el,ref}$	Reference efficiency for electricity generation

$n_{he,ref}$	Reference efficiency for heat generation
$P_{el,pu}$	Pump's electrical power consumption
$P_{el,exp}$	Expander's electrical power output
$P_{net,ORC}$	Net electrical output of ORC
Q_{CHP}	Total heat generated over system's lifetime
$\dot{Q}_h$	Total required heating load
r	Discount rate
t	Year index
$T_{con,in}$	Condenser inlet temperature
$T_{i,s}$	Indoor air temperature (setpoint)
T_o	Outdoor air temperature
U	Overall thermal transmittance
ϱ	Reflectance coefficient of the interior
τ	Permeability to solar radiation factor

Appendix A

Table A1. Experimental data on power generation of the ORC system at AUA.

$Freq_{pu}$ (Hz)	$Freq_{exp}$ (Hz)	$T_{con,in}$ (°C)	$P_{el,exp}$ (kW_{el})	$P_{el,pu}$ (kW_{el})	$P_{net,ORC}$ (kW_{el})
15	25	47.50	1.91	0.29	1.62
15	30	45.70	2.99	0.35	2.64
20	35	50.60	3.04	0.48	2.56
25	25	54.00	1.45	0.57	0.88
25	30	54.20	2.25	0.84	1.41
25	35	54.30	2.30	0.85	1.45
30	30	46.50	5.16	1.40	3.76
30	50	64.90	6.28	1.40	4.88
35	40	64.30	7.82	1.68	6.14
35	45	54.20	11.86	2.23	9.63
35	50	64.30	9.92	1.84	8.08
35	55	66.10	11.57	2.03	9.54
40	50	64.40	14.46	2.60	11.86
40	60	64.30	15.60	2.85	12.75
45	60	64.20	24.82	3.89	20.93

Table A2. Monthly estimations of thermal energy needs, operating hours, fuel supply, and electrical energy generation for the micro-CHP and BAU scenarios.

Month [Growing Period]	Thermal Energy Demand (kWh_{th})	Micro-CHP Operating Hours (h)	Diesel Supply (L)	Biomass Supply (kg)	ORC Electrical Energy Generation (kWh_{el})
March [GP1]	20,841.91	621.00	2386.73	4715.37	1875.77
April [GP1]	13,914.90	527.00	1593.47	3148.17	1252.34
May [GP1]	5205.69	421.00	596.13	1177.76	468.51
June [GP1]	871.84	206.00	99.84	197.25	78.47
September [GP2]	848.89	136.00	97.21	192.06	76.40
October [GP2]	6748.07	497.00	772.76	1526.71	607.33
November [GP2]	20,005.57	633.00	2290.95	4526.15	1800.50
December [GP2]	22,628.34	692.00	2591.30	5119.53	2036.55
TOTAL	91,065.20	3733.00	10,428.39	20,602.99	8195.87

Table A3. Micro-CHP bill of materials and purchase costs.

Component	Model	Qty.	Unit Cost (EUR)	Cost (EUR)
Refrigerant pump	EDURAGGR LBM 406 A120 L/5.5 KW	1 pc.	6060.00	6060.00
Expander	Hanbell ER-230	1 pc.	5488.00	5488.00
Evaporator	SWEP V250ASHx120/1P	1 pc.	1550.00	1550.00
Condenser	SWEP B250ASHx96/1P	1 pc.	1150.00	1150.00
Ball valve	GMC 1.3/8″	1 pc.	46.80	46.80
Oil ball valve	GMC 5/8″	1 pc.	19.94	19.94
Oil solenoid	GMC 5/8″	1 pc.	46.80	46.80
Filter drier oil	GMC 5/8″	1 pc.	9.66	9.66
Oil receiver	Vertical 10Lt	1 pc.	130.00	130.00
Liquid receiver	Horizontal 50Lt	1 pc.	260.00	260.00
Filter drier	GMC 1.3/8″ with cartridge	1 pc.	47.97	47.97
PHE water connections	Victaulic 3″ silicone gasket	4 pcs.	11.93	47.74
Safety valve	GMC SV1/34	1 pc.	17.55	17.55
Suction line	Copper tube 2.5/8″	2.10 m	47.97	100.74
Discharge line	Copper tube 2.1/8″	1.32 m	37.44	49.42
Liquid line	Copper tube 1.3/8″	3.17 m	18.34	58.15
Oil line	Copper tube 5/8″	3.00 m	11.05	33.15
Pipe insulation	Armaflex 9 mm Φ54	1.32 m	1.30	1.72
Rotalolock valve	1.3/8″ (tube) in 1.3/4″ (flare)	2 pcs.	14.82	29.64
Frame	1.90 × 0.80 × 0.81 m	1 pc.	2000.00	2000.00
Pump connections	Flange JIS 1.1/2″	2 pcs.	7.44	14.87
Liquid indicator	GMC 1.3/8″	1 pc.	17.55	17.55
Electric panel with PLC	0.6 × 0.6 × 0.25 m	1 pc.	6000.00	6000.00
TOTAL				23,179.70

References

1. Saragoça, A.; Silva, A.C.; Varanda, C.M.R.; Materatski, P.; Ortega, A.; Cordeiro, A.I.; Telo da Gama, J. Current Context of Cannabis sativa Cultivation and Parameters Influencing Its Development. *Agriculture* **2025**, *15*, 1635. [CrossRef]
2. Thetsane, R.M. Envisaging Challenges for the Emerging Medicinal Cannabis Sector in Lesotho. *J. Cannabis Res.* **2024**, *6*, 23. [CrossRef]
3. Dillis, C.; Polson, M.; Bodwitch, H.; Carah, J.; Power, M.E.; Sayre, N.F. Industrializing Cannabis? Socio-Ecological Implications of Legalization and Regulation in California. In *The Routledge Handbook of Post-Prohibition Cannabis Research*; Corva, D., Meisel, J.S., Eds.; Routledge: London, UK, 2021; pp. 221–230.
4. Chatzigianni, C.; Roussis, I.; Papadas, C.T.; Karidogianni, S.; Kouneli, V.; Kakabouki, I.; Folina, A.; Papastylianou, P.; Bilalis, D. Estimated Cost of Production for Medical Cannabis (*Cannabis sativa* L.) in Greece. *Bull. UASVM Hortic.* **2020**, *77*, 1843–5394. [CrossRef]
5. Hellenic Republic. *Law No. 4523—Gov. Gazette A' 41/07.03.2018*; Government Gazette, National Printing Office: Athens, Greece, 2018. (In Greek)
6. Folina, A.; Roussis, I.; Kouneli, V.; Kakabouki, I.; Karidogianni, S.; Bilalis, D. Opportunities for cultivation of medical cannabis (*Cannabis sativa* L.) in Greece. *Sci. Pap. Ser. A. Agron.* **2019**, *62*, 293–300.
7. Collado, C.E.; Hwang, S.J.; Hernández, R. Supplemental Greenhouse Lighting Increased the Water Use Efficiency, Crop Growth, and Cutting Production in *Cannabis sativa*. *Front. Plant Sci.* **2024**, *15*, 1371702. [CrossRef]
8. Dohlman, E.; Maguire, K.; Davis, W.V.; Husby, M.; Bovay, J.; Weber, C.; Lee, Y. *Trends, Insights, and Future Prospects for Production in Controlled Environment Agriculture and Agrivoltaics Systems*; Economic Information Bulletin, No. 264; U.S. Department of Agriculture, Economic Research Service: Washington, DC, USA, 2024. Available online: https://www.ers.usda.gov/sites/default/files/_laserfiche/publications/108221/EIB-264.pdf (accessed on 4 November 2025).
9. Azzaretti, C.; Carleton, B. *CEA Energy & Water Benchmarking Report: Establishing Preliminary Benchmarks for Controlled Environment Agriculture (CEA) Operations*; Resource Innovation Institute: Portland, OR, USA, 2023; Available online: https://resourceinnovation.org/wp-content/uploads/2023/08/RII-Benchmarking_2023.pdf (accessed on 4 November 2025).

10. Bafort, F.; Libault, A.; Maron, E.; Kohnen, S.; Ancion, N.; Jijakli, M.H. Operational Costs and Analysis of Agronomic Characteristics on Cannabidiol and Cannabigerol Hemp (*Cannabis sativa* L.) in Hydroponic Soilless Greenhouse and Field Cultivation. *Horticulturae* **2024**, *10*, 1271. [CrossRef]
11. Charles, A.P.R.; Gu, Z.; Archer, R.; Auwarter, C.; Hatterman-Valenti, H.; Rao, J.; Chen, B. Effect of High-Tunnel and Open-Field Production on the Yield, Cannabinoids, and Volatile Profiles in Industrial Hemp (*Cannabis sativa* L.) Inflorescence. *J. Agric. Food Chem.* **2024**, *72*, 12975–12987. [CrossRef] [PubMed]
12. Syed, A.; Hachem-Vermette, C. Climate Change Resilient Urban Prototypes—A Canadian Perspective on Net Zero Energy Design for Retail Amenities. *Energy Eng.* **2019**, *116*, 7–25. [CrossRef]
13. Marcelis, L.F.M.; Costa, J.M.; Heuvelink, E. Achieving Sustainable Greenhouse Production: Present Status, Recent Advances and Future Developments. In *Achieving Sustainable Greenhouse Production*; Marcelis, L.F.M., Heuvelink, E., Eds.; Burleigh Dodds Science Publishing: Cambridge, UK, 2019; pp. 1–14.
14. Mills, E. Comment on "Cannabis and the Environment: What Science Tells Us and What We Still Need to Know". *Environ. Sci. Technol. Lett.* **2021**, *8*, 483–485. [CrossRef]
15. Independent Electricity System Operator (IESO). *Energy Management Best Practices for Cannabis Greenhouses and Warehouses*; IESO: Toronto, ON, Canada, 2021.
16. Mills, E. The Emergence of Indoor Agriculture as a Driver of Global Energy Demand. *NPJ Sustain. Agric.* **2025**, *3*, 52. [CrossRef]
17. Kamminga, J.; Me, A. The Blind Men and the Elephant: Measuring the Environmental Impact of Cannabis. *J. Illicit Econ. Dev.* **2025**, *7*, 1–13. [CrossRef]
18. Massachusetts Department of Energy Resources. *Cannabis Energy Overview and Recommendations*; Massachusetts Cannabis Control Commission: Boston, MA, USA, 2018.
19. Mills, E. The Carbon Footprint of Indoor Cannabis Production. *Energy Policy* **2012**, *46*, 58–67. [CrossRef]
20. Cola, G.; Mariani, L.; Toscano, S.; Romano, D.; Ferrante, A. Comparison of Greenhouse Energy Requirements for Rose Cultivation in Europe and North Africa. *Agronomy* **2020**, *10*, 422. [CrossRef]
21. Sanford, S. *Reducing Greenhouse Energy Consumption—An Overview*; University of Wisconsin, Extension Service Report; University of Wisconsin: Madison, WI, USA, 2011.
22. Summers, H.M.; Sproul, E.; Quinn, J.C. The Greenhouse Gas Emissions of Indoor Cannabis Production in the United States. *Nat. Sustain.* **2021**, *4*, 644–650. [CrossRef]
23. Mills, E. Energy-Intensive Indoor Cultivation Drives the Cannabis Industry's Expanding Carbon Footprint. *One Earth* **2025**, *8*, 101179. [CrossRef]
24. Moreno, R.M. Dimensionamento e Análise Energética de uma Estufa de Cannabis Medicinal. Master's Thesis, Instituto Superior Técnico, Universidade de Lisboa, Lisbon, Portugal, 2020. (In Portuguese).
25. van Gruisen, D.; van Noort, F. *Towards Energy-Neutral Cannabis Cultivation: Exploring Energy Efficiency and Carbon Footprint Mitigation Strategies—A Desk Study on Regulated Medical Cannabis Production*; Kas als Energiebron/Wageningen University & Research: Wageningen, The Netherlands, 2024.
26. New Frontier Data. *The 2018 Cannabis Energy Report*; New Frontier Data: Washington, DC, USA, 2018.
27. Wells, M.J. Expert Opinion: How Humidity Works. *Cannabis Business Times*. 2017. Available online: https://www.cannabisbusinesstimes.com/home/article/15703674/moisture-matters (accessed on 4 November 2025).
28. National Cannabis Industry Association. *Environmental Sustainability in the Cannabis Industry: Impacts, Best Management Practices, and Policy Considerations*; NCIA: Washington, DC, USA, 2020.
29. Canova, A.; Chicco, G.; Genon, G.; Mancarella, P. Emission Characterization and Evaluation of Natural Gas-Fueled Cogeneration Microturbines and Internal Combustion Engines. *Energy Convers. Manag.* **2008**, *49*, 2900–2909. [CrossRef]
30. Seiler, J.C.; Pavlak, G.; Freihaut, J.D. Energy Dispatch Optimization at Controlled Environment Agriculture Sites with CHP: How Energy Utilization, Storage, and Market Exports Impact Operational Costs. *Energy Convers. Manag.* **2025**, *332*, 119743. [CrossRef]
31. Vermeulen, P.C.M.; van der Lans, C.J.M. Combined heat and power (CHP) as a possible method for reduction of the CO2 footprint of organic greenhouse horticulture. In Proceedings of the I International Conference on Organic Greenhouse Horticulture, Bleiswijk, The Netherlands, 11–14 October 2010; Dorais, M., Bishop, S.D., Eds.; ISHS: Leuven, Belgium, 2011; Volume 915, pp. 61–68.
32. Compernolle, T.; Witters, N.; Van Passel, S.; Thewys, T. Analyzing a Self-Managed CHP System for Greenhouse Cultivation as a Profitable Way to Reduce CO_2-Emissions. *Energy* **2011**, *36*, 1940–1947. [CrossRef]
33. Abbass, A. Hybrid Solar CHP Microgrid Optimization: Python Code Framework Using Real Equipment Data. *Adv. Eng. Lett.* **2025**, *4*, 164–174. [CrossRef]
34. Arsalis, A. A Comprehensive Review of Fuel Cell-Based Micro–Combined-Heat-and-Power Systems. *Renew. Sustain. Energy Rev.* **2019**, *105*, 391–414. [CrossRef]
35. Cardoso, D.; Nunes, D.; Faria, J.; Fael, P.; Gaspar, P.D. Intelligent Micro-Cogeneration Systems for Residential Grids: A Sustainable Solution for Efficient Energy Management. *Energies* **2023**, *16*, 5215. [CrossRef]

36. Kallio, S.; Siroux, M. Hybrid Renewable Energy Systems Based on Micro-Cogeneration. *Energy Rep.* **2022**, *8* (Suppl. S1), 762–769. [CrossRef]
37. Hammond, G.P.; Titley, A.A. Small-Scale Combined Heat and Power Systems: The Prospects for a Distributed Micro-Generator in the 'Net-Zero' Transition within the UK. *Energies* **2022**, *15*, 6049. [CrossRef]
38. Gul, E.; Baldinelli, G.; Bartocci, P. Energy Transition: Renewable Energy-Based Combined Heat and Power Optimization Model for Distributed Communities. *Energies* **2022**, *15*, 6740. [CrossRef]
39. Dobre, C.; Costin, M.; Constantin, M. A Review of Available Solutions for Implementation of Small–Medium Combined Heat and Power (CHP) Systems. *Inventions* **2024**, *9*, 82. [CrossRef]
40. Bellos, E. A Review of Organic Rankine Cycles with Partial Evaporation and Dual-Phase Expansion. *Sustain. Energy Technol. Assess.* **2024**, *72*, 104059. [CrossRef]
41. Lecompte, S.; Huisseune, H.; Van Den Broek, M.; Vanslambrouck, B.; De Paepe, M. Review of Organic Rankine Cycle (ORC) Architectures for Waste Heat Recovery. *Renew. Sustain. Energy Rev.* **2015**, *47*, 448–461. [CrossRef]
42. Karellas, S.; Leontaritis, A.D.; Panousis, G.; Bellos, E.; Kakaras, E. Energetic and Exergetic Analysis of Waste Heat Recovery Systems in the Cement Industry. *Energy* **2013**, *58*, 147–156. [CrossRef]
43. ORC World Map. Available online: https://orc-world-map.org/ (accessed on 4 November 2025).
44. Worley, J. *Greenhouses: Heating, Cooling and Ventilation*; Bulletin 792; University of Georgia, UGA Extension: Athens, GA, USA, 2014.
45. Cecilia, S.; Heuvelink, E.; Ooster, B.V. *Greenhouse Horticulture: Technology for Optimal Crop Production*, 2nd ed.; Brill | Wageningen Academic: Wageningen, The Netherlands, 2024; ISBN 9789004697041.
46. Bagg, M. Save Cash and Energy Costs via an LCC Model. *World Pumps* **2013**, *12*, 18–21. [CrossRef]
47. Cheng Hin, J.N.; Zmeureanu, R. Optimization of a Residential Solar Combisystem for Minimum Life Cycle Cost, Energy Use and Exergy Destroyed. *Sol. Energy* **2014**, *100*, 102–113. [CrossRef]
48. Gkoloni, N.; Golonis, C.; Kostopoulos, V. Integration of LCA and LCC for Decision Making in Biocomposite Production. *Proc. IOP Conf. Ser. Earth Environ. Sci.* **2022**, *1123*, 012071. [CrossRef]
49. Chaurasiya, P.K.; Azad, A.K.; Warudkar, V.; Ahmed, S. Advancement in Remote Sensing of Wind Energy. In *Advances in Clean Energy Technologies*; Azad, A.K., Ed.; Academic Press: Cambridge, MA, USA, 2021; pp. 207–233.
50. Spertino, F.; Di Leo, P.; Cocina, V. Economic Analysis of Investment in the Rooftop Photovoltaic Systems: A Long-Term Research in the Two Main Markets. *Renew. Sustain. Energy Rev.* **2013**, *28*, 531–540. [CrossRef]
51. Kumar, L.; Mamun, M.A.A.; Hasanuzzaman, M. Energy Economics. In *Energy for Sustainable Development*; Hasanuzzaman, M., Abd Rahim, N., Eds.; Academic Press: Cambridge, MA, USA, 2020; pp. 167–178.
52. Skiadopoulos, A.; Kosmadakis, G.; van Heule, X.; Lecompte, S.; De Paepe, M.; Manolakos, D. Hybrid Solar-Biomass Residential Micro-Combined Heat and Power Systems Driven by the Partially Evaporating Organic Rankine Cycle—A Case Study in Southern Europe. *Therm. Sci. Eng. Prog.* **2024**, *55*, 102906. [CrossRef]
53. Noussan, M. Allocation Factors in Combined Heat and Power Systems—Comparison of Different Methods in Real Applications. *Energy Convers. Manag.* **2018**, *173*, 516–526. [CrossRef]
54. European Environmental Agency. ENER19 Efficiency of Conventional Thermal Electricity and Heat Production. Available online: https://tinyurl.com/54mf2mfk (accessed on 4 November 2025).
55. Gholami, H.; Røstvik, H.N. Economic Analysis of BIPV Systems as a Building Envelope Material for Building Skins in Europe. *Energy* **2020**, *204*, 117931. [CrossRef]
56. *ISO 18134-1:2022*; Solid Biofuels—Determination of Moisture Content—Part 1: Reference Method. International Organization for Standardization: Geneva, Switzerland, 2022.
57. *ISO 17225-2:2021*; Solid Biofuels—Fuel Specifications and Classes—Part 2: Graded Wood Pellets. International Organization for Standardization: Geneva, Switzerland, 2021.
58. Falbo, L.; Algieri, A.; Morrone, P.; Perrone, D. Experimental Investigation into the Energy Performance of a Biomass Recuperative Organic Rankine Cycle (ORC) for Micro-Scale Applications in Design and Off-Design Conditions. *Energies* **2025**, *18*, 3201. [CrossRef]
59. Electricity Price Statistics. Available online: https://ec.europa.eu/eurostat/statistics-explained/index.php?title=Electricity_price_statistics (accessed on 4 November 2025).
60. Eurostat. Electricity Prices for Non-Household Consumers—Bi-Annual Data (from 2007 Onwards). Available online: https://ec.europa.eu/eurostat/databrowser/view/nrg_pc_205/default/table?lang=en (accessed on 4 November 2025).
61. Eurostat. Gas Prices for Non-Household Consumers—Bi-Annual Data (from 2007 Onwards). Available online: https://ec.europa.eu/eurostat/databrowser/view/NRG_PC_203__custom_110987/default/table?lang=en (accessed on 4 November 2025).

62. Intergovernmental Panel on Climate Change (IPCC). *2006 IPCC Guidelines for National Greenhouse Gas Inventories; Volume 2: Energy. Chapter 2: Stationary Combustion*; IGES: Hayama, Japan, 2006; Available online: http://www.ipcc-nggip.iges.or.jp/public/2006gl/pdf/2_Volume2/V2_2_Ch2_Stationary_Combustion.pdf (accessed on 4 November 2025).
63. The Green Tank. The Carbon Footprint of Electricity Production—January 2025. 2025. Available online: https://thegreentank.gr/en/2025/03/10/emissionswatch-jan25/ (accessed on 4 November 2025).
64. Huang, T.; Li, H.; Zhang, G.; Xu, F. Experimental Study on Biomass Heating System in the Greenhouse: A Case Study in Xiangtan, China. *Sustainability* **2020**, *12*, 5673. [CrossRef]
65. Vourdoubas, J. Economic and Environmental Assessment of the Use of Renewable Energies in Greenhouses: A Case Study in Crete–Greece. *J. Agric. Sci.* **2015**, *7*, 48–55. [CrossRef]
66. Chau, J.; Sowlati, T.; Sokhansanj, S.; Preto, F.; Melin, S.; Bi, X. Economic Sensitivity of Wood Biomass Utilization for Greenhouse Heating Application. *Appl. Energy* **2009**, *86*, 616–621. [CrossRef]
67. Tataraki, K.; Giannini, E.; Kavvadias, K.; Maroulis, Z. Cogeneration Economics for Greenhouses in Europe. *Energies* **2020**, *13*, 3373. [CrossRef]
68. van Beveren, P.J.M.; Bontsema, J.; van 't Ooster, A.; van Straten, G.; van Henten, E.J. Optimal Utilization of Energy Equipment in a Semi-Closed Greenhouse. *Comput. Electron. Agric.* **2020**, *179*, 105800. [CrossRef]
69. Nemś, A.; Nemś, M.; Świder, K. Analysis of the Possibilities of Using a Heat Pump for Greenhouse Heating in Polish Climatic Conditions—A Case Study. *Sustainability* **2018**, *10*, 3483. [CrossRef]
70. Harjunowibowo, D.; Omer, S.A.; Riffat, S.B. Experimental Investigation of a Ground-Source Heat Pump System for Greenhouse Heating–Cooling. *Int. J. Low-Carbon Technol.* **2021**, *16*, 1529–1541. [CrossRef]
71. Lee, C.G.; Kang, G.C.; Jang, J.K.; Yun, S.-W.; Moon, J.P.; Mun, H.-S.; Lagua, E.B. Efficacy of Hybrid Photovoltaic–Thermal and Geothermal Heat Pump System for Greenhouse Climate Control. *Energies* **2025**, *18*, 5386. [CrossRef]
72. Lee, C.-S.; Shin, H.; Park, C.; Park, M.-L.; Choi, Y. Economic Feasibility Analysis of Greenhouse–Fuel Cell Convergence Systems. *Sustainability* **2024**, *16*, 74. [CrossRef]

applied sciences

MDPI

Article

Effect of Real Gas Equations on Calculation Accuracy of Thermodynamic State in Hydrogen Storage Tank

Hao Luo [1,2,3,†], Qianqian Xin [4,†], Cenling Yao [1], Chenglong Li [1], Tianqi Yang [2,*], Xianhuan Wu [1,*], Richard Chahine [3] and Jinsheng Xiao [2,3]

1 School of Mechanical and Electrical Engineering, Wuhan Business University, Wuhan 430070, China; hao.luo@whut.edu.cn (H.L.); wuhanycl@163.com (C.Y.); lichenglong@whut.edu.cn (C.L.)

2 School of Automotive Engineering, Wuhan University of Technology, Wuhan 430070, China; jinsheng.xiao@uqtr.ca

3 Hydrogen Research Institute, Université du Québec à Trois-Rivières, Quebec, QC G8Z 4M3, Canada; richard.chahine@uqtr.ca

4 School of Mechanical Engineering, Wuhan Vocational College of Software and Engineering (Wuhan Open University), Wuhan 430070, China; xin11280429@whut.edu.cn

* Correspondence: tqyang@whut.edu.cn (T.Y.); wuxianhuan510@126.com (X.W.)

† These authors contributed equally to this work.

Abstract

The gas equation of state (EOS) serves as a critical tool for analyzing the thermal effects within the hydrogen storage tank during refueling processes. It quantifies the dynamic relationships among pressure, temperature and volume, playing a vital role in numerical simulations of hydrogen refueling, the development of refueling protocols, and ensuring refueling safety. This study first establishes a lumped-parameter thermodynamic model for the hydrogen refueling process, which combines a zero-dimensional gas model with a one-dimensional tank wall model (0D1D). The model's accuracy was validated against experimental data and will be used in combination with different EOSs to simulate hydrogen temperature and pressure. Subsequently, parameter values are derived for the van der Waals EOS and its modified forms—Redlich–Kwong, Soave, and Peng–Robinson. The accuracy of the modified forms is evaluated using the Joule–Thomson inversion curve. A polynomial EOS is formulated, and its parameters are numerically determined. Finally, the hydrogen temperatures and pressures calculated using the van der Waals EOS, Redlich–Kwong EOS, polynomial EOS, and the National Institute of Standards and Technology (NIST) database are compared. Within the initial and boundary conditions set in this study, the results indicate that among the modified forms for van der Waals EOS, the Redlich–Kwong EOS exhibits higher accuracy than the Soave and Peng–Robinson EOSs. Using the NIST-calculated hydrogen pressure as a benchmark, the relative error is 0.30% for the polynomial EOS, 1.83% for the Redlich–Kwong EOS, and 17.90% for the van der Waals EOS. Thus, the polynomial EOS exhibits higher accuracy, followed by the Redlich–Kwong EOS, while the van der Waals EOS demonstrates lower accuracy. This research provides a theoretical basis for selecting an appropriate EOS in numerical simulations of hydrogen refueling processes.

Keywords: hydrogen refueling; equation of state; Redlich-Kwong; van der Waals; high-pressure hydrogen storage

1. Introduction

Against the backdrop of the global energy sector's accelerated transition towards low-carbon and clean energy, hydrogen energy has emerged as a highly promising key component of the future energy system. Its significant advantages, including diverse sources, clean and pollution-free combustion products, and high energy density, have led to broad application prospects in transportation, power generation, industrial processes, and other fields, offering a new direction for addressing energy and environmental challenges [1].

Hydrogen storage technology is a core segment of the hydrogen industry chain, and its development level directly impacts the large-scale application and promotion of hydrogen energy. Current mainstream hydrogen storage methods include high-pressure gas storage, cryogenic liquid storage, and metal hydride storage. Among these, high-pressure gaseous hydrogen storage has become the most widely adopted mainstream method due to its simple structure, operational convenience, and relatively low energy consumption [2]. During high-pressure gaseous hydrogen refueling processes, the accurate determination of state parameters such as temperature and pressure within the storage tank is crucial. These parameters are key indicators for assessing the operational status of the hydrogen storage tank and are core elements for ensuring the safety and stability of the refueling process. However, under high-pressure conditions, the physical and chemical properties of hydrogen deviate significantly from those of an ideal gas. Factors such as intermolecular interactions and the volume of the molecules themselves exert a non-negligible influence on the gas state. The traditional ideal gas equation of state (EOS) cannot accurately describe the actual state of high-pressure hydrogen and is, therefore, unsuitable for directly calculating parameters such as temperature and pressure within the storage tank [3].

The gas EOS serves as a critical tool for analyzing the thermal effects inside hydrogen storage tanks during refueling. It quantifies the dynamic relationships between pressure, temperature, and volume, and is indispensable for the numerical simulation of hydrogen refueling, the development of refueling protocols, and safety assessments. Since van der Waals proposed the first EOS accounting for intermolecular forces (VDW EOS) in 1873, numerous researchers have conducted in-depth studies on gas EOS. The evolution of EOS has moved from general-purpose to specialized, high-precision models. It begins with concise but limited cubic equations (e.g., Van der Waals and the more advanced Peng–Robinson), progresses to more complex multiparameter formulations (e.g., Benedict–Webb–Rubin) for better accuracy, and culminates in modern Helmholtz free energy-based equations (e.g., NIST Leachman). These state-of-the-art models, offering great accuracy over wide ranges, are embedded in software like NIST REFPROP, signifying a shift from hand calculations to digital tools [4]. Chen et al. [3] proposed a simplified real gas EOS for hydrogen; compared against the National Institute of Standards and Technology (NIST) reference data, it showed maximum errors of 1.1% and 3.8% within the temperature ranges of 253 K to 393 K and 173 K to 393 K, respectively. Nasrifar et al. [5] employed eleven different EOSs to predict various properties of standard hydrogen and identified the most suitable models through comparison with experimental data. Bai-gang et al. [6] proposed a high-precision EOS where the error in hydrogen consumption remained within 0.5% under specific pressure and temperature ranges. Kontogeorgis et al. [4] discussed the capabilities, limitations, current status, and future challenges of EOS, including those with general applicability. Peng et al. [7] proposed a new simplified virial-type EOS—the Virial–Peng–Long (VPL) EOS—where the third to fifth virial coefficients are expressed using empirically fitted formulas. Bilgili et al. [8], through simulation analysis of different real gas and ideal gas EOSs, found that different EOSs significantly impact the calculated tank temperature and pressure, and the refueling time also varies slightly depending on the EOS used.

As the technical requirements for hydrogen storage continue to increase, selecting an appropriate gas EOS to calculate the thermodynamic state within hydrogen storage tanks accurately has become a pressing issue in the field of hydrogen energy. This paper aims to investigate the application of different gas EOSs in calculating the thermodynamic state of high-pressure hydrogen storage tanks. Through a comparative analysis of results from various EOSs, this study aims to provide a scientific basis for selecting the appropriate EOS for practical engineering applications.

2. Development and Validation of a Hydrogen Storage Tank Model

2.1. Development of a Hydrogen Storage Tank Model

To investigate the effect of different gas EOSs on the calculation accuracy of temperature and pressure inside the storage tank, this study first establishes a zero-dimensional gas and one-dimensional tank wall model (0D1D) for the hydrogen storage tank, as shown in Figure 1. Experimental research in Ref. [9] indicates that the temperature distribution within the tank is relatively uniform during the hydrogen refueling process. Therefore, it is reasonable to treat the hydrogen gas zone as a zero-dimensional lumped-parameter model during refueling. The tank wall is considered a multi-layer structure, with heat conduction occurring sequentially through each radial layer. The temperature distribution within each cylindrical layer is assumed to be uniform.

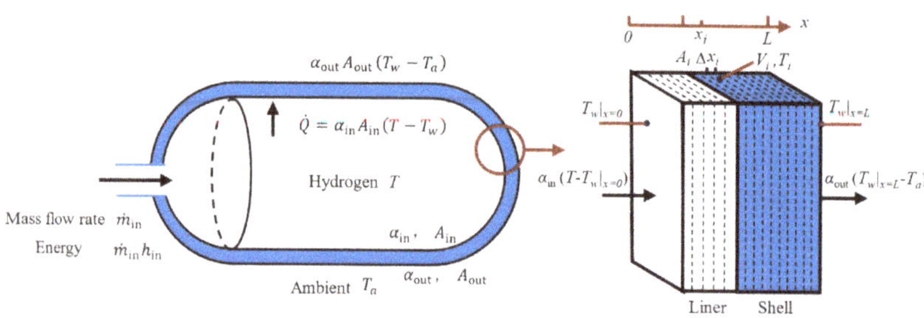

Figure 1. Zero-dimensional gas and one-dimensional tank wall model for the hydrogen storage tank.

During refueling process, the mass conservation of hydrogen can be expressed as [10]

$$\frac{\mathrm{d}m}{\mathrm{d}t} = \dot{m}_{\mathrm{in}} \quad (1)$$

where m represents the mass of hydrogen in the storage tank (kg); $\dot{m}_{\mathrm{in}}$ represents the mass flow rate of hydrogen (kg/s).

Heat exchange occurs between the hydrogen and the inner wall of the storage tank. The energy conservation of hydrogen can be expressed as [11]

$$\frac{\mathrm{d}(mu)}{\mathrm{d}t} = \dot{m}_{\mathrm{in}} h_{\mathrm{in}} - \alpha_{\mathrm{in}} A_{\mathrm{in}} (T - T_w|_{x=0}) \quad (2)$$

where u represents the specific internal energy of hydrogen; h_{in} represents the specific enthalpy of the inflow hydrogen (J/kg); T and $T_w|_{x=0}$ represent the temperature of the hydrogen and the temperature of the inner wall of the tank (K), respectively; A_{in} represents the surface area of the inner wall of the tank (m^2); α_{in} represents the heat transfer coefficient between the hydrogen and the inner wall of the tank ($\mathrm{W/m^2/K}$); $\alpha_{\mathrm{in}} = 0.14 \lambda \mathrm{Re}_{d_{\mathrm{in}}}{}^{0.67} / D_{\mathrm{in}}$, $\mathrm{Re}_{d_{\mathrm{in}}}$ is the Reynolds number; $\mathrm{Re}_{d_{\mathrm{in}}} = 4\dot{m}_{\mathrm{in}} / (\pi \mu d_{\mathrm{in}})$. λ and μ are the thermal conductivity and dynamic viscosity of the hydrogen; D_{in} and d_{in} are the inner diameter of the tank and the inlet injector.

Meanwhile, heat exchange occurs between the outer wall of the tank and the surrounding ambient air. The energy conservation of the tank wall can be expressed as [11]

$$\frac{\mathrm{d}(m_w c_w T_w)}{\mathrm{d}t} = \alpha_{\mathrm{in}} A_{\mathrm{in}}(T - T_w|_{x=0}) - \alpha_{\mathrm{out}} A_{\mathrm{out}}(T_w|_{x=L} - T_a) \tag{3}$$

where T_w represents the average temperature of the tank wall (K); c_w represents the specific heat capacity of the tank wall (J/kg/K); m_w represents the mass of the tank wall (kg); $T_w|_{x=L}$ and T_a represent the temperature of the outer wall of the tank and the ambient temperature (K), respectively; α_{out} represents the heat transfer coefficient between the outer wall of the tank and the ambient (W/m^2/K); $a_{\mathrm{out}} = \lambda_{\mathrm{air}}\left(0.4\mathrm{Re}_{D_{\mathrm{out}}}^{0.5} + 0.06\mathrm{Re}_{D_{\mathrm{out}}}^{2/3}\right)\mathrm{Pr}_{\mathrm{air}}^{0.4}/D_{\mathrm{out}}$, $\mathrm{Pr}_{\mathrm{air}} = \mu_{\mathrm{air}} c_{p_{\mathrm{air}}}/\lambda_{\mathrm{air}}$, $\mathrm{Re}_{D_{\mathrm{out}}} = \rho_{\mathrm{air}} v_{\mathrm{air}} D_{\mathrm{out}}/\mu_{\mathrm{air}}$. $\mathrm{Pr}_{\mathrm{air}}$ is the Prandtl number of the air; λ_{air}, μ_{air}, $c_{p_{\mathrm{air}}}$, ρ_{air} and v_{air} are the thermal conductivity, dynamic viscosity, specific heat capacity at constant pressure, density and velocity of the air, respectively; D_{out} is the outer diameter of the tank; A_{out} represents the surface area of the outer wall of the tank (m^2).

Heat exchange occurs internally inside the tank wall via thermal conduction. As shown in Figure 1, the middle part of the elliptical hydrogen storage vessel is a hollow cylinder, and both ends are hollow hemispheres. The wall thickness of the tank is much smaller than its inner diameter, so that the curvature effect can be neglected. Therefore, the one-dimensional radial problem in cylindrical or spherical coordinates can be simplified to a flat-plate model in Cartesian coordinates. After discretizing the tank wall into 15 concentric layers, each thin layer can be treated as a planar unit. This approach is the same as that used in Ref. [12]. The thermal conduction equations and the boundary conditions of the tank wall can be expressed as [12]

$$k\frac{\partial^2 T_w}{\partial x^2} = \rho_w c_w \frac{\partial T_w}{\partial t} \tag{4}$$

$$-k\frac{\partial T_w}{\partial x}|_{x=0} = \alpha_{\mathrm{in}}(T - T_w|_{x=0}) \tag{5}$$

$$-k\frac{\partial T_w}{\partial x}|_{x=L} = \alpha_{\mathrm{out}}(T_w|_{x=L} - T_a) \tag{6}$$

where L represents the thickness of the tank wall (m); ρ_w represents the density of the tank wall (kg/m^3); k represents the thermal conductivity of the tank wall material (W/m/K).

Once the hydrogen temperature inside the tank is calculated, the hydrogen pressure can be determined using real gas EOS. There are many types of real gas EOS, such as the Abel-Nobel EOS, the van der Waals EOS, and its modified forms like Redlich–Kwong, Soave, and Peng–Robinson. Another method is based on the database of the NIST, which can be expressed as

$$p = \frac{ZmRT}{V\mathrm{M}_{\mathrm{H}_2}} \tag{7}$$

where R represents the universal gas constant, with a value of 8.314 J/mol/K. $\mathrm{M}_{\mathrm{H}_2}$ represents the molar mass of hydrogen, with a value of 2.0159×10^{-3} kg/mol. p represents the hydrogen pressure inside the tank (Pa). V represents the volume of the tank (m^3). Z represents the compressibility factor. λ, μ, λ_{air}, μ_{air}, $c_{p_{\mathrm{air}}}$, ρ_{air} and Z can be obtained from the database of the NIST.

2.2. Validation of the Hydrogen Storage Tank Model

Based on the MATLAB/Simulink (Version: 9.4.0.813654 (R2018a)) software platform, the 0D1D numerical model of a storage tank for the hydrogen refueling process was established. To verify the accuracy of the 0D1D numerical model of the hydrogen storage tank, the physical parameters and experimental data of a 90.5 L hydrogen storage tank with a nominal working pressure (NWP) of 70 MPa from Ref. [13] were used for comparative validation. The physical parameters of the tank are shown in Table 1, and the initial and boundary conditions are presented in Table 2.

Table 1. Physical parameters of the hydrogen storage tank from the Ref. [13].

Parameter	Physical Definition	Value
L	Length of the horizontal section of the tank (m)	0.58
D_{in}	Inner diameter of the tank (m)	0.434
D_{out}	Outer diameter of the tank (m)	0.484
k_{wl}	Thermal conductivity of the tank wall liner (W/m/K)	0.4
k_{ws}	Thermal conductivity of the tank wall shell (W/m/K)	0.53
c_{wl}	Specific heat capacity of the tank wall liner (J/kg/K)	1917
c_{ws}	Specific heat capacity of the tank wall shell (J/kg/K)	942
ρ_{wl}	Density of the tank wall liner (kg/m^3)	954
ρ_{ws}	Density of the tank wall shell (kg/m^3)	1442
d_{in}	Diameter of the inlet injector in the tank (m)	0.006

Table 2. Initial and boundary conditions from the Ref. [13].

Parameter	Physical Definition	Value
T_0	Initial hydrogen temperature (°C)	1.9
T_{w0}	Initial tank wall temperature (°C)	1.9
T_a	Ambient temperature (°C)	1.9
T_{in}	Inlet temperature (°C)	7.9
p_0	Initial hydrogen pressure (MPa)	3
APRR	Average pressure ramp rate (refueling rate) (MPa/s)	0.07

In our previous work [14], this 0D1D model underwent extensive validation. The simulation results agree well with the experimental data, as illustrated in Figure 3 of Ref. [14], further substantiating the validity of the one-dimensional thermal conductivity model for the elliptical hydrogen storage vessel.

3. Thermodynamic States in Storage Tank Calculated by Different EOSs

During the hydrogen refueling process, the pressure of hydrogen inside the tank is high. Under these conditions, using the ideal gas EOS will lead to significant errors, and a real gas EOS should be employed instead. Many types of real gas EOS can be used for hydrogen. A detailed review of the main has been conducted, and the limitations and advantages of each EOS have been identified, as shown in Table 3. Moreover, we conducted further analysis on several representative EOSs among them.

Table 3. The main real gas EOSs used for hydrogen and their limitations and advantages.

Equation Name	Mathematical Expression	Advantages	Limitations
Van der Waals EOS (1873) [15]	$p = \frac{RT}{v-b} - \frac{a}{v^2}$	First real gas EOS; simple form; qualitatively explains deviations from ideal gas	Low accuracy; large errors for hydrogen
Redlich–Kwong EOS (1949) [16]	$p = \frac{RT}{v-b} - \frac{a}{\sqrt{T}v(v+b)}$	Improved applicability at high temperatures; relatively simple	Large deviations at high pressure and low temperature; poor accuracy for hydrogen

Table 3. *Cont.*

Equation Name	Mathematical Expression	Advantages	Limitations
Soave–Redlich–Kwong EOS (1972) [17]	$p = \frac{RT}{v-b} - \frac{a\alpha}{v(v+b)}$	Introduces $\alpha = 1.202e^{(-0.30288T_r)}$, improves vapor–liquid equilibrium calculations; widely used in engineering	Limited accuracy; not ideal for hydrogen under high pressure
Peng–Robinson EOS (1976) [18]	$p = \frac{RT}{v-b} - \frac{a}{v^2+2bv-b^2}$	Relatively accurate at medium and high pressures; widely used in gas storage and transportation	Deviations near the critical point; insufficient for light molecules like hydrogen
Benedict–Webb–Rubin EOS (BWR, 1940s) [19]	$P = RT\rho + \left(B_0RT - A_0 - \frac{C_0}{T^2}\right)\rho^2 + \ldots$	High accuracy over wide ranges; applicable for hydrogen experimental data fitting	Complex form with many parameters; inconvenient in engineering applications
Virial EOS [7]	$Z = 1 + \frac{B(T)}{v} + \frac{C(T)}{v^2} + \ldots$	Extremely accurate at low pressures; coefficients directly determined by experiments	Invalid at high pressures; cannot describe phase transitions
Lee–Kesler EOS (1975) [20]	$Z = Z^{(0)} + \omega\left(Z^{(1)} - Z^{(0)}\right)$	Widely used in engineering; simple and practical; suitable for light gases approximate calculation	Corresponding states approximation; limited accuracy for hydrogen under extreme conditions
NIST Leachman EOS (2009) [21]	Multiparameter form based on Helmholtz free energy	Developed specifically for hydrogen; very high accuracy; valid for 14–1000 K, $P \leq 2000$ MPa	Complex expression; requires software support
GERG-2008 EOS (Kunz & Wagner, 2008) [22]	Multiparameter Helmholtz energy form, developed for natural gas and mixtures	Applicable to hydrogen and hydrogen mixtures; high accuracy; widely used in transport engineering	Slightly less specialized for pure hydrogen compared to Leachman EOS
IAPWS Hydrogen EOS (2002) [23]	Multiparameter form based on Helmholtz free energy	International standard EOS; wide *T*–*P* coverage	Complex formulas; inconvenient for direct engineering use
NIST REFPROP Database [24]	Integrated multiparameter EOS (e.g., Leachman EOS for hydrogen)	Most widely used in research and engineering; supports pure hydrogen and mixtures; directly computes thermodynamic properties	Commercial software requiring license; no simple closed-form expressions

3.1. van der Waals Equation of State

A famous and influential real gas EOS is the van der Waals EOS:

$$p = \frac{RT}{v-b} - \frac{a}{v^2} \tag{8}$$

where a and b are the parameters, v is the molar volume of hydrogen (m^3/mol). At the critical point of the substance, where $T = T_c$, $p = p_c$ and $v = v_c$ (T_c, p_c, and v_c represent the critical temperature, critical pressure, and critical molar volume of the substance), the following conditions hold: $(\partial p/\partial v)_T = 0$ and $(\partial^2 p/\partial v^2)_T = 0$. Thus, we can obtain

$$\frac{2a}{v^3} - \frac{RT}{(v-b)^2} = 0 \tag{9}$$

$$\frac{2RT}{(v-b)^3} - \frac{6a}{v^4} = 0 \tag{10}$$

Then, at the critical point, solving Equations (8)–(10), it can be calculated that $p_c = a/27b^2$, $v_c = 3b$, $T_c = 8a/27Rb$. Combining critical thermodynamic parameters of hydrogen obtained experimentally, namely T_c = 33.3 K, p_c = 1.28 × 10^6 Pa, v_c = 6.5 × 10^{-5} m^3/mole [25], it can be calculated that a = 25.2 J·m^3/kilomole^2, b = 0.027 m^3/kilomole.

3.2. Modified Forms for van der Waals Equation of State

To further improve the accuracy of the gas EOS, many modified forms for van der Waals EOSs have been proposed. Among them, the modifications of Redlich–Kwong, Soave, and Peng–Robinson are widely used.

The modified form of Redlich–Kwong can be expressed as [8]

$$p = \frac{RT}{v-b} - \frac{a}{\sqrt{T}v(v+b)} \tag{11}$$

The modified form of Soave can be expressed as [8]

$$p = \frac{RT}{v-b} - \frac{a\alpha}{v(v+b)} \tag{12}$$

where $\alpha = 1.202e^{(-0.30288T_r)}$.

The modified form of Peng–Robinson can be expressed as [8]

$$p = \frac{RT}{v-b} - \frac{a}{v^2+2bv-b^2} \tag{13}$$

Referring to the method adopted for Equation (8), the expressions for parameters a and b in the Redlich–Kwong, Soave, and Peng–Robinson EOSs were solved. The results are shown in Table 4. By substituting the experimentally obtained critical temperature T_c = 33.3 K, critical pressure p_c = 1.28 × 10^6 Pa, and critical molar volume v_c = 6.5 × 10^{-5} m^3/mole of hydrogen into the expressions for parameters a and b in Table 4, the experimental values of a and b can be calculated.

Table 4. Parameter values in different modified forms for van der Waals equation of state.

Type	Parameter a			Parameter b			R^2
	Expression	Exp-Value	Fitted Value	Expression	Exp-Value	Fitted Value	
R-K	$0.42748R^2T_c^{2.5}/p_c$	1.477×10^{-1}	1.690×10^{-1}	$0.08662RT_c/p_c$	1.874×10^{-5}	1.874×10^{-5}	0.99
Soave	$0.427487T_c^2/p_c$	3.703×10^{-4}	3.239×10^{-2}	$0.08664RT_c/p_c$	1.874×10^{-5}	1.874×10^{-5}	0.99
P-R	$0.45724R^2T_c^2/p_c$	2.738×10^{-2}	2.249×10^{-2}	$0.07780RT_c/p_c$	1.683×10^{-5}	1.683×10^{-5}	0.99

The Joule–Thomson inversion curve is a highly sensitive test for EOS, particularly near the critical point. It can therefore be used to evaluate the parameters in different forms of the EOS. The Joule–Thomson inversion curve data for hydrogen are from the National Aeronautics and Space Administration (NASA) Technical Note D-6807 (1972) [26]:

$$p_r = \sum_{i=0}^{6} A_i T_r^i \tag{14}$$

where T_r, p_r, and v_r are the dimensionless form of reduced temperature, reduced pressure, and reduced molar specific volume ($T_r = T/T_c$, $p_r = p/p_c$, $v_r = v/v_c$). $A_0 = -15.5988252$, $A_1 = 26.0321395$, $A_2 = -9.7459013$, $A_3 = 2.4207304$, $A_4 = -0.49105816$, $A_5 = 0.05932495$, and $A_6 = -0.002913248$.

Set the range of T_r from 0.9 to 6.28 with a step size of 0.01. This range ensures that both the temperature and pressure of hydrogen calculated by Equation (14) are positive. The Joule–Thomson inversion curve can be calculated using Equation (14). By combining the experimentally obtained critical temperature, critical pressure, and critical molar volume of hydrogen, the corresponding value of v_r for the Joule–Thomson inversion curve can be calculated. Furthermore, since $T = T_r T_c$, $p = p_r p_c$, and $v = v_r v_c$, a three-dimensional spatial curve of (p, v, T) can ultimately be plotted, as shown by the black curve in Figure 2a, which can be regarded as standard data for the real gas EOS.

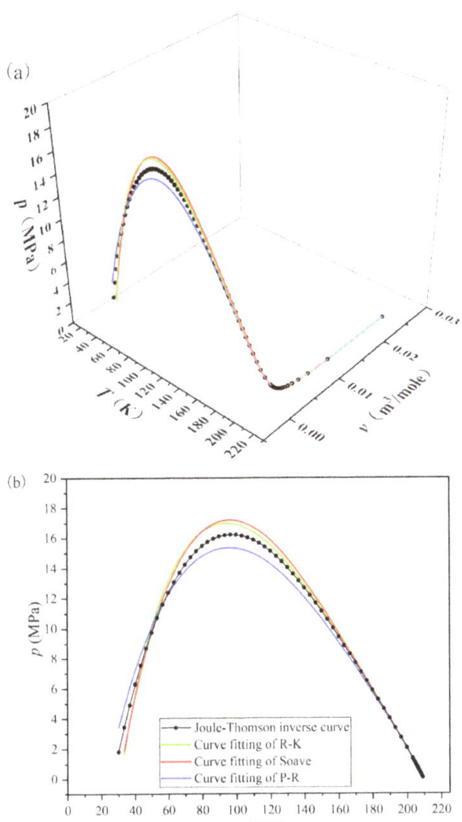

Figure 2. Fitting results for modified van der Waals EOSs using the Joule–Thomson inversion curve. (**a**) Fitted curve in three-dimensional space. (**b**) Projection of the three-dimensional fitted curve onto the p–T plane.

The Redlich–Kwong EOS, Soave EOS, and Peng–Robinson EOS were used to fit the black curve in Figure 2a. During the fitting process, the value of parameter b was fixed to the experimental value listed in Table 4 to facilitate comparison between the fitted value and the experimental value of parameter a. The parameter values determined by the fitting are shown as the fitted values in Table 4.

Table 4 shows that the R^2 obtained from fitting the three EOS are all greater than or equal to 0.99, indicating that the fitting process is accurate and reliable. The comparison between the fitted and experimental values of parameter a in Table 4 reveals that the relative error of parameter a in the Redlich–Kwong EOS is smaller. Figure 2b visually demonstrates that the fitted curve of the Redlich–Kwong EOS is closer to the standard Joule–Thomson inversion curve. Therefore, this study selects the Redlich–Kwong EOS for further research. Using the experimental values of parameters a and b, the specific form of the Redlich–Kwong EOS is determined as

$$p = \frac{RT}{v - 1.874 \times 10^{-5}} - \frac{1.477 \times 10^{-1}}{\sqrt{T}v(v + 1.874 \times 10^{-5})} \tag{15}$$

where p is the pressure of hydrogen (Pa), T is the temperature of hydrogen (K), v is the molar volume of hydrogen (m^3/mol), and R is the universal gas constant, with a value of 8.314 J/mole/K.

3.3. Polynomial Equation of State

Bourgeois and his collaborators argued that employing calculations based on the NIST database appears more suitable for simulating high-pressure hydrogen refueling processes than any other types of EOS [27]. However, data acquisition from the NIST database requires real-time table lookup operations, and the data are not presented in an explicit analytical form. In view of this, to enhance the convenience and efficiency of data usage, it is necessary to explore and develop a reliable equation that can directly represent NIST data in an explicit analytical form. This equation can be expressed as a polynomial, utilizing different parameter values to represent various thermophysical properties of hydrogen. The parameter values can be determined by fitting against data generated from the NIST database. This polynomial EOS will be a high-performance computational model specifically designed for hydrogen. Its customized architecture enables it to reproduce reference data within the target range accurately. Compared to the complex reference equations used in NIST REFPROP, its explicit polynomial structure offers a significant computational efficiency advantage, making it particularly suitable for numerical simulation scenarios that require high-frequency, high-precision thermodynamic calculations. First, the polynomial form of the EOS can be defined as

$$p = \sum_{i=0}^{N} \sum_{j=0}^{N-i} a_{ij} T^i \rho^j + d \tag{16}$$

where ρ represents the density of hydrogen (mol/L), T represents the temperature of hydrogen (K), and p represents the pressure of hydrogen (MPa). a_{ij} and d are the parameters. When N = 5, the specific form of the polynomial EOS (Equation (16)) is given by

$$\begin{aligned} p = \; & a_{00}T^0\rho^0 + a_{01}T^0\rho^1 + a_{02}T^0\rho^2 + a_{03}T^0\rho^3 + a_{04}T^0\rho^4 + \\ & a_{05}T^0\rho^5 + a_{10}T^1\rho^0 + a_{11}T^1\rho^1 + a_{12}T^1\rho^2 + a_{13}T^1\rho^3 + \\ & a_{14}T^1\rho^4 + a_{20}T^2\rho^0 + a_{21}T^2\rho^1 + a_{22}T^2\rho^2 + a_{23}T^2\rho^3 + \\ & a_{30}T^3\rho^0 + a_{31}T^3\rho^1 + a_{32}T^3\rho^2 + a_{40}T^4\rho^0 + a_{41}T^4\rho^1 + a_{50}T^5\rho^0 + d \end{aligned} \tag{17}$$

The hydrogen temperature varied from 223.15 K to 373.15 K in increments of 2 K, while the hydrogen pressure varied from 0.1 MPa to 100.1 MPa in increments of 2 MPa. Different combinations of temperature and pressure were employed, resulting in 3876 distinct hydrogen temperature–pressure data pairs. Calculations based on the NIST database were performed to obtain 3876 corresponding sets of (p, T, ρ) data. These data were used to fit the polynomial EOS (Equation (17)), and the parameter values are presented in Table 5.

Table 5. Parameter values in the polynomial equation of state.

Parameter Values (R^2 = 0.998)				
a_{00}: 0	a_{01}: -1.185×10^2	a_{02}: 3.744	a_{03}: -8.399×10^{-2}	a_{04}: 1.157×10^{-3}
a_{05}: -6.779×10^{-6}	a_{10}: -3.512×10	a_{11}: 1.325	a_{12}: -2.807×10^{-2}	a_{13}: 3.811×10^{-4}
a_{14}: -2.353×10^{-6}	a_{20}: 2.094×10^{-1}	a_{21}: -5.506×10^{-3}	a_{22}: 7.174×10^{-5}	a_{23}: -4.348×10^{-7}
a_{30}: -6.217×10^{-4}	a_{31}: 1.026×10^{-5}	a_{32}: -6.150×10^{-8}	a_{40}: 9.188×10^{-7}	a_{41}: -7.177×10^{-9}
a_{50}: -5.408×10^{-10}	d: 2.346×10^3			

3.4. Pressure in Hydrogen Storage Tank Calculated by Different Equations of State

Based on the validated 0D1D model described above, simulation studies were conducted using the van der Waals EOS, Redlich–Kwong EOS, the polynomial EOS, and the NIST database-based method. For all cases, the refueling time was set to 3 min, which is the recommended refueling time for light-duty fuel cell vehicles according to the SAE J2601 hydrogen refueling protocol. The calculation results indicate that the choice of EOS has a minor impact on the computed hydrogen temperature but a significant effect on the computed hydrogen pressure.

Figure 3 compares the hydrogen pressure calculated using different forms of the EOS. Figure 3a shows that when using the final hydrogen pressure calculated based on the NIST database as the benchmark, the relative error of the final hydrogen pressure calculated by the polynomial EOS is 0.30%, while it is 1.83% for the Redlich–Kwong EOS, and 17.90% for the van der Waals EOS. This indicates that the polynomial EOS aligns most closely with the NIST-based calculations, whereas the van der Waals EOS exhibits lower accuracy. The Redlich–Kwong EOS, a modified form of the van der Waals EOS, demonstrates higher accuracy than the van der Waals EOS, suggesting that the modifications have improved upon the original van der Waals EOS to some extent.

Figure 3b shows a similar trend to Figure 3a, but the final hydrogen pressure in Figure 3b is higher. The relative error between the van der Waals EOS and the NIST data is 34.57%, which is greater than the error observed in Figure 3a. This indicates that as the hydrogen pressure inside the tank increases, the deviation in the pressure calculated using the van der Waals EOS becomes more significant. In summary, due to the high hydrogen pressures encountered during the refueling process, using the NIST database-based calculations or the polynomial EOS can help reduce computational errors. Although the polynomial EOS offers high accuracy, it involves more coefficients and a relatively complex form. The Redlich–Kwong EOS provides relatively high accuracy with fewer parameters and a simpler form.

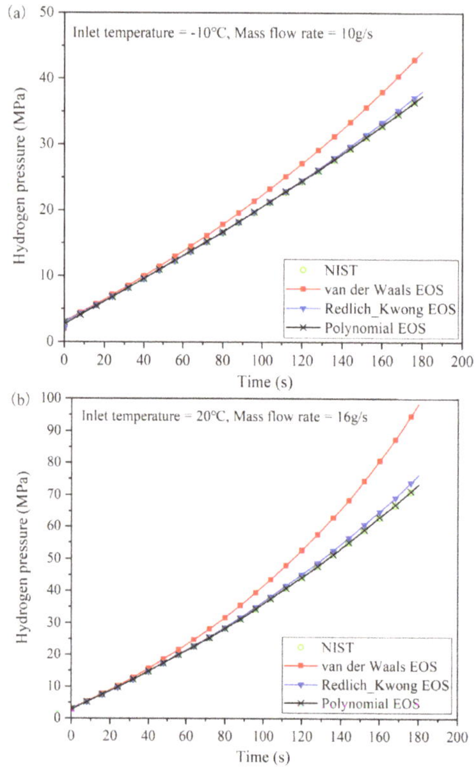

Figure 3. Hydrogen pressure calculated using different EOSs under varying inlet temperatures and mass flow rates. (**a**) Inlet temperature of −10 °C, mass flow rate of 10 g/s. (**b**) Inlet temperature of 20 °C, mass flow rate of 16 g/s.

4. Conclusions

This study established and validated a zero-dimensional gas and one-dimensional tank wall model (0D1D) for a hydrogen storage tank. The parameter values in the van der Waals equation of state (EOS) and its modified forms (Redlich–Kwong, Soave, and Peng–Robinson) were calculated using mathematical methods. The accuracy of the modified forms for van der Waals EOS was examined using the Joule–Thomson inversion curve. A polynomial EOS was constructed, and the values of its parameters were determined. The hydrogen temperature and pressure calculated by the van der Waals EOS, the Redlich–Kwong EOS, and the polynomial EOS were compared against those calculated based on the NIST database. Within the scope of the initial conditions and boundary conditions set in this study, the results are as follows:

(1) A lumped-parameter thermodynamic model integrating a zero-dimensional gas model and a one-dimensional tank wall model (0D1D) was developed for simulating the hydrogen refueling process. The model demonstrates high accuracy, as validated through comparative analysis with experimental data.

(2) Among the modified forms for van der Waals EOS (Redlich–Kwong, Soave, and Peng–Robinson), the fitting curve of the Redlich–Kwong EOS more closely approximates the Joule–Thomson inversion curve, implying that the Redlich–Kwong EOS exhibits higher accuracy than the Soave and Peng–Robinson EOSs.

(3) Using the final hydrogen pressure calculated by the NIST database as the benchmark, the relative error of the polynomial EOS is 0.30%, followed by 1.83% for the Redlich–Kwong EOS, and 17.90% for the van der Waals EOS. This indicates that the polynomial EOS exhibits higher accuracy, followed by the Redlich–Kwong EOS, while the van der Waals EOS demonstrates lower accuracy.

This research provides a theoretical basis for selecting an appropriate EOS in the numerical simulation of the hydrogen refueling process. Moreover, the van der Waals EOS and its modified forms (Redlich–Kwong, Soave, Peng–Robinson) in this article are all mean-field EOSs. However, hydrogen in practical settings may operate near metastable or near-spinodal conditions, where mean-field EOS become unreliable and fluctuations materially affect dynamics and control phase behavior [28]. Future work will incorporate intermolecular interactions details and thermal hydrodynamic fluctuations to extend the applicability of the current model.

Author Contributions: Conceptualization, R.C. and J.X.; methodology, T.Y. and C.L.; software, X.W.; validation, Q.X.; formal analysis, C.L.; investigation, Q.X.; resources, C.Y. and J.X.; data curation, Q.X.; writing—original draft preparation, H.L. and Q.X.; writing—review and editing, H.L. and T.Y.; visualization, X.W.; supervision, C.Y.; project administration, H.L. and X.W.; funding acquisition, H.L. and X.W. All authors have read and agreed to the published version of the manuscript.

Funding: This research is funded by the Ministry of Education of the People's Republic of China (Fund name: 2024 China University–Industry Research Innovation Fund. Fund number: 2024MZ015).

Institutional Review Board Statement: Not applicable.

Informed Consent Statement: Not applicable.

Data Availability Statement: The data are contained within this article.

Conflicts of Interest: The authors declare no conflicts of interest.

Abbreviations

The following abbreviations are used in this manuscript:

EOS	Equation of state
NWP	Nominal working pressure
NIST	National Institute of Standards and Technology
NASA	National Aeronautics and Space Administration
PRR	Pressure ramp rate
SAE	Society of Automotive Engineers
0D1D	Zero-dimensional gas and one-dimensional tank wall

References

1. Li, Q.; Zhang, Q.; Zhang, L.; Lang, J.; Yuan, W.; An, G.; Lei, T.; Yan, J. A comprehensive review of advances and challenges of hydrogen production, purification, compression, transportation, storage and utilization technology. *Renew. Sustain. Energy Rev.* **2026**, *226*, 116196. [CrossRef]
2. Al Kareem, S.S.A.; Hassan, Q.; Fakhruldeen, H.F.; Hanoon, T.M.; Jabbar, F.I.; Algburi, S.; Khalaf, D.H. A review on physical and chemical hydrogen storage methods for sustainable energy applications. *Unconv. Resour.* **2025**, *8*, 100235. [CrossRef]
3. Chen, H.; Zheng, J.; Xu, P.; Li, L.; Liu, Y.; Bie, H. Study on real-gas equations of high pressure hydrogen. *Int. J. Hydrogen Energy* **2010**, *35*, 3100–3104. [CrossRef]
4. Kontogeorgis, G.M.; Liang, X.; Arya, A.; Tsivintzelis, I. Equations of state in three centuries. Are we closer to arriving to a single model for all applications? *Chem. Eng. Sci. X* **2020**, *7*, 100060. [CrossRef]
5. Nasrifar, K. Comparative study of eleven equations of state in predicting the thermodynamic properties of hydrogen. *Int. J. Hydrogen Energy* **2010**, *35*, 3802–3811. [CrossRef]
6. Bai-gang, S.; Dong-sheng, Z.; Fu-shui, L. A new equation of state for hydrogen gas. *Int. J. Hydrogen Energy* **2012**, *37*, 932–935. [CrossRef]

7. Peng, Y.; Long, X. A new simplified virial equation of state for high temperature and high pressure gas. *AIP Adv.* **2022**, *12*, 015119. [CrossRef]
8. Bilgili, M.; Yumşakdemir, R.F. Effects of real gas equations on the fast-filling process of compressed hydrogen storage tank. *Int. J. Hydrogen Energy* **2024**, *53*, 816–828. [CrossRef]
9. Zheng, J.; Guo, J.; Yang, J.; Zhao, Y.; Zhao, L.; Pan, X.; Ma, J.; Zhang, L. Experimental and numerical study on temperature rise within a 70 MPa type III cylinder during fast refueling. *Int. J. Hydrogen Energy* **2013**, *38*, 10956–10962. [CrossRef]
10. Xiao, J.; Bénard, P.; Chahine, R. Charge-discharge cycle thermodynamics for compression hydrogen storage system. *Int. J. Hydrogen Energy* **2016**, *41*, 5531–5539. [CrossRef]
11. Xiao, J.; Wang, X.; Zhou, X.; Bénard, P.; Chahine, R. A dual zone thermodynamic model for refueling hydrogen vehicles. *Int. J. Hydrogen Energy* **2019**, *44*, 8780–8790. [CrossRef]
12. Rothuizen, E.; Mérida, W.; Rokni, M.; Wistoft-Ibsen, M. Optimization of hydrogen vehicle refueling via dynamic simulation. *Int. J. Hydrogen Energy* **2013**, *38*, 4221–4231. [CrossRef]
13. Bourgeois, T.; Ammouri, F.; Weber, M.; Knapik, C. Evaluating the temperature inside a tank during a filling with highly-pressurized gas. *Int. J. Hydrogen Energy* **2015**, *40*, 11748–11755. [CrossRef]
14. Luo, H.; Xiao, J.; Bénard, P.; Chahine, R.; Yang, T. Effects of filling strategies on hydrogen refueling performance. *Int. J. Hydrogen Energy* **2024**, *51*, 664–675. [CrossRef]
15. Waals, J.D.v.d. Molekulartheorie eines Körpers, der aus zwei verschiedenen Stoffen besteht. *Z. Phys. Chem.* **1890**, *5U*, 133–173. [CrossRef]
16. Redlich, O.; Kwong, J.N.S. On the Thermodynamics of Solutions. V. An Equation of State. Fugacities of Gaseous Solutions. *Chem. Rev.* **1949**, *44*, 233–244. [CrossRef] [PubMed]
17. Soave, G. Equilibrium constants from a modified Redlich-Kwong equation of state. *Chem. Eng. Sci.* **1972**, *27*, 1197–1203. [CrossRef]
18. Peng, D.-Y.; Robinson, D.B. A New Two-Constant Equation of State. *Ind. Eng. Chem. Fundam.* **1976**, *15*, 59–64. [CrossRef]
19. Benedict, M.; Webb, G.B.; Rubin, L.C. An Empirical Equation for Thermodynamic Properties of Light Hydrocarbons and Their Mixtures, I. Methane, Ethane, Propane and n-Butane. *J. Chem. Phys.* **1940**, *8*, 334–345. [CrossRef]
20. Lee, B.I.; Kesler, M.G. A generalized thermodynamic correlation based on three-parameter corresponding states. *AIChE J.* **1975**, *21*, 510–527. [CrossRef]
21. Leachman, J.; Jacobsen, R.; Penoncello, S.; Lemmon, E.W. Fundamental Equations of State for Parahydrogen, Normal Hydrogen, and Orthohydrogen. *J. Phys. Chem. Ref. Data* **2009**, *38*, 721–748. [CrossRef]
22. Kunz, O.; Wagner, W. The GERG-2008 Wide-Range Equation of State for Natural Gases and Other Mixtures: An Expansion of GERG-2004. *J. Chem. Eng. Data* **2012**, *57*, 3032–3091. [CrossRef]
23. Wagner, W.; Pruß, A. The IAPWS Formulation 1995 for the Thermodynamic Properties of Ordinary Water Substance for General and Scientific Use. *J. Phys. Chem. Ref. Data* **2002**, *31*, 387–535. [CrossRef]
24. Lemmon, E.W.; Bell, I.H.; Huber, M.L.; McLinden, M.O. *NIST Standard Reference Database 23: Reference Fluid Thermodynamic and Transport Properties-REFPROP, Version 10.0*; National Institute of Standards and Technology: Gaithersburg, MD, USA, 2018. [CrossRef]
25. Sears, F.W.; Salinger, G.L. *Thermodynamics, Kinetic Theory, and Statistical Thermodynamics*; Addison-Wesley Publishing Company: Boston, MA, USA, 1975; Available online: http://narosa.com/books_display.asp?catgcode=978-81-85015-71-2 (accessed on 10 September 2025).
26. Hendricks, R.C.; Peller, I.C.; Baron, A.K. *Joule-Thomson Inversion Curves and Related Coefficients for Several Simple Fluids*; NASA Technical Note D-6807; National Aeronautics and Space Administration: Washington, DC, USA, 1972; Available online: https://ntrs.nasa.gov/api/citations/19720020315/downloads/19720020315.pdf (accessed on 10 September 2025).
27. Bourgeois, T.; Ammouri, F.; Baraldi, D.; Moretto, P. The temperature evolution in compressed gas filling processes: A review. *Int. J. Hydrogen Energy* **2018**, *43*, 2268–2292. [CrossRef]
28. Lulli, M.; Biferale, L.; Falcucci, G.; Sbragaglia, M.; Yang, D.; Shan, X. Metastable and unstable hydrodynamics in multiphase lattice Boltzmann. *Phys. Rev. E* **2024**, *109*, 045304. [CrossRef]

applied sciences

MDPI

Article

Power Smoothing in a Wave Energy Conversion Using Energy Storage Systems: Benefits of Forecasting-Enhanced Filtering for Reduction in Energy Storage Requirements

Marcos Blanco [1], Luis Mazorra [2], Isabel Villalba [2], Gustavo Navarro [1], Jorge Nájera [1] and Marcos Lafoz [1,*]

1 Centro de Investigaciones Energéticas, Medioambientales y Tecnológicas (CIEMAT), Avda. Complutense, 40, 28040 Madrid, Spain

2 Electric Engineering Department, Universidad de Las Palmas de Gran Canaria (ULPGC), C/ Juan de Quesada 30, 35001 Las Palmas de Gran Canaria, Spain

* Correspondence: marcos.lafoz@ciemat.es

Featured Application: This work focuses on the integration of wave energy converters (WECs) with energy storage systems (ESS) for grid-connected renewable energy applications. By leveraging ESS, the approach reduces the impact of power fluctuations inherent to ocean waves, enhancing the reliability of energy delivery. The proposed forecasting-enhanced filtering strategies can be applied to WEC farms with ESS and allow for optimized (reduced) ESS sizing while maintaining equivalent levels of power smoothing. It is worth mentioning that this approach is inherently scalable and its effectiveness is not limited by the size of the farm. In fact, as the number of WEC units increases, the natural cancellation of power fluctuations between units can further reduce storage requirements, even without enhanced forecasting strategies, making the method applicable to larger-scale deployments.

Abstract

This paper presents a power smoothing strategy for wave energy converters (WECs) by means of energy storage systems (ESS) with integrated forecasting filtering algorithms applied to their control. The oscillatory nature of wave energy leads to high variability in power output, posing significant challenges for grid integration. A case study in Tenerife, Spain, was modeled in MATLAB-Simulink (release r2020b) to evaluate the impact of prediction-enhanced smoothing filters on ESS sizing. Various forecasting algorithms were assessed, including Bayesian Neural Networks, ARMA models, and persistence models. The simulation results demonstrate that the use of forecasting algorithms substantially reduces energy storage requirements while maintaining grid stability. Specifically, the application of Bayesian Neural Networks reduced the required ESS energy by up to 36.52% compared to traditional filters. In a perfect prediction scenario, reductions of up to 53.91% were achieved. These results highlight the importance of combining appropriate filtering strategies with advanced forecasting techniques to improve the technical and economic viability of wave energy projects. The paper concludes with a parametric analysis of moving average filter windows and prediction horizons, identifying the optimal combinations for different sea conditions. In summary, this study provides practical information into reducing the storage capacity required for power smoothing in wave energy systems, thereby contributing to the mitigation of grid integration challenges that may arise with the large-scale deployment of marine renewable energy.

Keywords: wave energy; energy storage systems; forecasting filtering; Bayesian neural networks; grid integration

1. Introduction

The global potential of wave energy remains highly significant and largely untapped, with an estimated 2.1 TWh available in open waters [1]. This is one of the main reasons for the growing interest in the field, both at academic and industrial levels. In fact, wave energy represents a key component in strategies for the energy transition and sustainable development, as highlighted in the EU Blue Economy Report 2025 [2]. Several studies have examined the influence of technical, economic, and political aspects [3,4] in enabling wave energy to become a feasible option within the global energy generation mix in the near future. Such interest has led to substantial funding efforts; for instance, European R&D programs in ocean energy were estimated to have reached €0.85 billion by 2018 [5]. By 2050, ocean energy is projected to contribute up to 100 GW of installed capacity, equivalent to approximately 10% of the European electricity consumption in 2019 [6]. Beyond electricity, it is also worth mentioning that the combination of water resources and wave energy is a practical reality. The proximity between these resources has prompted studies on seawater desalination processes powered directly by wave energy, demonstrating the potential for integrating marine renewable energy into water management applications [7–9]. These potentials is particularly relevant within the European decarbonization framework, as achieving the 2050 climate targets will require zero-emission electricity generation.

To reach this objective, demand management, energy storage, and the deployment of all flexible renewable energy sources—among them ocean energy—will be necessary [10]. In this context, the Strategic Research and Innovation Agenda (SRIA) on ocean energy [11] has identified several priority topics and associated actions, such as "Demonstrate delivery of grid-compliant power including short-term energy storage solutions to smooth power output when needed." This priority is directly linked to the grid integration challenges arising from the fluctuating power output of Wave Energy Converters (WECs). Due to the oscillatory nature of ocean waves, the power generated typically exhibits high peak-to-average ratios, with oscillation periods on the order of several seconds, and may include zero crossings during each cycle [12–14]. Although a proper spatial distribution of WECs in a wave farm can reduce power oscillations, this measure alone is insufficient to fully overcome grid integration issues [15–17]. Such issues stem from the fact that oscillations in power generation lead to transient imbalances between supply and demand in the electrical grid, which then trigger corrective mechanisms [18–20]. On the one hand, these oscillations cause deviations in grid frequency due to the instantaneous mismatch between generated and consumed power. These deviations must remain within the limits defined by each grid code. Moreover, for a given power imbalance, frequency oscillations tend to be more severe in weaker electrical grids with lower system inertia [21]. Island systems appear to be particularly suitable for wave energy deployment, given their proximity to the resource and the typically higher electricity prices compared to continental systems [22]. However, these island grids are often electrically weak due to their smaller scale, which exacerbates the impact of power oscillations [23]. On the other hand, the activation of frequency regulation mechanisms enhances conventional generators (e.g., thermal or hydroelectric plants) to operate more frequently, increasing mechanical fatigue and component wear [19,24]. In such contexts, rapid-response energy storage solutions can play a crucial role by absorbing short-term imbalances, thereby supporting grid stability and reducing mechanical stress on conventional generators. Therefore, to address these issues, the integration of energy storage systems—either standalone or hybrid configurations—into wave energy generation systems has been proposed [25]. Beyond the technical role of the energy storage systems, these systems are increasingly recognized as key enablers of the global energy transition, with non-negligible implications for sustainability and environmental performance. The deployment of these technologies must therefore be assessed not only in terms of operational efficiency, but also with respect to their lifecycle impacts, including resource extraction, manufacturing, and recycling [26,27].

These storage systems can mitigate grid integration challenges by implementing power smoothing strategies, typically based on filtering algorithms [28,29]. Preliminary studies have shown that using forecasting filtering algorithms can reduce the required storage capacity, and that fast-response storage technologies such as supercapacitors [30,31] or flywheels (possibly complemented by battery systems to handle intra-day variations in hybrid storage systems) are particularly well-suited for this purpose [32]. Additionally, fast-response storage can alleviate battery aging and reduce the wear of conventional power plants involved in frequency regulation [33]. While several studies have addressed the development of forecasting algorithms for conventional renewables [34,35], similar efforts exist for wave energy systems as well. However, in this domain, forecasting strategies typically aim at enhancing energy-extraction control algorithms by forecasting force-related variables [36–38] over short time horizons (few seconds). Some studies have addressed wave resource forecasting using Long Short-Term Memory (LSTM) neural networks combined with modal decomposition techniques [39], have focused on predicting power output through adjusted power matrices and Bi-directional LSTM (BiLSTM) architectures combined with Convolutional Neural Networks (CNNs) [40] or focused on predicting sea conditions in order to plan actions on offshore systems [41].

The algorithm proposed in this work, however, is based on Bayesian Neural Networks and focuses on forecasting the power output, with time horizons ranging from a few seconds up to the typical wave period (8–14 s). Regarding the application of forecasting algorithms in renewable energies such as solar and wind, although primarily aimed at improving grid integration, the focus is typically on longer time horizons, centered on resource forecasting for participation in electricity markets, plant planning, and operation, where "short-term" horizons are generally defined as periods below 4 h [42,43]. On the other hand, while there are also applications of long-term resource forecasting in wave energy [44], the main focus has been on maximizing the energy extracted from waves, given the non-causal nature of these systems, which operate over very short horizons of only a few seconds (i.e., below 4 s) [45–47]. Additionally, some studies have explored power smoothing in wave energy systems; however, these are mainly preliminary works addressing different storage technologies and proof-of-concept analyses of the capability of ESS to smooth power output, considering batteries [48], supercapacitors [31], or hybrid systems combining both [49]. It is worth noting that a few studies [50] have applied forecasting algorithms to wave energy power smoothing, although in those cases the objective was mainly to limit the generated power to reduce oscillations while controlling the ESS state of charge. In contrast, the present work specifically develops a forecasting-based algorithm for power smoothing and investigates the ability of forecasting strategies to reduce ESS sizing requirements within this control framework, thereby justifying the integration of forecasting and demonstrating its practical design advantages.

The methodology builds on previous work conducted to (1) design a suitable WEC for a specific location (northern coast of Tenerife, Canary Islands, Spain) [51]; (2) optimize the spatial layout of WECs within a wave farm to minimize power oscillations [52]; and (3) evaluate the grid impact of the wave farm's integration, including an assessment of the maximum admissible power oscillations for the electrical system [23]. Based on these foundations and using the same case study, a power prediction algorithm is developed and integrated into a power smoothing strategy. A comparative analysis is carried out involving simpler computational approaches—such as persistence and ARMA models—in order to justify the use of neural networks. Finally, results are presented comparing scenarios with and without forecasting algorithms, showing the corresponding power and energy requirements for the storage systems as well as the achieved power smoothing performance.

2. Methodology

The mathematical model used to analyze the case study has been developed using MATLAB-Simulink software (release r2020b). The model consists of the following components:

- A WEC model (prime mover + PTO), which allows for the evaluation of the electrical power output profile of a single WEC under a given sea state. This model accounts for the relative position of the WEC with respect to other devices and for the directional characteristics of the incident waves. This enables the aggregation of individual power profiles to determine the total output of a wave energy farm (see Section 2.1).
- An energy storage system model, coupled with a power smoothing control algorithm. In this case study, the storage system is modeled as an "agnostic" energy storage unit, and a moving average filter is used as the smoothing algorithm (see Section 2.2).
- A forecasting algorithm, which complements the power smoothing control algorithm of the storage system by enabling the estimation of the real-time moving average filter (MAF) using past power output samples and predicted values as future samples (see Section 2.3).

2.1. Description of the WEC and Wave Farm Model

This work is framed within a case study focused on the integration of a wave energy farm (WF) into the electrical grid of an island power system. The wave energy installation is assumed to be located on the northern coast of Tenerife and is electrically connected to the island's grid. Several projects have been developed in this region [53–55] due to its significant wave energy potential [56], estimated within the range of 20–70 kW/m. The main characteristics of the WF are summarized as follows:

- Ocean wave data are obtained as described in Section 2.1.1, considering the WF location at latitude 28.4° N and longitude 17° W.
- The primary WEC device is a point absorber with a peak power output of 160 kW. The Power Take-Off (PTO) system consists of a direct-drive linear electric generator rated at 160 kW.
- The WF is composed of 15 groups of WEC units arranged in 9 rows following a triangular layout.

The complete mathematical model of the wave farm (denoted as WFM) is parameterized according to the characteristics of the WEC device (Section 2.1.2) and the spatial layout of the WF (Section 2.1.3). The model outputs the electric power profile based on the ocean resource conditions provided as input (Section 2.1.1).

2.1.1. Wave Resource Conditions

As previously mentioned, this case study is based on Tenerife Island, one of the seven main islands of the Canary Islands (Spain). The metocean data used are derived from the numerical dataset corresponding to the so-called SIMAR points [57]—specifically point 1012013 (see location in Figure 1a and wave contingency diagram in Figure 1b).

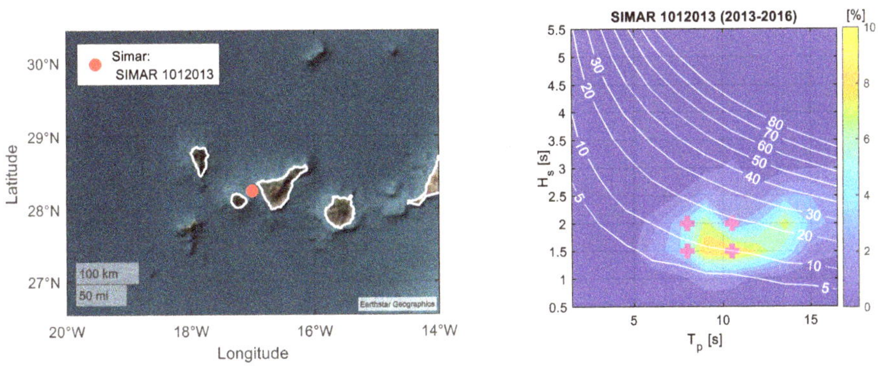

(a) (b)

Figure 1. (**a**) Location of the WF (source: http://www.puertos.es/es-es/oceanografia/Paginas/portus.aspx—accessed in 1 September 2025); (**b**) wave contingency diagram showing the annual occurrence of various sea states, characterized by peak period (T_p) and significant wave height (H_s). The four sea states selected and detailed in Table 1 are highlighted with magenta cross markers, and the corresponding wave power per unit of wave front (Pj [14]) is indicated by contour lines in white.

Table 1. Wave conditions considered in this study.

No. Case	H_s [m]	T_p [s]	α_W [°]	P_j [kW/m]
CASE 1	1.5000	8.0000	0	7.63
CASE 2	1.5000	10.5000	0	10.02
CASE 3	2.0000	8.0000	0	13.57
CASE 4	2.0000	10.5000	0	17.81
CASE 5	1.5000	8.0000	15	7.63
CASE 6	1.5000	10.5000	15	10.02
CASE 7	2.0000	8.0000	15	13.57
CASE 8	2.0000	10.5000	15	17.81
CASE 9	1.5000	8.0000	30	7.63
CASE 10	1.5000	10.5000	30	10.02
CASE 11	2.0000	8.0000	30	13.57
CASE 12	2.0000	10.5000	30	17.81
CASE 13	1.5000	8.0000	45	7.63
CASE 14	1.5000	10.5000	45	10.02
CASE 15	2.0000	8.0000	45	13.57
CASE 16	2.0000	10.5000	45	17.81

The relevant information available includes the significant wave height (H_s [14]), the peak period of the wave energy spectrum (T_p [14]), and the main wave propagation direction (α_W). Figure 1b presents the annual occurrence of various sea states, characterized by their peak period (Tp) and significant wave height (Hs). The four sea states selected—based on T_p and H_s—in Table 1, correspond to those most significant in terms of occurrence, wave energy, and suitability with the WEC device (which is detailed in the following subsection), providing a representative basis for the analysis of wave energy potential and storage requirements. Based on several years of data, 16 representative sea states for the present analysis are selected, and a synthetic wave elevation profile is generated across the spatial domain of the WEC farm. This profile accounts for H_s, T_p, and α_W, as well as the frequency spectrum and directional spreading of the wave energy [14]. The 16 selected cases correspond to the permutations of two values for H_s, two for T_p, and four for α_W, as summarized in Table 1. Additionally, to provide context and facilitate interpretation, Table 1 presents the wave energy per unit wave front (P_j [14]).

2.1.2. WEC Model

The WEC model corresponds to a two-body point absorber whose dimensions result from an optimization procedure aimed at maximizing the extracted energy for the selected location, following the methodology described in [51]. The dimensions of the device are shown in Figure 2. In [51], a differential evolution optimization algorithm was employed to determine the dimensions that maximize the energy captured by the WEC's prime mover (i.e., the components interacting directly with the waves, in this case, the two bodies of the point absorber). The optimization considered the site-specific wave conditions and the design constraints of the power take-off system—PTO—(i.e., the device converting mechanical energy from the prime mover into electrical energy, here a linear generator), including limits on speed, stroke, force, and nominal power. This process resulted in a WEC configuration with a peak power of 160 kW, corresponding to the rated power of the selected PTO.

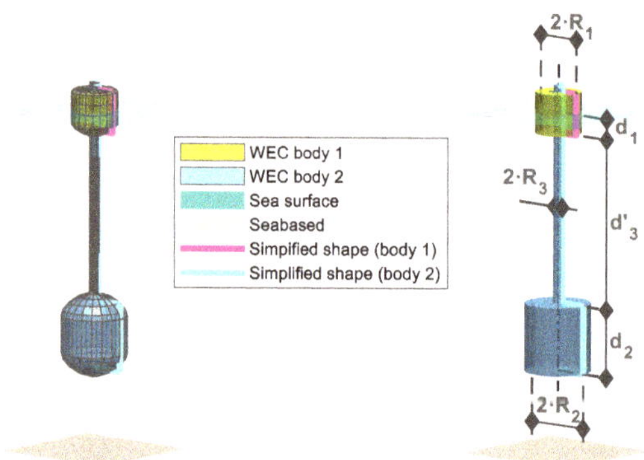

Dimension	Value [m]
$2{\cdot}R_1$	7.4
$2{\cdot}R_2$	16.76
$2{\cdot}R_3$	1.5
d_1	2.03
d_2	31.2
d_3	0.1

Figure 2. Diagram of the modeled point absorber.

The dynamic model is parameterized using the hydrodynamic coefficients associated with these dimensions, obtained through the NEMOH software (v3.0.0) [58]. The simulation is performed in MATLAB-Simulink (release r2020b) using the WECSIM library (v7.0.0) [59], which is based on a state-space realization of the hydrodynamic parameters in the frequency domain, as described in [60], and the implementation of a dynamic model similar to that in [61,62].

The model includes a loss model for the linear generator [63], and the energy extraction control algorithm is implemented in the time domain using a Gain Scheduling approach [64]. This approach involves determining, through simulations, the optimal damping factor that the linear generator must apply to maximize energy extraction under different sea states. A lookup table is then used in the time-domain simulation to apply the appropriate damping factor based on the current sea state.

2.1.3. Wave Farm Model

The wave farm model builds upon the results from the WEC model described in Section 2.1.2. A spacing between WEC units is assumed to be sufficient to neglect hydrodynamic interaction among them, while considering the position of each device and the spatial variability of the wave resource.

The layout of each device has been defined to minimize the power oscillations of the total farm output, resulting in a farm composed of 50 units arranged in a triangular (staggered) pattern according to the results presented in [52] (see Figure 3).

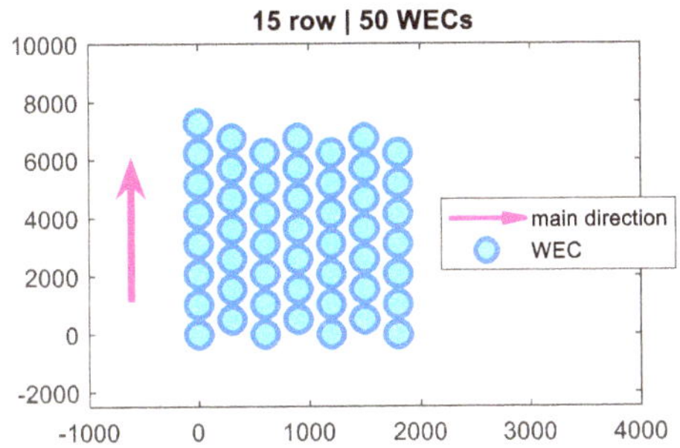

Figure 3. Spatial layout scheme of the WEC devices.

2.2. Description of the Energy Storage System Model

The wave farm (WF) is assumed to include an energy storage system (ESS) designed to smooth the oscillating power output generated by the WF. Both the connection layout and control scheme of the ESS are shown in Figure 4.

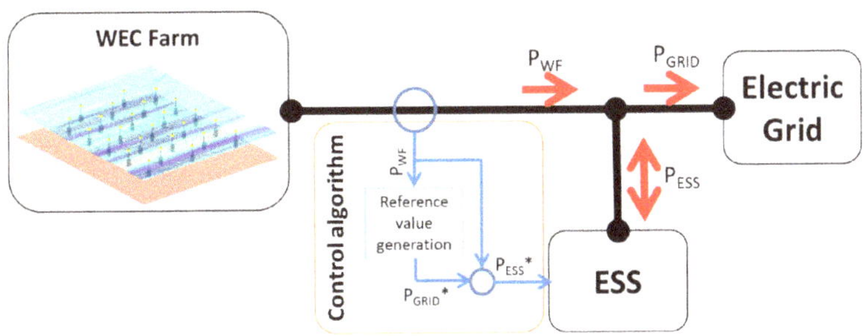

Figure 4. Connection and control scheme of the energy storage system (ESS).

The ESS is connected at the point of common coupling (PCC) of the WF and it is assumed to be installed onshore to reduce maintenance requirements. The ESS control reference is generated according to the scheme presented in Figure 4. In this scheme, the target power to be injected into the grid ($P_{GRID}*$) is obtained by filtering the power generated by the WF (P_{WF}), while the ESS target power ($P_{ESS}*$) is defined as the difference between the generated power and the filtered power to be injected into the grid, as expressed in Equation (1). The moving average filter (MAF) operates within a fixed time window of a number of samples of Ns_MAF, incorporating both measured (real) samples (number of real samples: $Ns\_MAF - Ns\_PRED$) and predicted samples (number of predicted samples: $nPRED$) provided by different forecasting algorithms (described in Section 2.3), as formulated in Equation (2). MAF evaluates the average value of a certain variable in a Ns_MAF number of samples.

$$P_{ESS} = P_{GRID} - P_{WF} \tag{1}$$

$$\begin{gathered} P_{ESS} = P_{ESS}* = P_{GRID} * - P_{WF} = MAF(P_{WF}) - P_{WF}; \\ P_{GRID} * (t_0) = MAF(P_{WF}) = \frac{\sum past\ samples + \sum predicted\ samples}{total\ no.\ samples} \\ = \frac{\sum_{i=0}^{i=Ns\_MAF-Ns\_PRED} P_{WF}(t_o - i\cdot\Delta t) + \sum_{i=1}^{i=Ns\_PRED} P_{WF}(t_o + i\cdot\Delta t)}{Ns\_MAF} \end{gathered} \tag{2}$$

where P_{ESS} is the power exchanged by the storage system, $P_{GRID}{}^*$ is the smoothed power delivered to the grid, P_{WF} is the instantaneous wave farm power, Ns_MAF is the total number of samples in the moving average filter window, and Ns_PRED is the number of forecasted samples included in the filter (i.e., the number of real or non-predicted samples is

$Ns\_MAF - Ns\_PRED$). In addition, MAF(x) denotes the application of the moving average filter to the variable x, x(t_i) represents the value of the variable x at time instant t_i (e.g., $P_{WF}(t_0 + i \cdot \Delta t)$ corresponds to the value of the wave farm power at $t_0 + i \cdot \Delta t$), Δt is the sampling time, t_0 is the evaluation instant, and i is the sample index

In order to define window sizes, the variables *nMAF* and *NPred* are introduced too, which correspond to window sizes expressed in seconds for the filter and prediction, respectively, i.e., $nMAF = Ns\_MAF^{*}\Delta t$ and $nPred = Ns\_Pred^{*}\Delta t$.

The target grid-injected power is filtered using a moving average filter (MAF) with a window size (*nMAF*) of 30 s and a sampling period (Δt) of 0.1 s (i.e., a *nMAF* of 30 s correspond, in number of samples, with a *Ns_MAF* of 300 samples), in accordance with the parameters defined in the previous study presented in [23]. In that study, the impact of injecting oscillating power into the island grid was analyzed, a MAF-based solution was proposed, and its performance was evaluated using the aforementioned parameters.

From the ESS power demand profiles (P_{ESS}), the required power and energy capacities of the ESS (P_{rated} and E_{rated}, respectively) can be calculated for each specific case, according to Equations (3) and (4), evaluated in each simulation period.

$$P_{rated} = \max(\text{abs}(P_{ESS}(t))); \ \forall \text{ t of a complete simulation} \tag{3}$$

$$\begin{gathered} E_{ESS}(t) = \int_0^t P_{ESS}(t)\, dt; \\ E_{rated} = \max(E_{ESS}(t)) - \min(E_{ESS}(t)); \ \forall \text{ t of a complete simulation} \end{gathered} \tag{4}$$

In this study, the ESS is considered fully agnostic, with sufficient capacity such that its state of charge (SoC) does not affect its behavior. The calculations of P_{rated} and E_{rated} based on Equations (3) and (4) were performed assuming an ideal, technology-agnostic energy storage system (ESS) in order to compare the different filtering and forecasting strategies. This approach isolates the effects of short-horizon forecasts on the required storage capacity for a given level of power smoothing. However, it should be noted that differences may arise when applying the methodology to real-world systems or more detailed ESS models, but these are not expected to alter the main qualitative conclusions of the manuscript.

The analysis of storage technologies and SoC control algorithms is beyond the scope of this work, which is instead focused on implementing a forecasting algorithm aimed at reducing the capacity requirements of the ESS.

2.3. Forecasting Algorithms

This section will describe the forecasting models used in this article to evaluate the predicted samples of the power generated by the WF (P_{WF}), highlighting their application in system energy management. Storage models will draw on these predictions to enhance energy management. A range of statistical forecasting models will be used to identify the most appropriate model for each case study. The development of any sophisticated methodology is useless if certain naïve or reference models demonstrate superior performance to that proposed. The objective is to compare two naïve models with increasingly complex prediction models. Therefore, in this study, it is proposed to work with two ARMA-based models and one NN-based model. The more complex models should offer, a priori, better accuracy (as Figure 8 shown in Section 4.2). The general forecasting function that will be used to relate the input variables to the expected results is as shown in Equation (5).

$$\hat{y}(t+h) = F[y(t), \ \ldots, \ y(t-i)] \tag{5}$$

where $\hat{y}(t+h)$ is the P_{WF} calculated for time horizon h, in our case $h = 0.1, 1, \ldots, 8$ s, $y(t-i)$ is P_{WF} from dataset of each case of study for 'i' past values.

2.3.1. Naïve Models

In this study, two elementary models were examined for their forecasting capabilities across a range of time horizons. It was observed that both models utilised past values data, employing normalized data series for analysis. The first is the simple persistence (Pers) model [65] defined by Equation (6).

$$\hat{y}(t+h) = F[y(t), \ldots, y(t-i)] \tag{6}$$

This model is based on the assumption that the forecasted data for each time horizon 'h' is solely dependent on the previous value, implying that conditions remain unchanged between time 't' and time '$t + h$'. The next approach, known as the Smart-Persistence (Pers Smart) model, offers a simple enhancement to the persistence model. It predicts wave performance for each time horizon 'h' by averaging the 'h' preceding values [66] The Smart-Persistence model is mathematically defined in Equation (7).

$$\hat{y}(t+h) = mean(y(t), \ldots, y(t-h)) \tag{7}$$

2.3.2. Linear Models

For time series estimation, statistical linear models have been extensively developed. This study follows the approach outlined in [67,68] for climatic variables, utilizing a methodology based on linear regression models. Linear models operate using a combination of past values and errors. The autoregressive moving average (ARMA) time series model is based on two fundamental components: the autoregressive (AR) model and the moving average (MA) model. These models represent a linear combination of a specific number of past series values and errors, as defined by Equation (8).

$$y(t \hat{+} h) = \sum_{i=0}^{p-1} [\Phi_{i+1} y(t-1)] + \sum_{j=0}^{q-1} [\Theta_{j+1} \epsilon_{t-j}] \tag{8}$$

The forecasted variable for horizon 'h' is represented by $\hat{y}(t+h)$. The values of $y(t-i)$ correspond to past data points from the time series, selected to establish a linear relationship with the forecasted value, while ϵ_t represents the error term. The set of autoregressive parameters, $\{\Phi_i\}_{i=1,2,\ldots,p}$, is derived from the sample data during training for the AR regression using past values. Meanwhile, the parameters $\{\Theta_i\}_{i=1,2,\ldots,q}$ correspond to the series of errors ϵ_t in the MA regression. The parameter 'p' defines the order of the AR model, indicating the number of past values used for solar radiation forecasting. Similarly, 'q' represents the order of the MA model. Both model parameters are determined through least squares regression, which compares past data used as input with the future data to be predicted.

One of the most critical decisions during the training process is determining the model's complexity. For an ARMA model, the optimal values of p and q must be established during the training process, with the model being represented as ARMApq. In this case, the optimal order p is obtained by calculating the Partial Autocorrelation Function (PACF) and the Bayesian Information Criterion (BIC). Similarly, the optimal order q for the MA model is determined by calculating the Autocorrelation Function (ACF) and defining the number of error terms used in the prediction. In many cases, ARMA models with low-order parameters have been shown to produce very good results [69]. For example, in this study, an ARMA with optimal p and q values are considered (ARMApq), and also considers an ARMA model with p = 1 and q = 2 (ARMA12) for sake of comparison.

2.3.3. Artificial Neural Networks

Artificial neural networks (NN) [70] consist of interconnected units called neurons that receive inputs either from other neurons or directly from an external data source. These networks are designed to learn the relationship between input data and corresponding outputs. Each connection between neurons has an associated weight, which is adjusted during the training process. The optimal weights are determined by optimising a cost function.

The NN structure used in this article consists of an input layer, at least one hidden layer and an output layer, with no feedback or lateral connections between them. The hidden layer consists of several non-linear neurons. Each neuron processes input variables, weighted accordingly, and sums them to produce an output. This sum is then passed through a non-linear activation function or transfer function to control the output amplitude. The activation function used is the hyperbolic tangent function, Equation (9), while the output layer uses a linear activation function.

$$f(x) = \tanh(x) = \frac{e^x - e^{-x}}{e^x + e^{-x}} \tag{9}$$

The relation between input variables and forecasted data is established by Equation (10). The weights of the network are initially set by random initialization and then optimized during the training phase. This optimization is achieved by minimizing an error function (cost function) using the back-propagation algorithm. The most used error function is the mean squared error, which measures the difference between the network's output $\hat{y}(t+h)$ and the desired target output, Equation (11).

$$\hat{y}(t+h) = \alpha_0 + \sum_{j=1}^{h} \alpha_j \cdot f\left[\beta_{0,j} + \sum_{i=1}^{p} \beta_{i,j} \cdot y(t-i)\right] \tag{10}$$

$$E(w) = \frac{1}{2}\sum_{i=1}^{N} (\hat{y}(t+h) - y_i)^2 \tag{11}$$

The accuracy of artificial neural networks (NN) in approximating continuous functions is influenced by the structure of the network. If the network fits the noise in the function, it may perform well during training but poorly on new data, a problem known as overfitting. To avoid this, regularization techniques are used to control model complexity [70–72]. In this study, Bayesian regularization was used to control the complexity of the model. The Bayesian approach involves a probability density function over the weight space, where the optimal neural network parameters correspond to the maximum of this density. This framework introduces two new hyperparameters, α and β, which regulate the complexity of the model. Consequently, the cost function is defined as in Equation (12).

$$S(w) = \frac{\beta}{2}\sum_{i=1}^{N} (\hat{y}(t+h) - t_i)^2 + \frac{\alpha}{2}\sum_{i=1}^{m} {w_i}^2 \tag{12}$$

To determine the optimal values of the hyperparameters α and β, as well as the optimal weight vector w_{MP}, an iterative procedure is required [72,73]. Initially, the hyperparameters are obtained through the numerical implementation of the Bayesian framework using the computed values from the first iteration. These values are then used to compute the first group of optimal weights using the standard training algorithm, specifically the scaled conjugate gradient method, which minimizes $S(w)$.

The Bayesian framework offers several approaches to assess model complexity. One of these techniques, known as Automatic Relevance Determination (ARD), allocates a distinct regularization coefficient to each input, facilitating the identification of the most significant inputs during training. Furthermore, it calculates the model's probability (model evidence), which serves as a criterion for model selection [73,74]. In this study, we trained various NN with different numbers of hidden units, and the network with the highest model evidence was chosen as the best-performing one.

3. Case Study

As previously commented, power oscillation injected in the grid would have a direct impact in electric grid due to power oscillations cause frequency oscillations and the electric frequency cannot have variations beyond a narrow margin (a few hundreds of millihertz). And among other options of reducing the power oscillation in WEC farms [75], electric energy storage (ESS) is one of the better options.

To better contextualize the scope of the work, a representative case study has been selected and introduced at this section based on a previous studies [23,51,52], describing the new analysis to be carried out over it. The Tenerife Island grid was chosen due to its weak grid characteristics, which makes it particularly sensitive to power oscillations from renewable generation. In this context, the study illustrates the motivation for smoothing wave power generation and remarks on the role of ESS-based smoothing strategies. Specifically, the case study focuses on two main aspects: (1) power smoothing capacity in order to reduce the electric frequency oscillations in weak grids, (2) and rated power and energy requirements of the ESS depending on the adopted control algorithm. This provides a practical framework for comparing filtering and forecasting strategies and defines the parametric analysis whose results are shown in Section 4.

3.1. Application Example: Forecasting in Moving Average Filtering

By way of example, some results derived from the study case presented in [23] are shown in Figure 5. This study case framed on Tenerife Island (SPAIN) is considering the connection of a wave power generation farm of 120 MW (the modal value of the minimum power generation in Tenerife Island is 452.5 MW, which implies a maximum penetration of 25%). Figure 5a shows an example of the oscillatory electric power generated with a WEC farm in a particular sea state; and Figure 5b shows the electric frequency oscillations produced as a consequence.

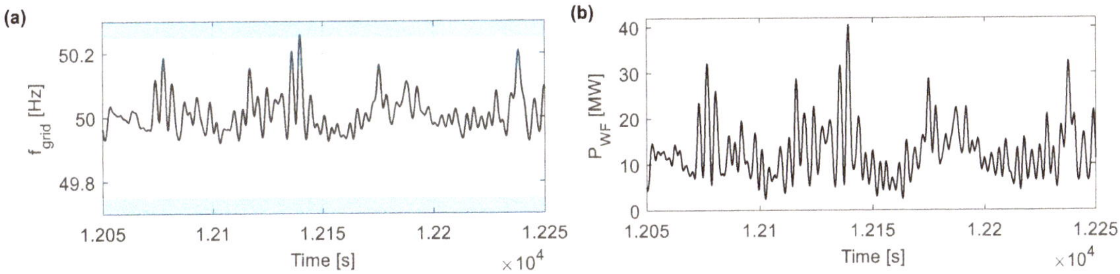

Figure 5. (**a**) electric frequency oscillations in the electric grid. These results correspond with an example of the complete study case presented in [23]—the frequency limits of ±150 mHz and ±205 mHz are highlighted in light blue and dark blue, respectively. (**b**) Electric power generated by the WEC farm.

As previously commented in Section 2.2, the power oscillations can be smoothed by means of ESS installed in the Point of Common Coupling (PCC). The power smoothing grade should reduce/avoid the number of events where the frequency oscillations are out of bounds (±150 mHz according to Spanish grid code [76]). The ESS smooths the power generation oscillations, so that the power injected into the network is a certain reference value (see Figure 4). The reference value is defined in order to reduce the frequency oscillations, and it can be calculated in different ways, resulting in different control algorithms. In this example, two control algorithms are proposed:

1. Moving average filter.
2. Moving average filter considering future (predicted/forecasted) values. In this example (Section 3.1), it is considered perfect prediction.

Figure 6 shows a comparison of the two ways to calculate reference values compared to the original wave power generation level. The area between these two curves (marked with a coloured area) represents the energy storage capacity required to transform the original generated power into the smoothed one proposed by the control algorithm. The figure clearly shows that the control algorithm selected has a great impact in the energy storage requirements due to the delay introduced by the MAF—of 30 s of window size and without prediction (nMAF = 30; nPred = 0)—increase the area between these two curves.

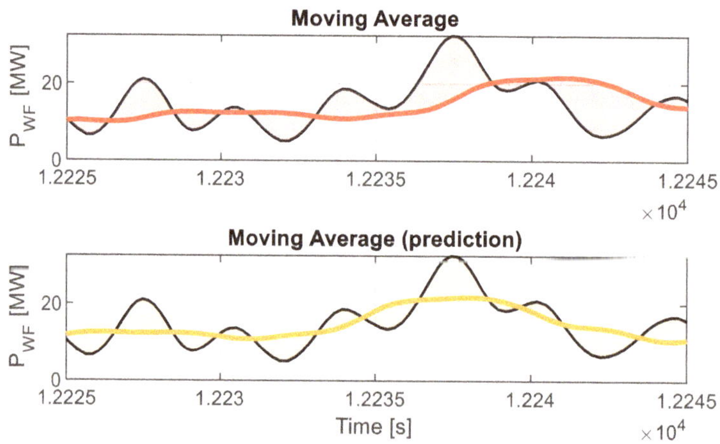

Figure 6. Example of the EES reference values calculated with the four described control algorithms.

Table 2 shows the results of the two control algorithm options in this example, including the ESS energy and power requirements evaluated by means of Equations (3) and (4), the number of over-frequency and under-frequency events, and the variance of the wave power generations and the electric frequency time profiles evaluated by means of Equation (5).

Table 2. Summary of the comparison between the four control algorithm options.

Control Algorithm	σ (P_{WEC})	σ (f_{grid})	No. Alert Events [$\Delta(f_{grid})$ > 150 mHz]	No. Emerg. Events [$\Delta(f_{grid})$ > 250 mHz]	ESS Rated Power [KW] (P_{ESS})	ESS Rated Energy [kW·h] (E_{ESS})
base case (no ESS)	4854.688	0.05011	340	14	0	0
MAF	3117.836	0.03170	7	0	2647	2.257
MAF with Pred.	3116.857	0.03169	7	0	2239.80	0.900

In Table 2, the frequency evaluation is derived from the case study presented in [23]. In this study, the considered frequency limits are ±150 mHz around the nominal 50 Hz for alert events, and ±250 mHz for emergency events [76]. The frequency evaluation was performed using a dynamic single-bus inertial model of the Tenerife island grid (Spain), which accounts for both the generation profiles and the system dynamics, including frequency restoration mechanisms, as described in the referenced study [23].

The table shows that the control algorithms that use predicted values require ESS with less energy and power rated values. However, the smoothing levels are similar for the version with and without prediction. For example, the control algorithm that uses moving average filter needs 250% more of ESS rated energy than a control algorithm using prediction to achieve the same levels of power smoothing.

In summary, the electric frequency oscillations caused by the oscillating wave power generations can be smoothed by energy storage systems till the required level by the electric grid. However, the smoothing control algorithm of the ESS has a great impact on the utilization of its rated characteristics; the same smoothing effect requires different levels of rated ESS power and energy, depending on the control algorithm.

It should be noted that in this example, no forecasting algorithm is used, and the prediction is considered perfect by using a future measured value of the complete profile. The results with the different proposed forecasting algorithms (Section 2.3) are shown in the next Section 4.2, with more modest reductions in ESS capacity requirements.

3.2. Description of the Case Study

Section 4 presents the results for 16 representative sea states selected for the location introduced in Section 2.1.1 (see Table 1). The corresponding one-hour power generation profiles are displayed in Figure 7.

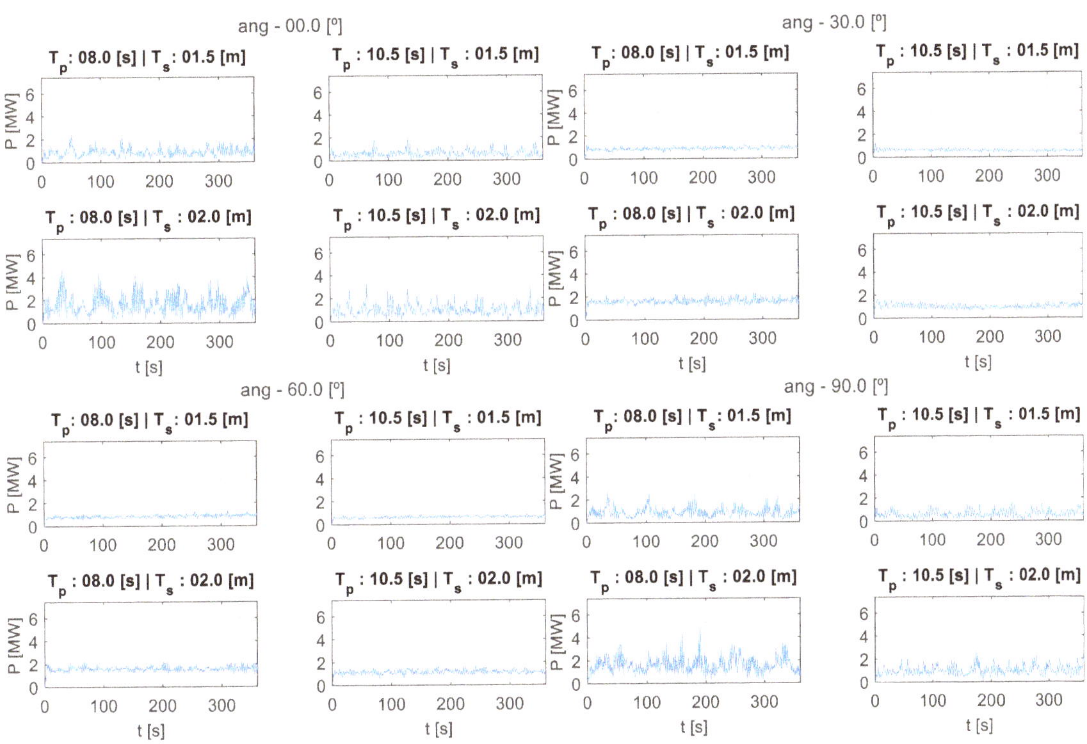

Figure 7. Hourly power generation profiles for the 16 sea state scenarios considered for the wave farm (WF) power generation.

The analysis compares results obtained using the forecasting algorithms described in Section 2.3 and listed in Table 3. Forecasted values are used in MAF to define the ESS power command of the power smoothing control. As in [23], the main objective is to minimize the impact of the WF on the grid by reducing power oscillations. A key performance metric is the variance of the power delivered to the grid, P_{GRID}, by the combined ESS + WF system. This variance is calculated using Equation (13) and the standard deviation is calculated according Equation (14).

$$\sigma^2{}_P = \frac{1}{N}\sum_{i=1}^{N}(P_{ESS}(t_i) - mean(P_{ESS}(t))); \forall \text{ t of a complete simulation} \quad (13)$$

$$\sigma_P = \sqrt{\sigma^2{}_P}; \quad (14)$$

Table 3. Forecasting algorithms considered in the study.

Type	Acronym	Prediction Values Acronym	Description/Complete Name
Base Case	noPred	noPred	No forecasting algorithm in MAF (extra case)
Naïve models (Section 2.3.1)	Pers	PresPredict	Simple persistence
	PersSmart	PersSmartPredict	Smart persistence
Linear models (Section 2.3.2)	ARMApq	ARMAPQpredict	ARMA with optimal p and q values
	ARMA12	ARMA12predict	Simple ARMA modelo (p = 1; q = 2)
Artificial Neural Networks (Section 2.3.3)	NN	NNpredict	Artifical Neural Network (Bayesian Framework)
Limit Case	PredPerf	PredPerf	Prefect prediction values in MAF (extra case)

In addition, the study also evaluates the ESS energy and power requirements based on Equations (3) and (4).

A comparison is carried out between the proposed Bayesian neural network-based forecasting method and classical approaches (see Section 2.3), which are summarized in Table 3. For benchmarking purposes, two additional scenarios, one assuming perfect prediction (*Perf*) as a limit case and another with no prediction (*noPred*) as base case, are included. This serves to establish an upper performance bound for the forecasting-based control strategies.

The control strategies are parametrized using two values: the moving average window size—expressed in number of samples (*Ns_MAF*) or in seconds (*nMAF*)—and the forecast window size (*Ns_Pred* expressed in number of predicted samples, or *nPred* expressed in seconds), as defined in Equation (2). For clarity, all parameter values are expressed both in number of samples and in equivalent seconds, considering the sampling time $\Delta t = 0$, which remains constant throughout this work. For instance, a total moving average window of *Ns_MAF* = 300 corresponds to *nMAF* = 30 s, and a forecasting horizon of *Ns_PRED* = 80 corresponds to *nPRED* = 8 s. This notation has been applied consistently throughout the manuscript, including the captions of figures and tables. The combinations analysed are derived from the following vectors:

- Total window size: *Ns_MAF*: [0, 10, 20, . . . , 320] samples; *nMAF*: [0, 1, 2, . . . , 32] seconds;
- Forecast window size: *Ns_Pred*: [0, 10, . . . , 80] samples; *nPred*: [0, 1, . . . , 8] seconds.

The configuration includes, on one hand, a no-storage scenario (*nMAF* = 0) as a baseline for evaluating the effectiveness of ESS integration. On the other hand, all scenarios with *nPred* = 0 represent filtering strategies without forecasting and serve to establish a lower bound for forecasting improvement.

Based on the example shown in Section 3.1, the expected results (detailed in Section 4) can be summarized as follows:

- The reduction in power variance is primarily driven by the total filtering window size *nMAF*, with limited influence from the forecasting window.
- The decrease in ESS capacity requirements, particularly energy, is strongly influenced by the forecast window size *nPred*.

4. Results

4.1. Training of Neural Networks and Tuning of Forecasting Models

In this article, we have used historical data from the power series to train the forecasting models, both linear and neural networks. One of the most important steps to achieve good forecasting and avoid overfitting is to study the complexity of the models. In this case, the steps described in [77] ere followed. The data set is divided into a training set and a test set. The training set is used to train the forecasting model, and the test set allows us to study the accuracy of the model when presented different data from the training data, thus avoiding overfitting.

In the case of the linear models (Section 2.3.2), the complexity of the model consists of the number of inputs for the AR model 'p' and the number of error terms for the MA model 'q'. The partial autocorrelation function sample (PACF), the autocorrelation function sample (ACF) and the Bayesian information criterion (BIC) are used to study the complexity of the model.

PACF establishes a correlation between two values of a series with a time difference between them, so it is used to find the relevant past values. The maximum order value of p is set within the 95% range of the PACF sample. Similarly, an ACF sample is used to set the maximum order of q in the MA model. Once the maximum order of p and q has been determined, training is performed with all possible combinations of p and q from 1 to the maximum. Finally, the complexity of the model is decided among all these simulations by calculating the BIC and the model error %rRMSE (Root Mean Square Error) with the test data set. For each case study and each prediction horizon, the optimal p and q were calculated according to this procedure. The prediction error with a simple ARMA(2,1) model was also calculated for comparison.

In the case of neural networks, the complexity of the model is studied by choosing the number of inputs (past data of the series) and the number of hidden units. The Bayesian framework allows us to study the number of inputs with the ARD technique and the number of hidden units with the log of the evidence [73,74]. Both techniques control the overfitting of the model. ARD assigns a different hyperparameter to each set of weights associated with an input. At the end of training, weights whose hyperparameter has a high value are close to zero and the input is considered irrelevant to the model. After training with different numbers of inputs and hidden units, these techniques made it possible to choose the number of model parameters. The past values used in the prediction are those corresponding to all the steps within the previous one second (steps of 0.1 s) and the values of the previous 9 s (steps of 1 s).

Figure 8 shows the errors of the different forecasting models. The figure shows a comparison of the accuracy of all models for each case study and for all forecasting time horizons. As expected, it can be seen that the prediction error increases as the time horizon increases. In general, in all cases and for all prediction horizons, the neural networks proposed in this study provide more suitable forecasting and present the lower errors. The linear models do not provide a very significant improvement over the use of a naive model.

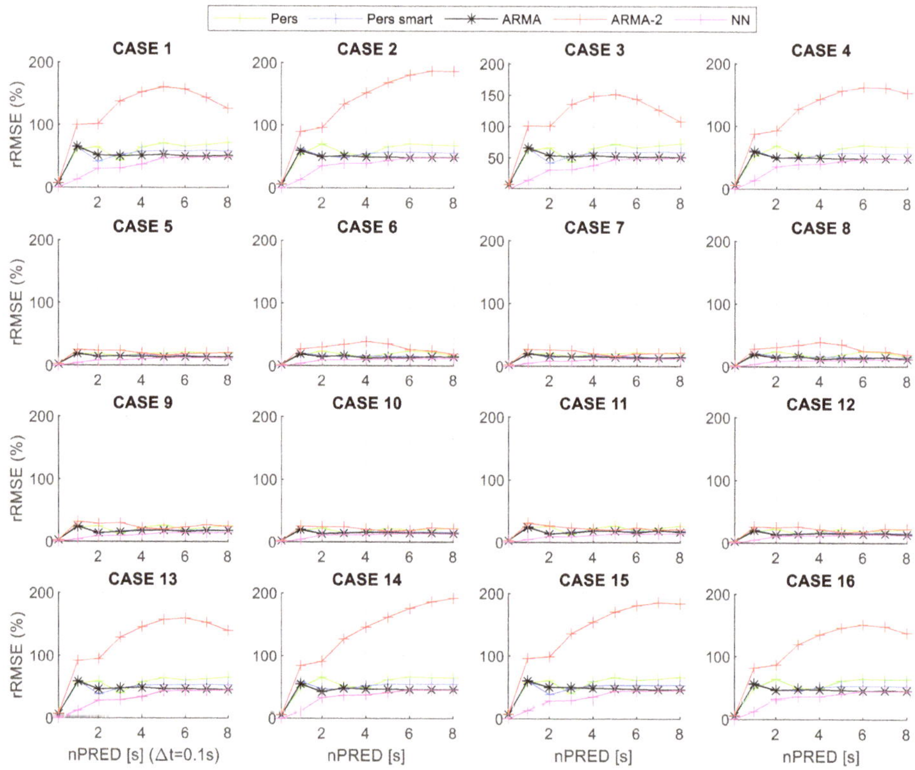

Figure 8. Forecasting model results for different time horizons (*nPred* in seconds with a sample time $\Delta t = 0.1$ s) in all study cases (rRMSE).

4.2. Results of Applying the Forecast-Enhanced Filtering Strategy to the Case Study

Following the training (in the case of the Bayesian network -NN-) or parameter fitting (for the ARMA-based models), a set of valid forecasting systems was obtained for analysing all the scenarios described in Section 3. In total, the evaluation covers 16 representative sea states, 6 forecasting algorithms (including perfect forecasting), 31 total MAF window sizes (including the case without MAF), and 9 forecasting window lengths (including the case without forecasted values), resulting in a total of 26,784 configurations.

The evaluation is based on three key performance metrics—P_{rated}, E_{rated}, and the standard deviation of the grid-injected power, σ_P—along with their respective normalized ratios compared to the no-forecasting case:

1. Required rated power for the ESS (P_{rated})—Equation (3).
2. Required rated energy for the ESS (E_{rated})—Equation (4).
3. Standard deviation of grid-injected power after filtering via MAF and ESS (σ_P)—Equation (14).
4. Ratio of P_{rated} compared to the no-forecasting case.
5. Ratio of E_{rated} compared to the no-forecasting case.
6. Ratio of σ_P compared to the no-forecasting case.

To illustrate the methodology, results for Case 1 are presented in detail in Figure 9, showing the six performance metrics across the five forecasting algorithms and the perfect forecasting scenario (*PredPerf*). Each set of plots presents the metrics versus the MAF window size (*nMAF*), grouped by forecasting window length (*nPRED*). Additionally, the optimal (*nMAF*, *nPRED*) pair that minimizes energy storage requirements is highlighted. Figures for all 16 sea states are provided in Appendix A.

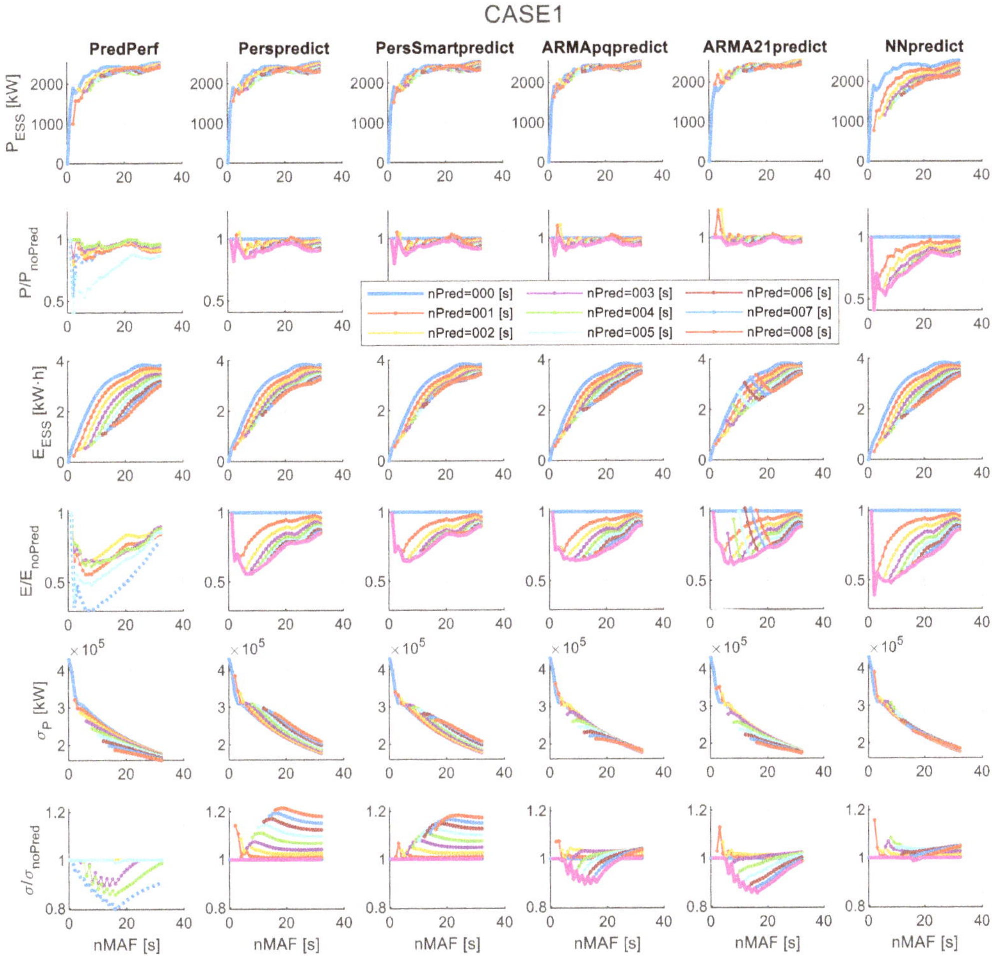

Figure 9. Results for Case 1 across different MAF sizes and forecasting algorithms.

These plots show that, regardless of the forecasting algorithm, incorporating forecasted values consistently reduces energy storage requirements compared to no forecasting (see the E/E_{nopred} plots, always below 1). The rated power requirement P_{rated} remains similar to the no-forecasting case (see P/P_{nopred}), except for the NN model, where deviations are observed. Regarding σ_P, the standard deviation tends to remain close to, or slightly above, the values of the no-forecasting case for ARMA and NN algorithms (see σ/σ_{nopred}, with values close to or below 1, except for low *nMAF* values). This behaviour suggests that MAF design can be oriented toward σ_P limitation, largely independent of forecasting window and algorithm, which can then be optimized for energy savings.

Figure 10 shows an aggregated analysis across the 16 sea states, focusing on the proposed Bayesian network forecasting model (NN). Each subplot displays the Pareto front (in red) for σ_P versus E_{rated}, along with the optimal (*nMAF*, *nPRED*) combinations. In Figure 11, results are grouped by wave incidence angle (in sets of four cases), and the number of times each (*nMAF*, *nPRED*) pair appears in the Pareto fronts is counted. The bottom plot of Figure 11 summarizes the most frequently optimal parameter combinations. Figure 12 extends this analysis by presenting the same summary for all six forecasting algorithms, including perfect forecasting.

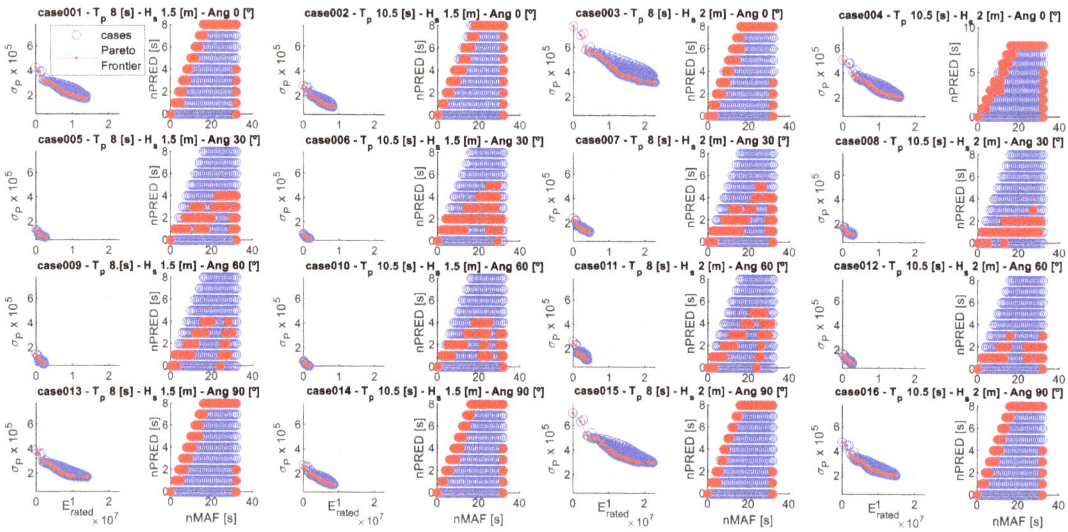

Figure 10. Summary of results for the 16 sea states using the NN model. Pareto fronts and optimal (*nMAF*, *nPRED*) combinations are highlighted.

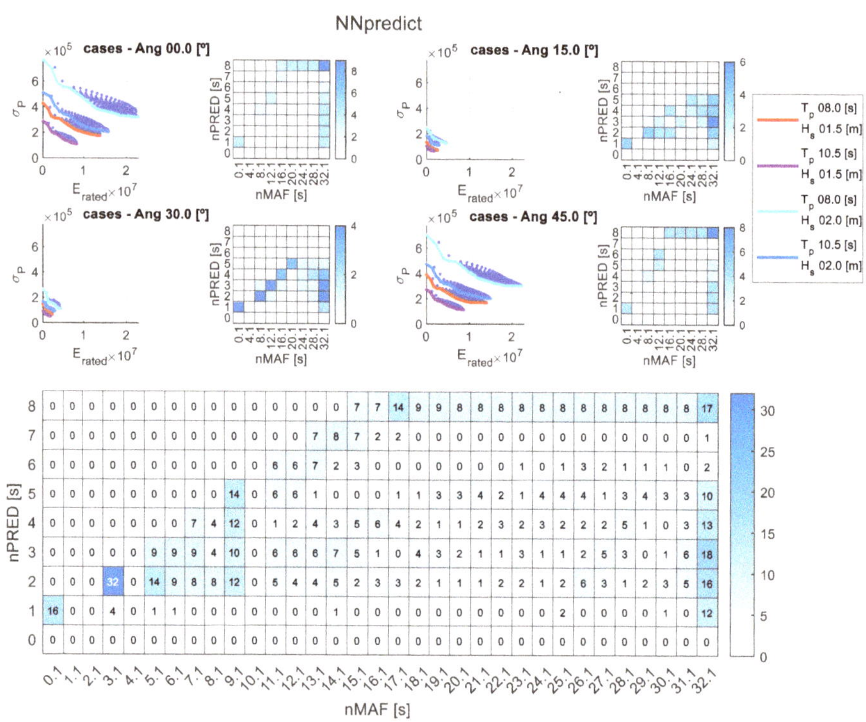

Figure 11. Grouped analysis of Pareto-optimal parameter combinations (*nMAF*, *nPRED*) across the 16 sea states using NN.

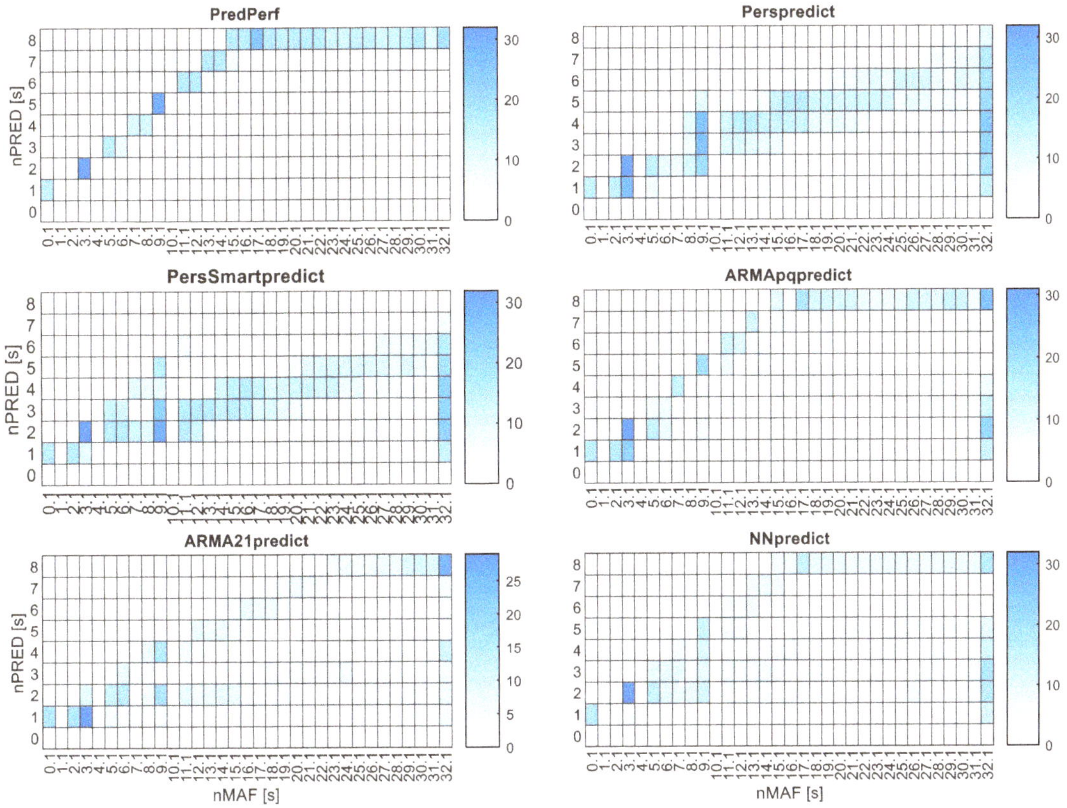

Figure 12. Pareto-optimal parameter combinations ($nMAF$, $nPRED$) across all forecasting algorithms and the perfect forecasting case.

From Figure 10, we observe that larger forecasting windows ($nPRED$) yield better results for sea states 1–4 and 13–16, which are characterized by higher σ_P. For the remaining cases (5–12), which benefit from spatial smoothing due to WEC positioning (see Figure 7), prediction remains beneficial, though maximum $nPRED$ values are not necessarily optimal. When considering all 16 sea states (Figure 11), maximum values for both $nMAF$ and $nPRED$ tend to appear frequently in the Pareto-optimal combinations. Figure 12 confirms that forecast-enhanced filtering outperforms filtering without forecasting across all algorithms—forecasted values consistently contribute to reductions in both σ_P and E_{rated}. In the perfect forecasting scenario, the maximum $nPRED$ value is always selected.

Figure 13 compares Pareto fronts across the six forecasting algorithms, again highlighting improvements relative to filtering without forecasting. In Figure 13, we observe consistent performance gains from all forecasting algorithms over the no-forecasting baseline (black line). Improvements are especially pronounced for sea states with high σ_P (cases 1–4 and 13–16). NN often outperforms simpler models, although ARMApq may yield better results in high-E_{rated} scenarios where filtering smoothness takes precedence over fidelity. Perfect forecasting results are also shown (in magenta), representing the theoretical upper bound for performance improvement.

Although the primary objective of this work is to demonstrate the advantages of incorporating forecasting algorithms into the smoothing of power output generated by a wave farm (WF), and to present the implementation and results of applying real forecasting algorithms in moving average filters, rather than to provide a detailed sizing methodology for prediction and filtering systems, it is still possible to report the performance of the objective variable E_{rated} with respect to the parametrization values of the forecasting and filtering algorithms (i.e., optimal combinations of *nMAF* and *nPRED*). Furthermore, one may assess which of these combinations meet the minimum requirements in terms of reducing the standard deviation of power (σ_P), with the aim of selecting the design parameters for a specific filter configuration.

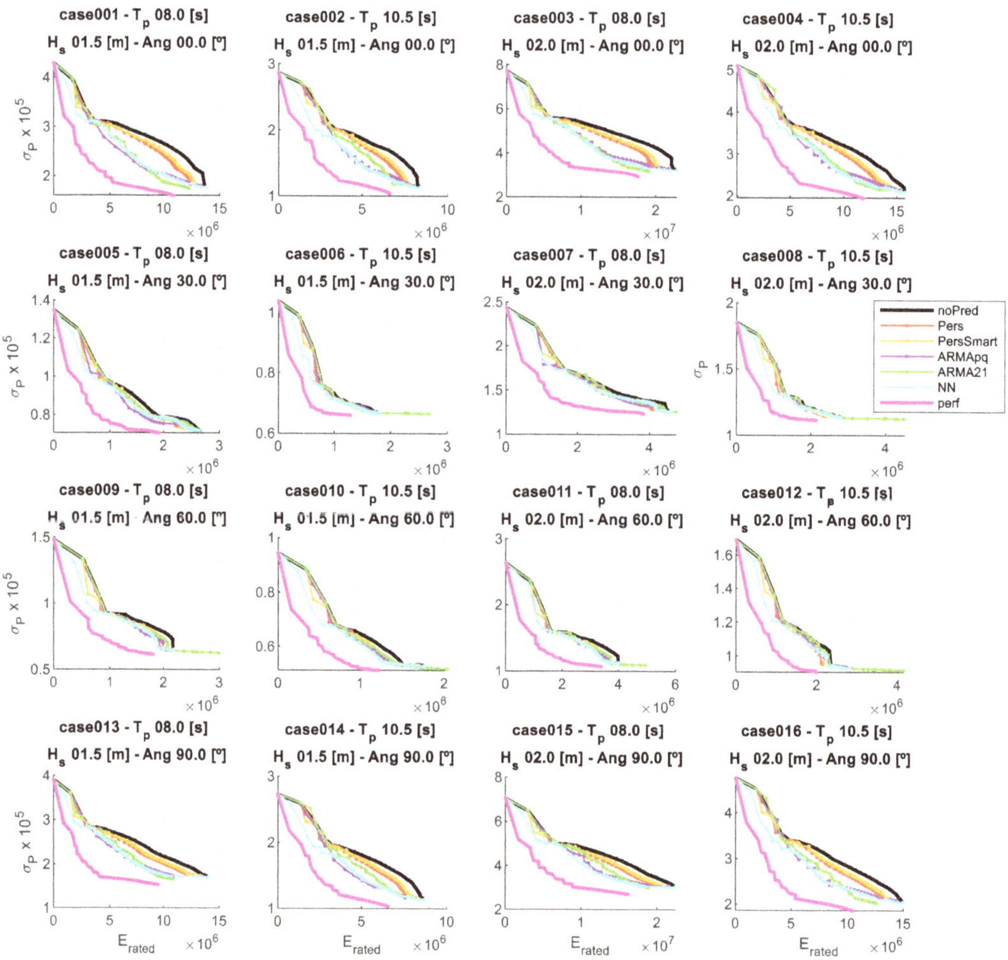

Figure 13. Pareto fronts of σ_P and E_{rated} for all forecasting algorithms across the 16 sea states.

Based on the reference study in [23], a σ_P threshold of 5.12×10^6 W was established to ensure grid frequency stability when integrating the studied WF (see Section 2.1) into the isolated electrical system of Tenerife. This threshold is used to identify configurations *nMAF-nPRED* capable of limiting grid disturbances.

Results are shown in Figure 14. Heatmap graphs of E_{rated} and σ_P as functions of (*nMAF*, *nPRED*) for all forecasting algorithms, including the perfect forecasting scenario. They indicate that higher *nPRED* values systematically reduce E_{rated}, regardless of forecasting algorithm. For NN, optimal performance is achieved with *nPRED* = 8 s and *nMAF* = 16 s, yielding acceptable energy requirements while satisfying the σ_P constraint. This configuration also outperforms all other algorithms except perfect forecasting.

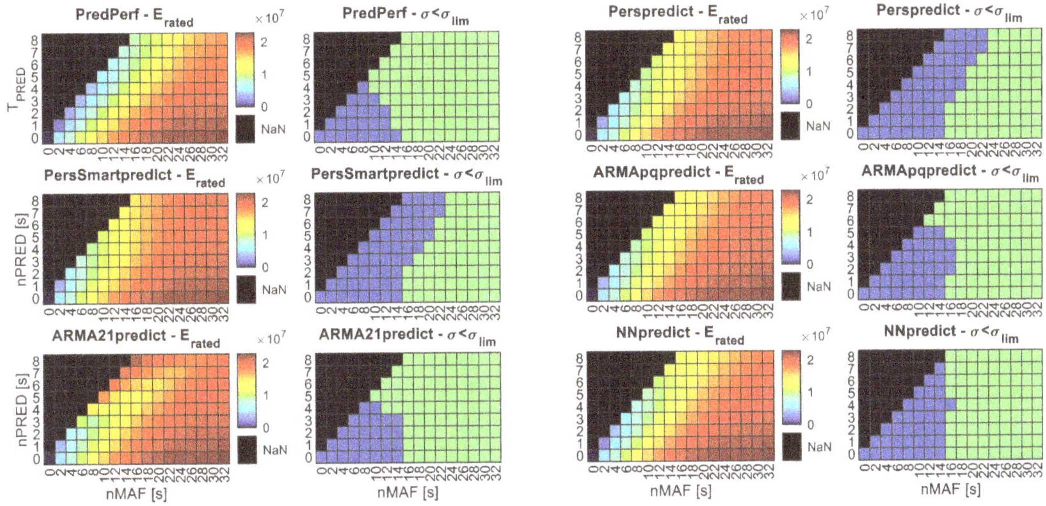

Figure 14. Heatmap graphs of E_{rated} and σ_P as functions of ($nMAF$, $nPRED$) for all forecasting algorithms, including the perfect forecasting scenario.

Focusing on the results obtained for the NN algorithm (see Figure 15 for details of sea state CASE 1) it can be observed that, in general, improvements are achieved in reducing the energy storage system requirements (E_{rated}) and even the power capacity (P_{rated}), albeit at the expense of a slight degradation in the standard deviation of the injected power (σ_P), which worsens by less than 7% compared to the cases without forecasting.

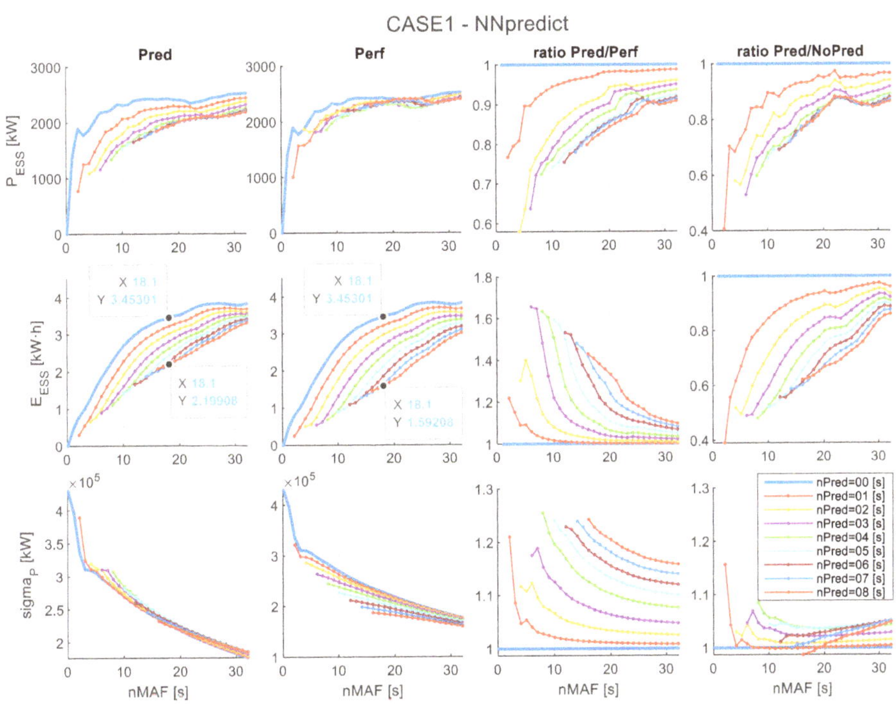

Figure 15. Summary of σ_P and E_{rated} results for different ($nMAF$, $nPRED$) values using the NN forecasting model.

If we select the filter values of 8 s for *nPRED* and 16 s for *nMAF*, the results for σ_P and E_{rated} can be shown for each of the 16 cases (compared against perfect forecasting and the no-forecasting scenario), as well as a time-domain example of the filtering using the chosen filter with NN and without forecasting (see Figure 16).

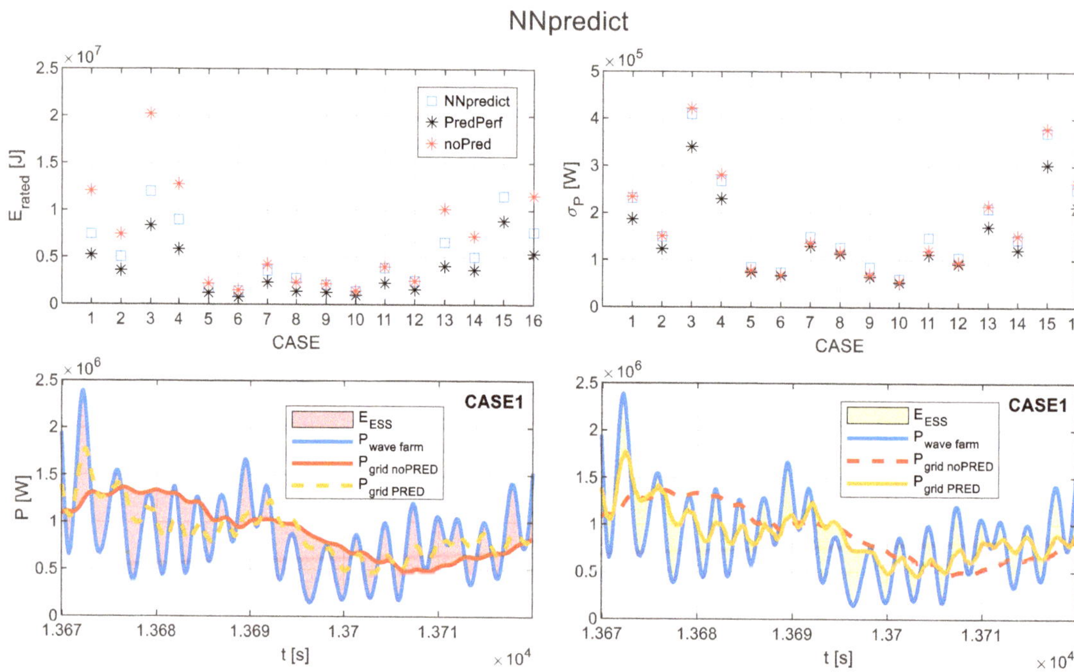

Figure 16. Summary of results obtained for σ_P and E_{rated} for the filter configuration combination of *nMAF* and *nPRED* at 16 and 8 s, respectively, and for the NN forecasting algorithm.

In Figure 16, it can be observed that E_{rated} achieves better values compared to the no-forecasting case, while the improvement in σ_P is less pronounced. Regarding the time-domain profiles shown, the results are very similar to those already presented in the example of Figure 6; the reduction in the phase lag between the generated power (P_{WF}) and the filtered power (P_{GRID}) obtained by using forecasting algorithms leads to a smaller area enclosed between the P_{WF} and P_{GRID} profiles when forecasting is applied.

In summary, it can be observed that the power used and the smoothing capacity achieved are not strongly influenced by forecasting (the size of the prediction window *nPRED*), but there is an improvement in the required energy (E_{rated}). Moreover, in the case of using NN, it can be seen that the best combination of total window size and predicted window size is 16 and 8 s, respectively.

With this combination, we obtain results demonstrating that including forecasting leads to an improvement in the required storage size (compared to the case with the same total window size but without forecasting), reducing the storage requirements to achieve the same smoothing results of power oscillations (reaching reductions of up to 36.52%). Additionally, the results with NN outperform those obtained with simplified forecasting algorithms such as Pers.

On the other hand, it is also demonstrated that within the forecasting algorithms' capabilities, there is room for improvement, since results achieved with perfect forecasting could further reduce storage requirements (achieving reductions of up to 53.91%).

Finally, it is worth mentioning the limitations of the study. One of the main limitations is that the assessment of the power and energy requirements were carried out assuming an ideal, technology-agnostic energy storage system (ESS), as described in Section 2. While this assumption allows for a clear comparison of filtering and short-horizon forecasting strategies, it does not capture several engineering constraints that would be present in real-world deployments. In practice, the dynamic behavior of the state of charge (SoC), ramp-rate limits, round-trip efficiency, charge/discharge asymmetry, and potential ageing of storage devices could influence both the sizing requirements and the achievable smoothing performance. These factors would generally tend to reduce the numerical gains reported here and/or could lead to modifications in the control algorithms to take them into account. Nevertheless, even when considering these limitations, the main qualitative conclusion remains valid: short-horizon forecasts consistently reduce the energy required to achieve a given level of power smoothing. By presenting the results under an idealized ESS framework, the study provides a general benchmark that can be adapted to specific storage technologies in future applications. Other limitations to consider are that it focuses on the power smoothing capacity of the ESS rather than on the effects of power oscillations on the electrical grid; that the prediction algorithms are executed in simulation, with their real-time execution to be purchased in the future; and that the high-frequency effects introduced by power electronics have not been considered. All these limitations open the door to further development in future work.

5. Conclusions

The use of energy storage systems (ESS) in wave energy generation is key to reducing power oscillations and improving the stability of the electric grid. In particular, in low-inertia electric systems, such as islanded grids, the introduction of power smoothing mechanisms is essential to avoid frequency oscillations. The combination of moving average filtering techniques with advanced forecasting algorithms, based on Bayesian Neural Networks, allows for significant optimization of energy storage requirements, reducing the need for power and energy without compromising the grid stability.

The results obtained in the case study have shown that using prediction in ESS control algorithms reduces the energy storage system required by 36.52% compared to a configuration without prediction. In the case of a perfect prediction scenario, the maximum reduction achieved is 53.9%. This improvement is very relevant because it has led to lower storage implementation costs, making wave energy projects more economically viable.

Another key aspect identified in this paper is the importance of selecting properly the parameters of the moving average filter and the prediction window. It has been demonstrated that using a prediction window of 8 s and a total filter window of 16 s optimizes the balance between reducing power oscillations and minimizing the ESS energy requirements. This result highlights the need to adjust these parameters according to the specific characteristics of the wave conditions and the electrical grid where the solution is implemented.

Furthermore, the comparison between different forecasting algorithms has confirmed that Bayesian Neural Networks present a better performance than simpler methods, such as persistence models and ARMA. However, the results indicate that there is still room for improvement in the predictive capacity of these algorithms since their optimal performance remains below the perfect prediction scenarios. This suggests that future research should focus on improving the accuracy of forecasting models, exploring hybrid approaches that combine different machine learning techniques to achieve more precise power generation estimations.

Therefore, smoothing the oscillating power generated by wave energy farms (given the oscillating nature of the energy source: waves) reduces the impact of these types of systems on the electric grid. This ultimately improves grid stability and can increase the penetration limits of this renewable energy in electric grids. Furthermore, given the need for these storage systems for large-scale deployment of wave energy, optimizing the sizing of storage systems using control algorithms can be very useful. It should be noted that this approach is inherently scalable and is not limited by the size of the farm. As the number of WEC units increases, the natural cancellation of power oscillations between units can further reduce storage requirements, even without improved forecasting strategies, making the method applicable to larger-scale deployments.

The implementation of these strategies represents an important step towards greater and more efficient grid integration of renewable energies, contributing to the energy transition and the achievement of decarbonization goals.

Author Contributions: Conceptualization, M.B., M.L. and L.M.; methodology, M.B., G.N. and L.M.; software, M.B., G.N. and L.M.; validation, MB., I.V. and J.N.; formal analysis, M.B. and M.L.; investigation, M.B. and J.N.; resources, I.V. and M.L.; data curation, M.B., I.V., J.N. and G.N.; writing—original draft preparation, M.B., I.V. and L.M.; writing—review and editing, M.B., I.V., L.M. and M.L.; visualization, M.B. and G.N.; supervision, M.B. and M.L.; project administration, J.N. and M.B.; funding acquisition, M.L., G.N., J.N. and M.B. All authors have read and agreed to the published version of the manuscript.

Funding: This research, developed under the Projects STORIES (ID: 101036910), has received funding from European Union's Horizon 2020 research and innovation program under H2020-EU.1.4—EXCELLENT SCIENCE—Research Infrastructures (LC-GD-9-1-2020) and Project HYBRIDHYDRO (TED2021-132794A-C22), which has received funding from MCIN/AEI/10.13039/501100011033 and from the European Union "Next Generation EU"/PRTR.

Data Availability Statement: The data presented in this study are available on request from the corresponding author.

Conflicts of Interest: The authors declare no conflicts of interest.

Appendix A

Results for the 16 representative wave cases (see Table 1), 6 forecasting algorithms (including perfect forecasting), 31 total window sizes of the MAF filter (including the case without MAF), and 9 window sizes with predicted values (including the case without predicted values), resulting in a total of 26,784 cases to analyze.

The variables used for the analysis are as follows:

- Required power in the ESS (P_{rated})—Equation (3).
- Required energy in the ESS (E_{rated})—Equation (4).
- Standard deviation of the power injected into the electrical grid after filtering through MAF and ESS (σ_p)—Equation (14).
- Ratio of required power in the ESS relative to the case without forecasting.
- Ratio of required energy in the ESS is relative to the case without forecasting.
- Ratio of the standard deviation of injected power relative to the case without forecasting.

The results of each of the 16 cases are presented in a set of different graphs. Each set of graphs shows each of the 6 variables to be analyzed. These figures also display the best combination between predicted values and window size, aiming to minimize the amount of storage.

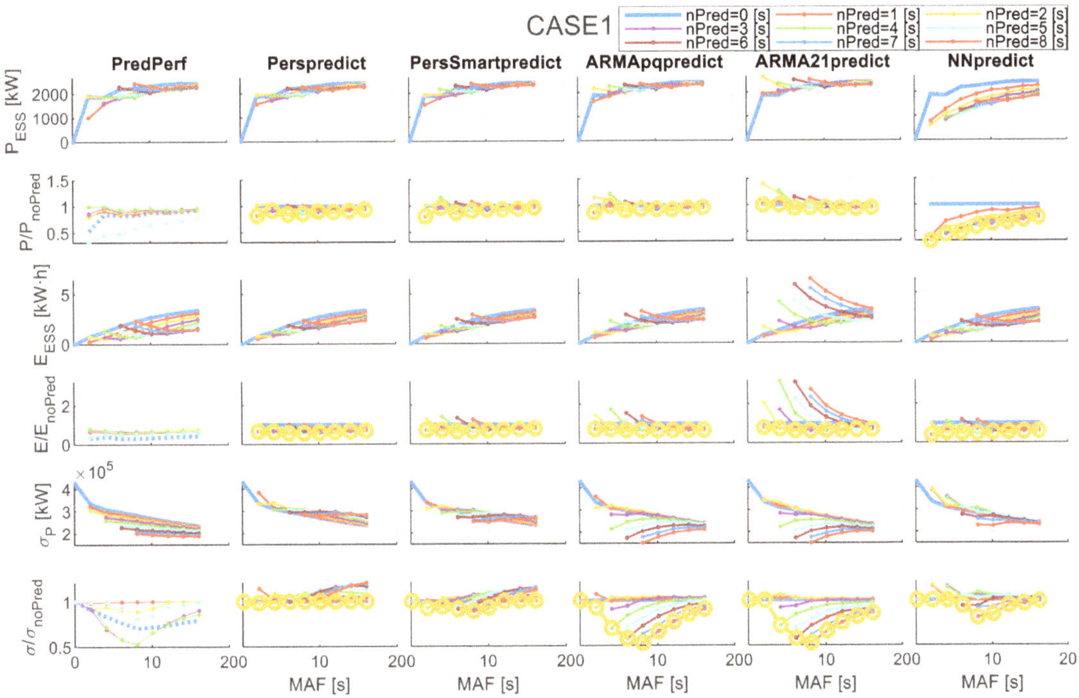

Figure A1. Results for CASE 1 with different MAFs and forecast algorithms.

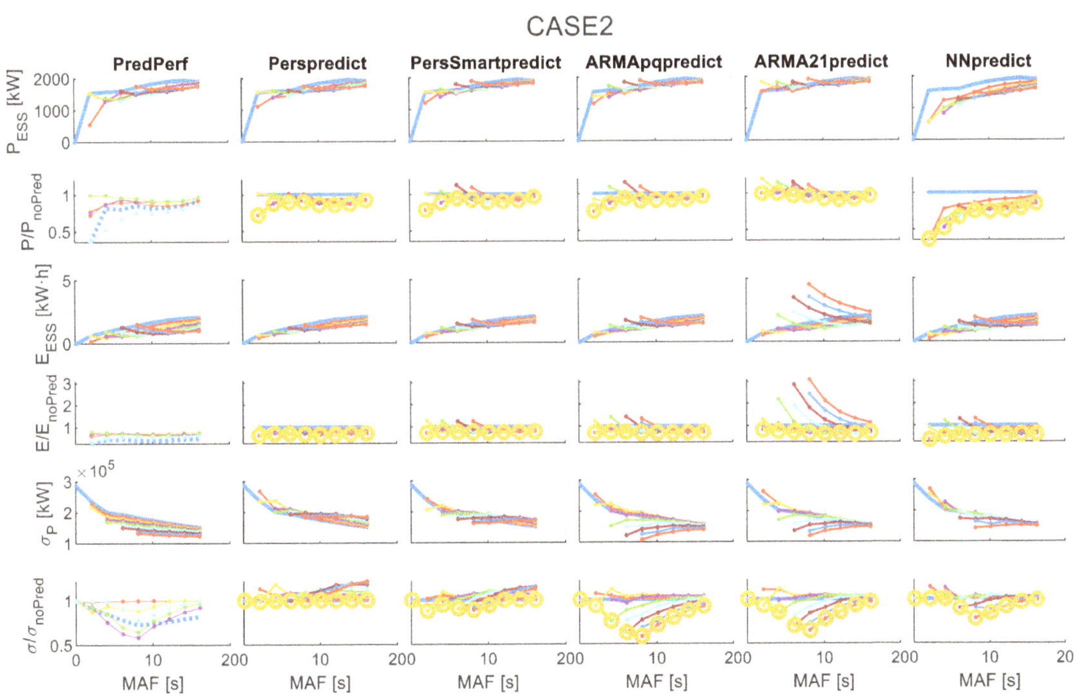

Figure A2. Results for CASE 2 with different MAFs and forecast algorithms.

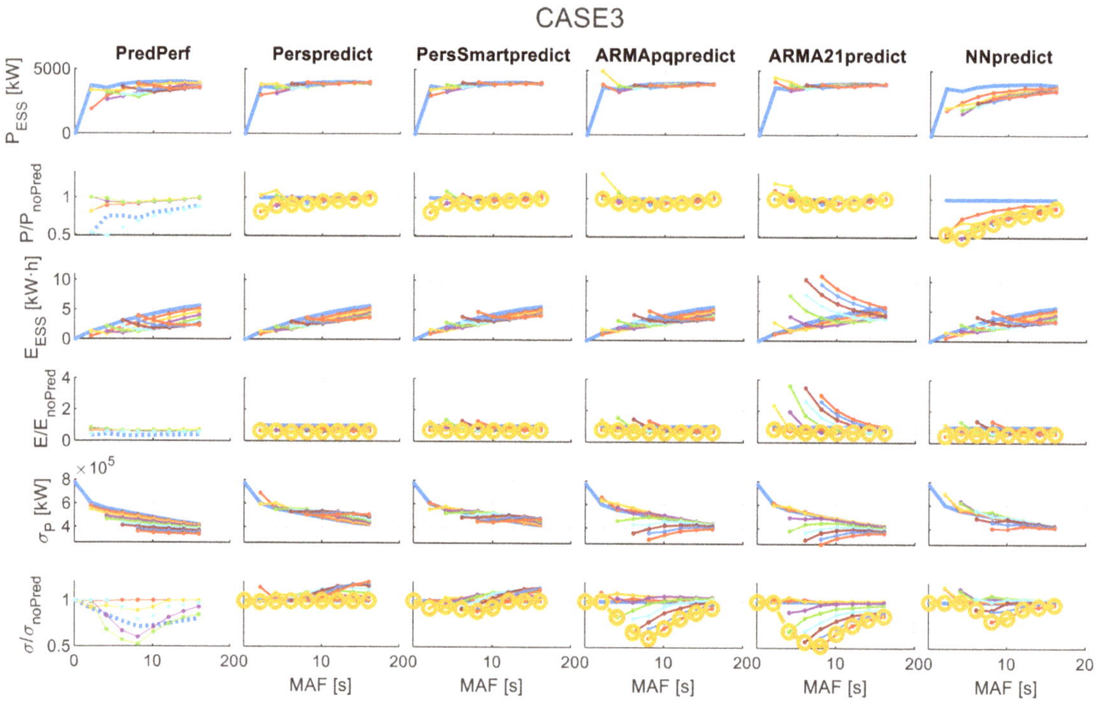

Figure A3. Results for CASE 3 with different MAFs and forecast algorithms.

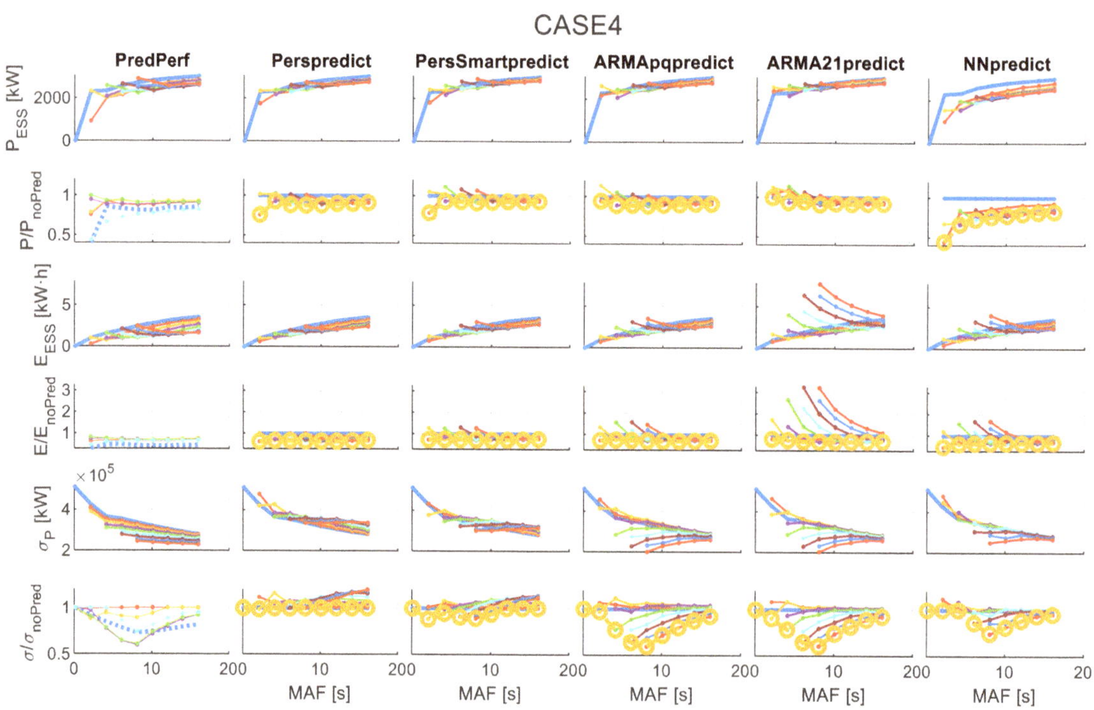

Figure A4. Results for CASE 4 with different MAFs and forecast algorithms.

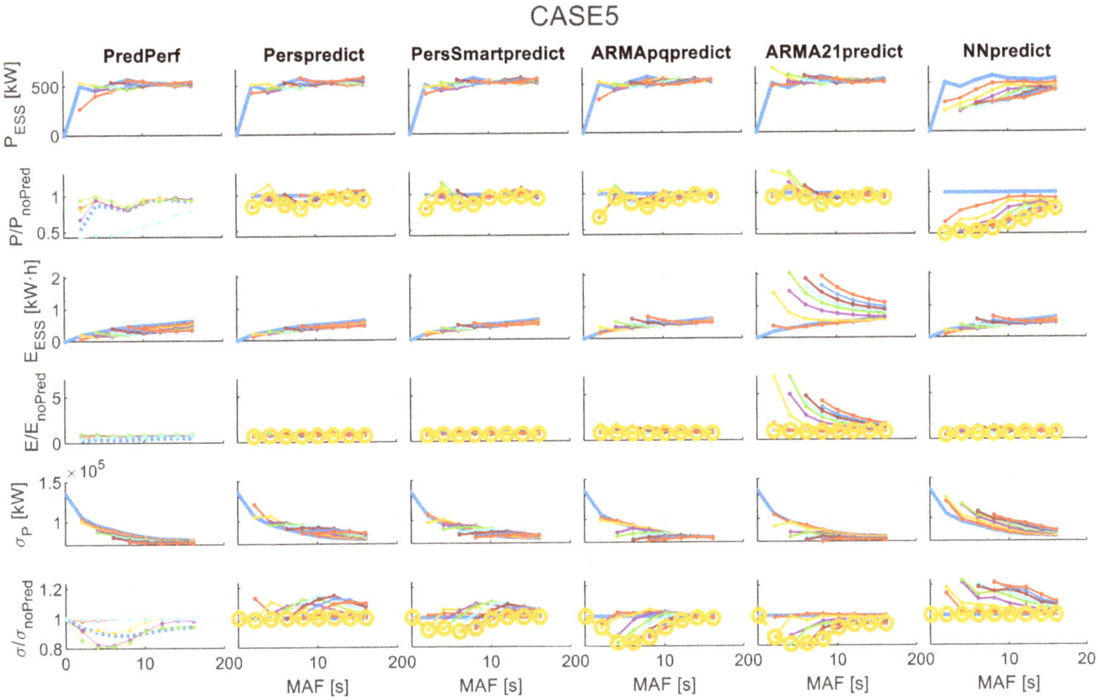

Figure A5. Results for CASE 5 with different MAFs and forecast algorithms.

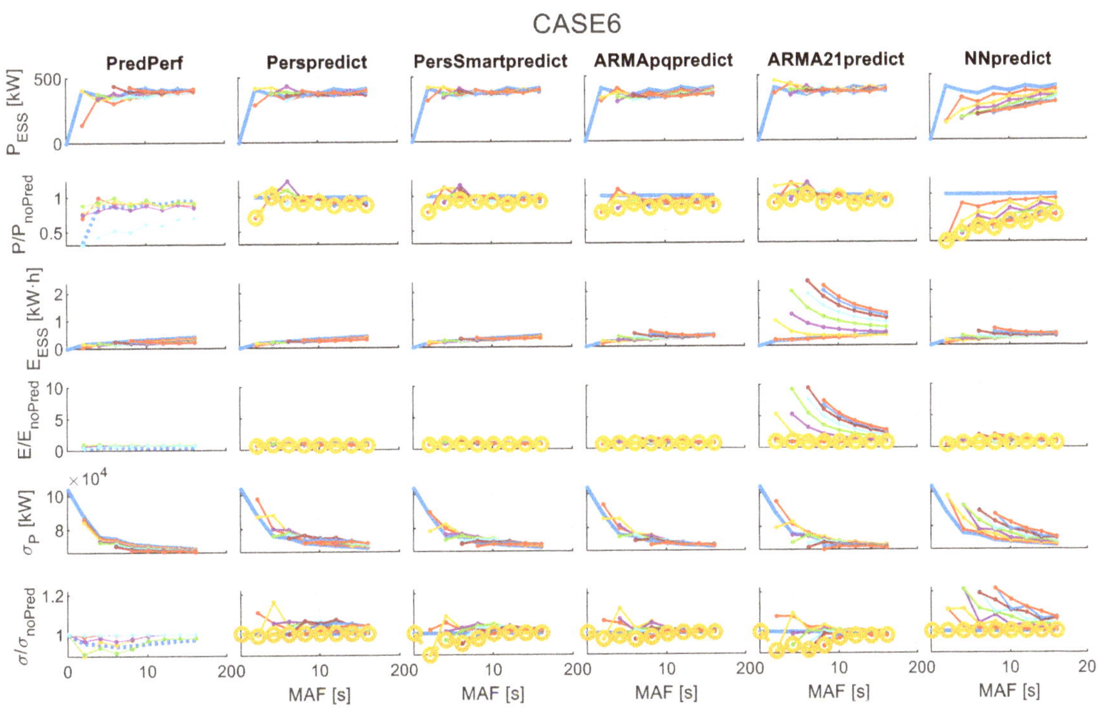

Figure A6. Results for CASE 6 with different MAFs and forecast algorithms.

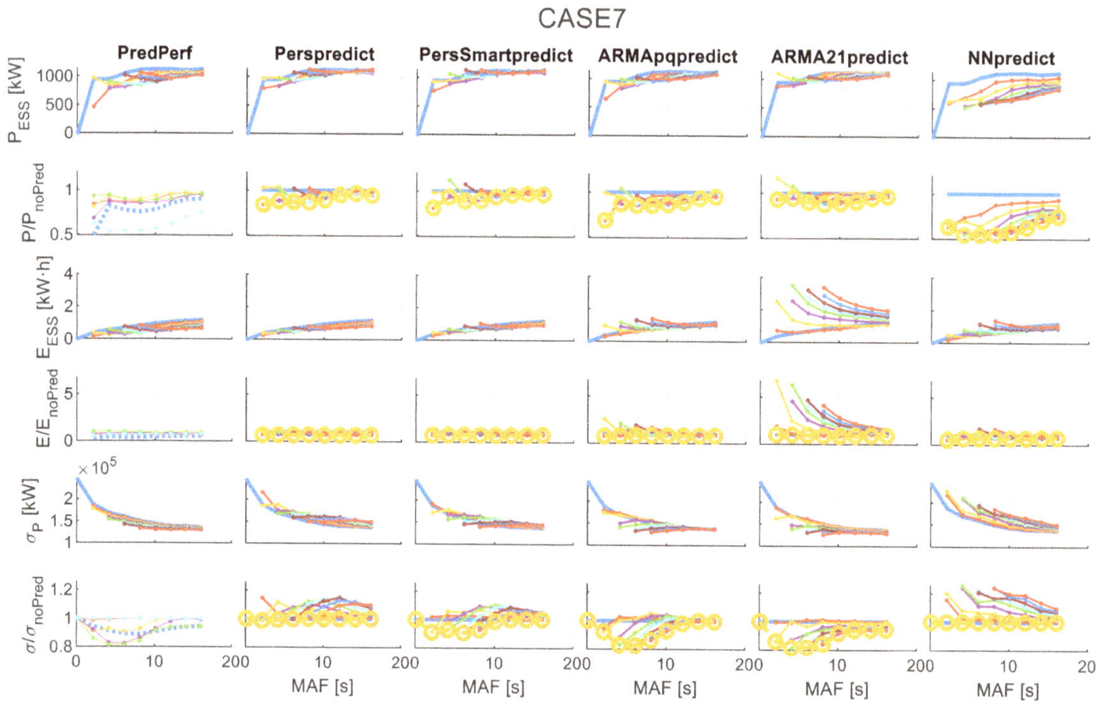

Figure A7. Results for CASE 7 with different MAFs and forecast algorithms.

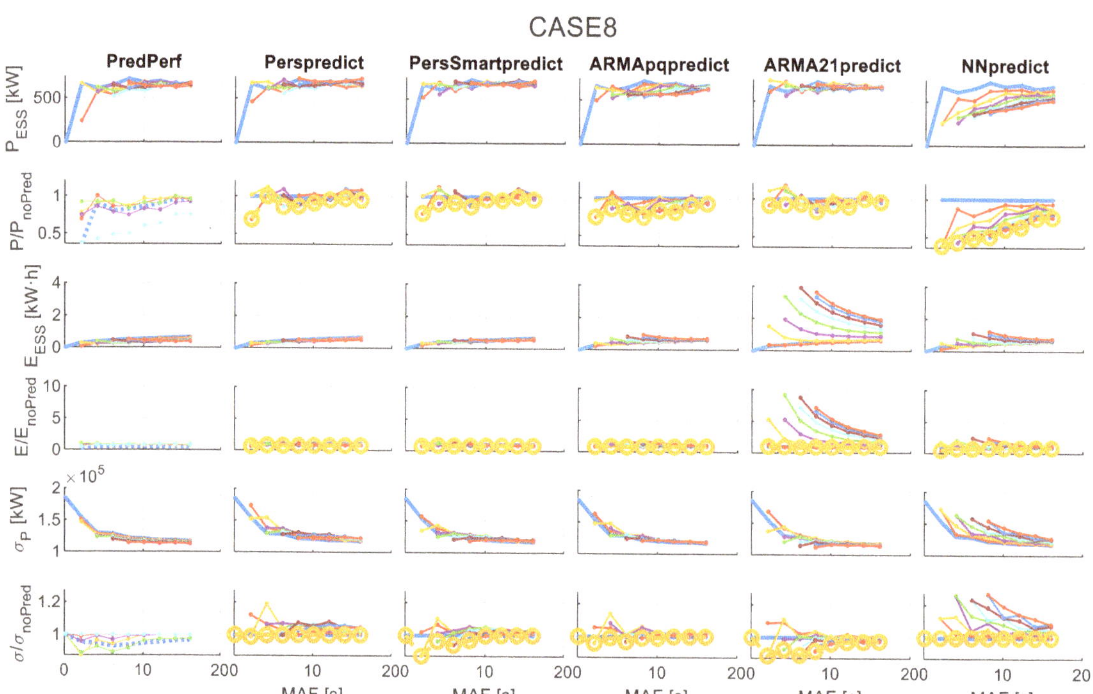

Figure A8. Results for CASE 8 with different MAFs and forecast algorithms.

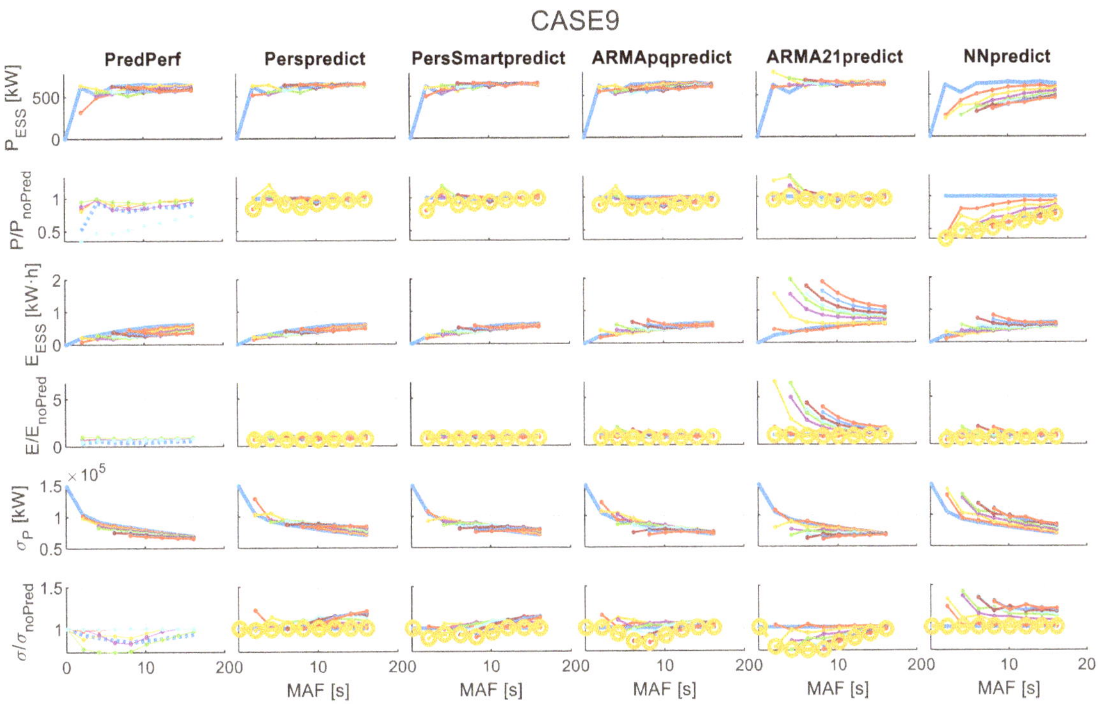

Figure A9. Results for CASE 9 with different MAFs and forecast algorithms.

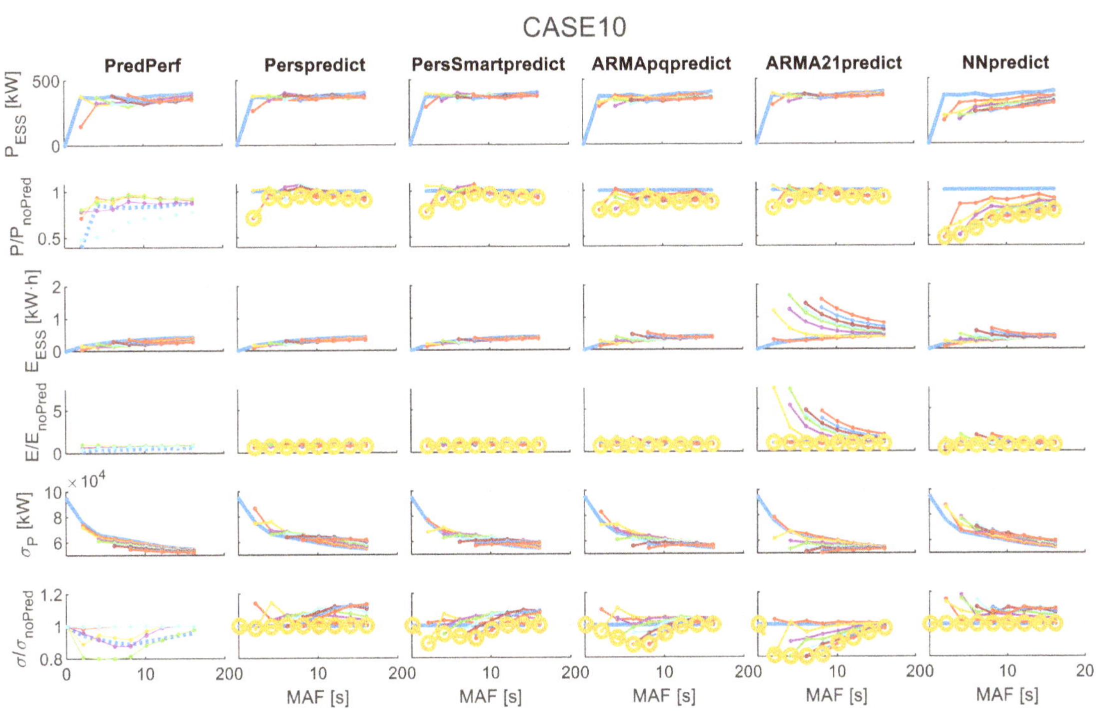

Figure A10. Results for CASE 10 with different MAFs and forecast algorithms.

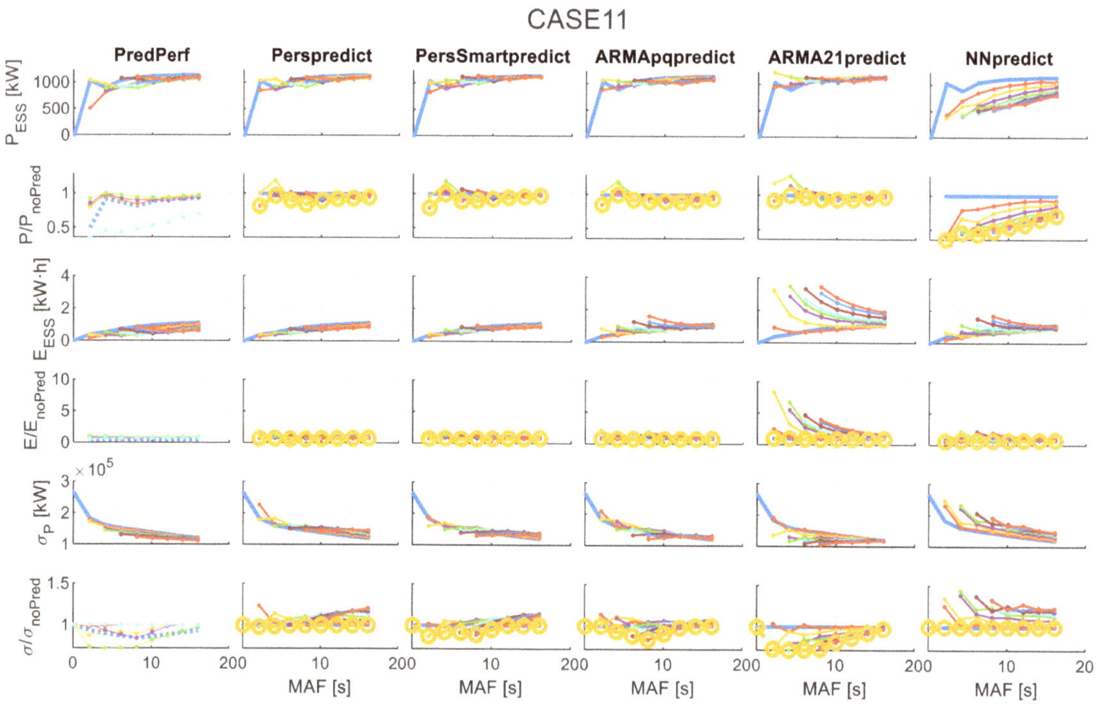

Figure A11. Results for CASE 11 with different MAFs and forecast algorithms.

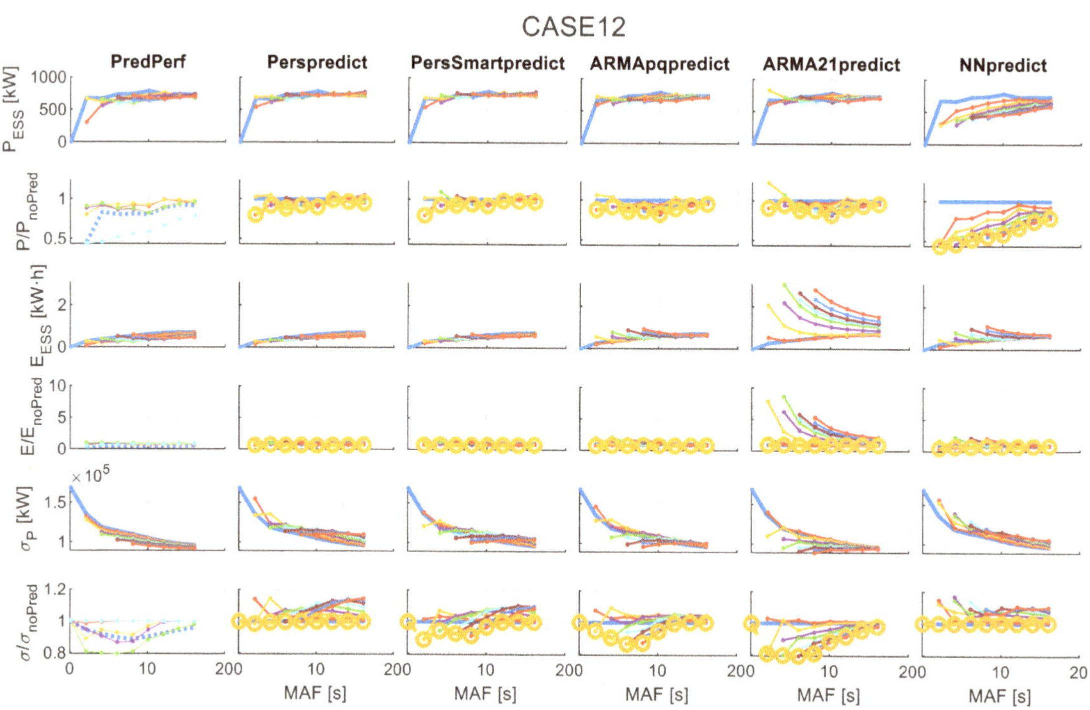

Figure A12. Results for CASE 12 with different MAFs and forecast algorithms.

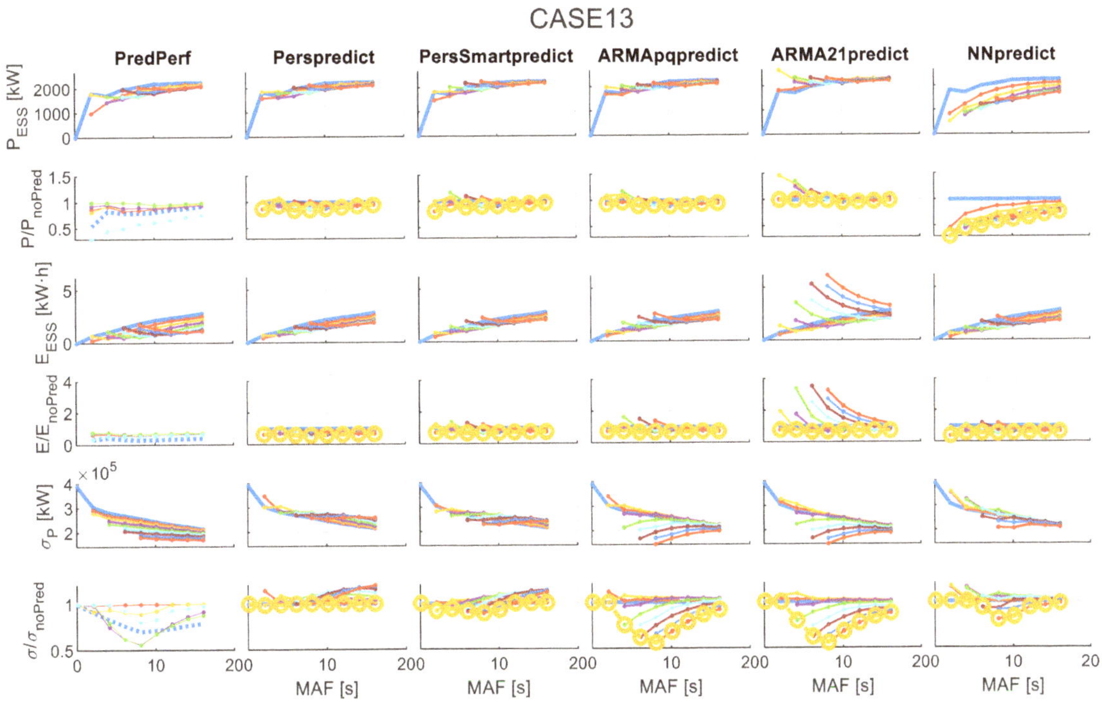

Figure A13. Results for CASE 13 with different MAFs and forecast algorithms.

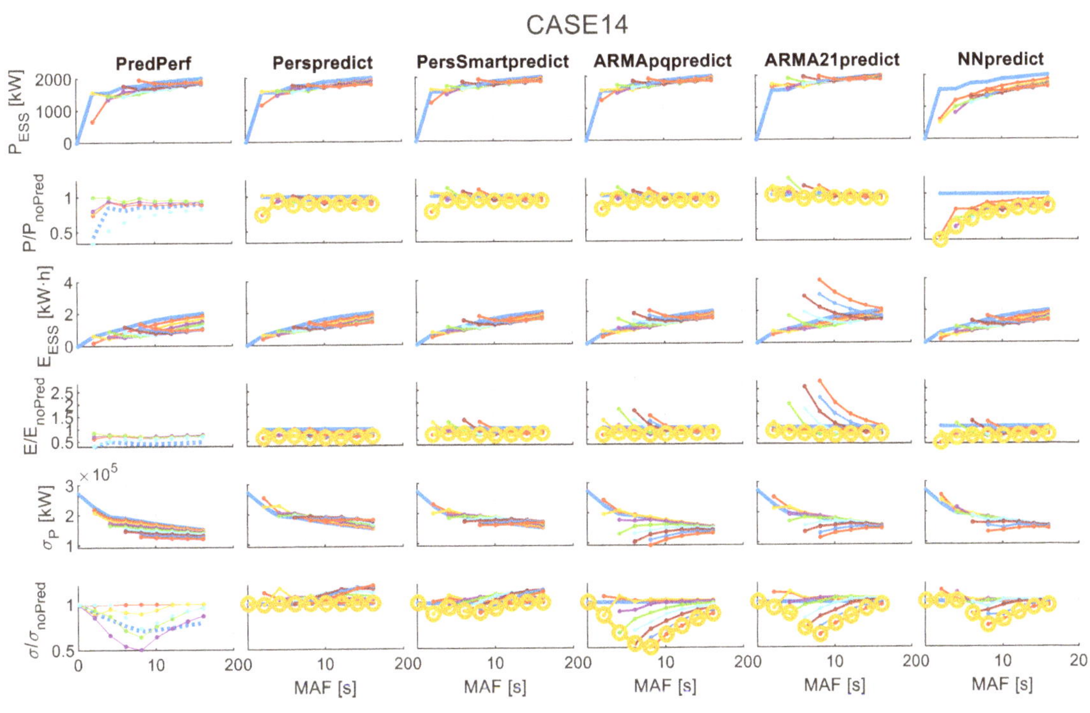

Figure A14. Results for CASE 14 with different MAFs and forecast algorithms.

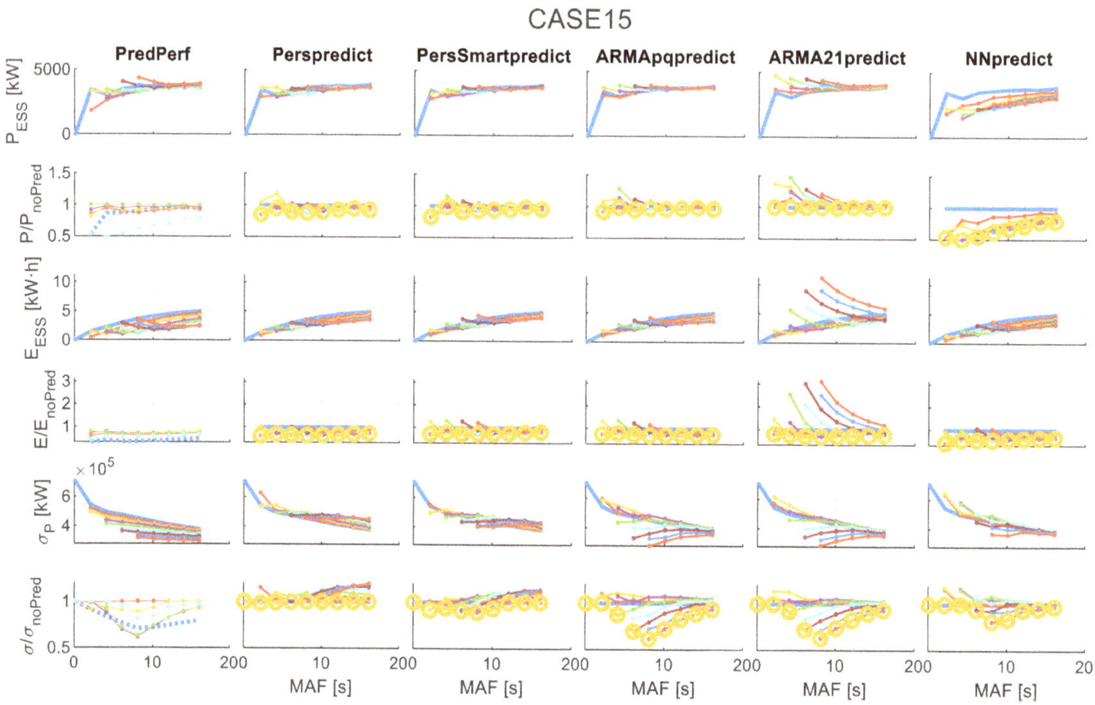

Figure A15. Results for CASE 15 with different MAFs and forecast algorithms.

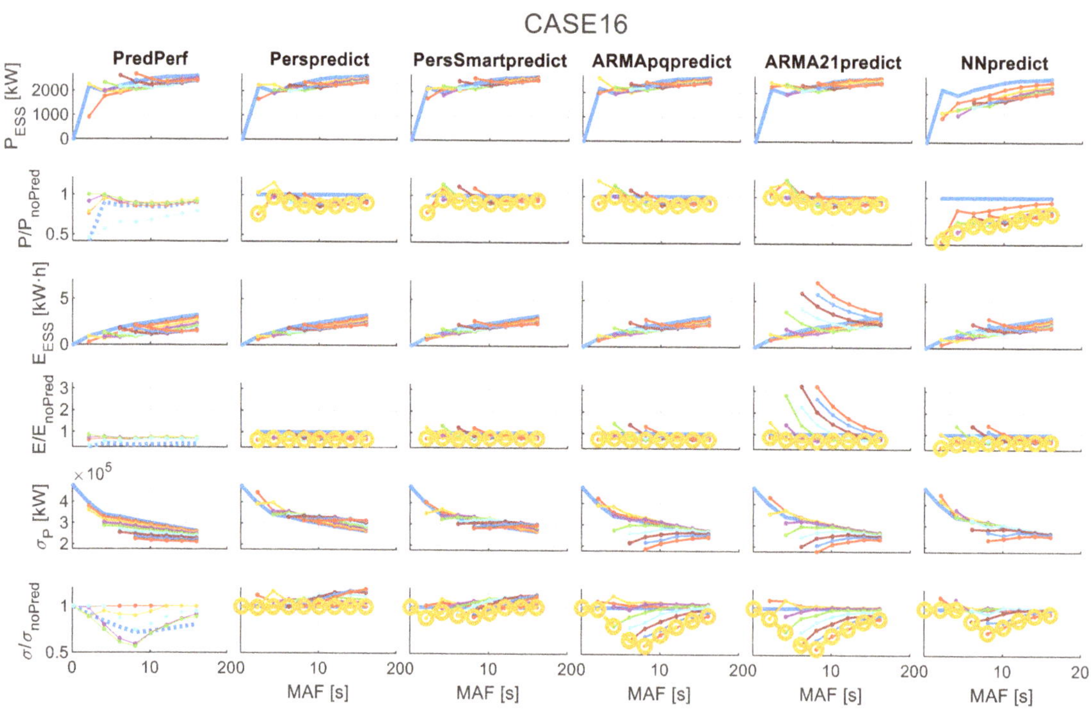

Figure A16. Results for CASE 16 with different MAFs and forecast algorithms.

References

1. Gunn, K.; Stock-Williams, C. Quantifying the Global Wave Power Resource. *Renew. Energy* **2012**, *44*, 296–304. [CrossRef]
2. Europea, C.; de Asuntos Marítimos y Pesca, D.G.; Borriello, A.; Calvo Santos, A.; Feyen, L.; Ghiani, M.; Guillén, J.; McGovern, L.; Petrucco, G.; Pistocchi, A.; et al. *The EU Blue Economy Report 2025*; Oficina de Publicaciones de la Unión Europea: Luxembourg, 2025.
3. Apolonia, M.; Fofack-Garcia, R.; Noble, D.R.; Hodges, J.; da Fonseca, F.X. Legal and Political Barriers and Enablers to the Deployment of Marine Renewable Energy. *Energies* **2021**, *14*, 4896. [CrossRef]
4. Astariz, S.; Iglesias, G. The Economics of Wave Energy: A Review. *Renew. Sustain. Energy Rev.* **2015**, *45*, 397–408. [CrossRef]
5. Magagna, D. *Ocean Energy Technology Development Report*; European Commission JRC: Brussels, Belgium, 2019.
6. Magagna, D. *Ocean Energy Technology Market Report*; European Commission JRC: Brussels, Belgium, 2019.
7. Mi, J.; Wu, X.; Capper, J.; Li, X.; Shalaby, A.; Wang, R.; Lin, S.; Hajj, M.; Zuo, L. Experimental Investigation of a Reverse Osmosis Desalination System Directly Powered by Wave Energy. *Appl. Energy* **2023**, *343*, 121194. [CrossRef]
8. Chakravarthi, T.V.S.K.; Chaudhuri, A.; Samad, A. Wave Powered Desalination Systems—Recent Developments. In Proceedings of the IEEE 2024 International Conference on Sustainable Energy: Energy Transition and Net-Zero Climate Future (ICUE), Pattaya City, Thailand, 21–23 October 2024; pp. 1–7.
9. Hasan, I.; Karayaka, H.B.; Kandil, T.; Yu, Y.-H. Analysis of Ocean Wave Energy-Driven Desalination Methods for Sustainable Water Production. In Proceedings of the 2025 IEEE Green Technologies Conference (GreenTech), Wichita, KS, USA, 26–28 March 2025; pp. 26–30.
10. Ocean Energy TWG. *SET-Plan Ocean Energy Implementation Plan*; European Commission: Brussels, Belgium, 2018.
11. Ruiz-Minguela, P.; Berque, J.; Villate, J.L.; Pirttimaa, L.; Gruet, R.; Noble, D.; Jeffrey, H. *Strategic Research and Innovation Agenda for Ocean Energy*; ETIP Ocean: Brussels, Belgium, 2024.
12. Tedeschi, E.; Molinas, M. Wave-to-Wave Buoys Control for Improved Power Extraction under Electro-Mechanical Constraints. In Proceedings of the 2010 IEEE International Conference on Sustainable Energy Technologies (ICSET), Kandy, Sri Lanka, 6–9 December 2010; pp. 1–6.
13. Salter, S.H.; Taylor, J.R.M.; Caldwell, N.J. Power Conversion Mechanisms for Wave Energy. *Proc. Inst. Mech. Eng. Part M J. Eng. Marit. Environ.* **2002**, *216*, 1–27. [CrossRef]
14. Falnes, J. *Ocean Waves and Oscillating Systems, Linear Interactions Including Wave-Energy Extraction*; Cambridge University Press: Cambridge, UK, 2002; ISBN 9780511754630.
15. Blanco, M.; Navarro, G.; Najera, J.; Lafoz, M.; Sarasua, J.I.; García, H.; Martínez-Lucas, G.; Pérez-Díaz, J.I.; Villalba, I. Wave Farms Integration in a 100% Renewable Isolated Small Power System—Frequency Stability and Grid Compliance Analysis. In Proceedings of the 15th European Wave and Tidal Energy Conference, Bilbao, Spain, 3–7 September 2023; Volume 15. [CrossRef]
16. Göteman, M.; Engström, J.; Eriksson, M.; Isberg, J. Optimizing Wave Energy Parks with over 1000 Interacting Point-Absorbers Using an Approximate Analytical Method. *Int. J. Mar. Energy* **2015**, *10*, 113–126. [CrossRef]
17. Blavette, A.; O'Sullivan, D.L.; Lewis, A.W.; Egan, M.G. Impact of a Wave Farm on Its Local Grid: Voltage Limits, Flicker Level and Power Fluctuations. In Proceedings of the 2012 OCEANS—Yeosu, Yeosu, Republic of Korea, 21–24 May 2012; pp. 1–9.
18. ENTSO-E. *Continental Europe Operation Handbook*; ENTSO-E: Brussels, Belgium, 2009.
19. ENTSO-E. *Network Code on Load-Frequency Control and Reserves*; ENTSO-E: Brussels, Belgium, 2013; Volume 6.
20. Robles, E.; Haro-Larrode, M.; Santos-Mugica, M.; Etxegarai, A.; Tedeschi, E. Comparative Analysis of European Grid Codes Relevant to Offshore Renewable Energy Installations. *Renew. Sustain. Energy Rev.* **2019**, *102*, 171–185. [CrossRef]
21. Grainger, J.J.; Stevenson, W.D. *Power System Analysis*; McGraw-Hill: New York, NY, USA, 1994; ISBN 0070612935.
22. LiVecchi, A.; Copping, A.; Jenne, D.; Gorton, A.; Preus, R.; Gill, G.; Robichaud, R.; Green, R.; Geerlofs, S.; Gore, S.; et al. *Exploring Opportunities for Marine Renewable Energy in Maritime Markets*; U.S. Department of Energy (DOE): Washington, DC, USA, 2019.
23. Villalba, I.; Blanco, M.; Pérez-Díaz, J.I.; Fernández, D.; Díaz, F.; Lafoz, M. Wave Farms Grid Code Compliance in Isolated Small Power Systems. *IET Renew. Power Gener.* **2019**, *13*, 171–179. [CrossRef]
24. Sarasua, J.I.; Martínez-Lucas, G.; García-Pereira, H.; Navarro-Soriano, G.; Molina-García, Á.; Fernández-Guillamón, A. Hybrid Frequency Control Strategies Based on Hydro-power, Wind, and Energy Storage Systems: Application to 100% Renewable Scenarios. *IET Renew. Power Gener.* **2022**, *16*, 1107–1120. [CrossRef]
25. Pelosi, D.; Gallorini, F.; Alessandri, G.; Barelli, L. A Hybrid Energy Storage System Integrated with a Wave Energy Converter: Data-Driven Stochastic Power Management for Output Power Smoothing. *Energies* **2024**, *17*, 1167. [CrossRef]
26. Bielewski, M.; Pfrang, A.; Bobba, S.; Kronberga, A.; Georgakaki, A.; Letout, S.; Kuokkanen, A.; Mountraki, A.; Ela, I.; Shtjefni, D.; et al. *Clean Energy Technology Observatory: Batteries for Energy Storage in the European Union—2022 Status Report on Technology Development, Trends, Value Chains and Markets*; Publications Office of the European Union: Brussels, Belgium, 2022.
27. European Commision Com(2020) 798/3 Final—Regulation of the European Parliament and of the Council Concerning Batteries and Waste Batteries, Repealing Directive 2006/66/EC and Amending Regulation (EU) No 2019/1020. 2020. Available online: https://eur-lex.europa.eu/legal-content/EN/TXT/?uri=CELEX%3A52020PC0798 (accessed on 1 September 2025).

28. Lafoz, M.; Blanco, M.; Beloqui, L.; Navarro, G.; Moreno-Torres, P. Dimensioning Methodology for Energy Storage Devices and Wave Energy Converters Supplying Isolated Loads. *IET Renew. Power Gener.* **2016**, *10*, 1468–1476. [CrossRef]
29. Moreno-Torres Concha, P.; Lafoz Pastor, M.; Navarro Soriano, G.; Blanco Aguado, M.; García-Tabarés Rodríguez, L. System for the Conditioning of the Electrical Power Generated in a Wave Generation System. Spanish Patent ES2547029A1, 8 July 2015.
30. Murray, D.B.; Hayes, J.G.; O'Sullivan, D.L.; Egan, M.G. Supercapacitor Testing for Power Smoothing in a Variable Speed Offshore Wave Energy Converter. *IEEE J. Ocean. Eng.* **2012**, *37*, 301–308. [CrossRef]
31. Elamin, A.Y.; Wahyudie, A. Optimal Control of a Supercapacitor Energy Storage System for Smoothing Wave Output Power. In Proceedings of the IEEE 2022 International Conference on Electrical and Computing Technologies and Applications (ICECTA), Ras al-Khaimah, United Arab Emirates, 23–25 November 2022; pp. 186–189.
32. Torres, J.; Blanco, M.; Lafoz, M.; Navarro, G.; Nájera, J.; Santos-Herran, M. Dimensioning Methodology of Energy Storage Systems for Power Smoothing in a Wave Energy Conversion Plant Considering Efficiency Maps and Filtering Control Techniques. *Energies* **2020**, *13*, 3380. [CrossRef]
33. Basheer, Y.; Pérez-Díaz, J.I.; Blanco, M.; Fraile-Ardanuy, J.; Nájera, J.; Navarro, G.; Sarasúa, J.I. Active Power Control of a Hybrid Power Plant Comprising Hydropower, Battery and Supercapacitor to Provide Frequency Control. In Proceedings of the 9th Hybrid Power Plants & Systems Workshop, Åland Islands, Finland, 3–4 June 2025.
34. Szostek, K.; Mazur, D.; Drałus, G.; Kusznier, J. Analysis of the Effectiveness of ARIMA, SARIMA, and SVR Models in Time Series Forecasting: A Case Study of Wind Farm Energy Production. *Energies* **2024**, *17*, 4803. [CrossRef]
35. Bahri, M.; Vahidnia, S.; Ghignone, L.; Mohtashamkhani, M. Wind Speed Prediction Using a SEEMD-LSTM Model. In Proceedings of the IEEE 2023 International Conference on Engineering and Emerging Technologies (ICEET), Istanbul, Turkiye, 27–28 October 2023; pp. 1–6.
36. Fusco, F.; Ringwood, J.V. A Study of the Prediction Requirements in Real-Time Control of Wave Energy Converters. *Sustain. Energy IEEE Trans.* **2012**, *3*, 176–184. [CrossRef]
37. Kaiser, M.S.; Iida, T.; Taniguchi, T.; Katayama, T.; Yoshimura, R.; Irifune, K. Optimal Prediction Horizon Length in Model Predictive Control to Maximise Energy Absorption by a Point Absorber Wave Energy Converter. *Ocean Eng.* **2025**, *329*, 121130. [CrossRef]
38. Li, D.; Wang, T.; Tao, J.; Sharma, S.; Borthwick, A.G.L.; Dong, X.; Shi, H. Model Predictive Control of a Single-Buoy Wave Energy Converter with Coupled Constraints and Model Adaptation. *Ocean Eng.* **2025**, *315*, 119887. [CrossRef]
39. Wang, J.; Zhang, D.; Huang, Q.; Cui, Z. Multiple-Step Accurate Prediction of Wave Energy: A Hybrid Model Based on Quadratic Decomposition, SSA and LSTM. *Int. J. Green Energy* **2025**, *22*, 100–123. [CrossRef]
40. Zhang, Y.; Liu, S.; Shen, Q.; Zhang, L.; Li, Y.; Hou, Z.; Chen, R. Short-Term Prediction Model of Wave Energy Converter Generation Power Based on CNN-BiLSTM-DELA Integration. *Electronics* **2024**, *13*, 4163. [CrossRef]
41. Elkhrachy, I.; Alhamami, A.; Alyami, S.H.; Alviz-Meza, A. Novel Ocean Wave Height and Energy Spectrum Forecasting Approaches: An Application of Semi-Analytical and Machine Learning Models. *Water* **2023**, *15*, 3254. [CrossRef]
42. Teixeira, R.; Cerveira, A.; Pires, E.J.S.; Baptista, J. Advancing Renewable Energy Forecasting: A Comprehensive Review of Renewable Energy Forecasting Methods. *Energies* **2024**, *17*, 3480. [CrossRef]
43. Sengupta, M.; Habte, A.; Wilbert, S.; Gueymard, C.; Remund, J.; Lorenz, E.; van Sark, W.; Jensen, A.R. *Best Practices Handbook for the Collection and Use of Solar Resource Data for Solar Energy Applications*, 4th ed.; NREL: Golden, CO, USA, 2024.
44. Raj, N.; Prakash, R. Assessment and Prediction of Significant Wave Height Using Hybrid CNN-BiLSTM Deep Learning Model for Sustainable Wave Energy in Australia. *Sustain. Horiz.* **2024**, *11*, 100098. [CrossRef]
45. Demonte Gonzalez, T.; Anderlini, E.; Yassin, H.; Parker, G. Nonlinear Model Predictive Control of Heaving Wave Energy Converter with Nonlinear Froude–Krylov Forces. *Energies* **2024**, *17*, 5112. [CrossRef]
46. Elamin, A.Y.; Wahyudie, A.; Hashfi, T.B.; Shareef, H.; Errouissi, R.; Laghari, M.S.; Mubin, M.B.; Mekhilef, S. Real-Time Model Predictive Control Framework for a Point Absorber Wave Energy Converter with Excitation Force Estimation and Prediction. *IEEE Access* **2024**, *12*, 4078–4098. [CrossRef]
47. Zhang, M.; Yu, S.-R.; Zhao, G.-W.; Dai, S.-S.; He, F.; Yuan, Z.-M. Model Predictive Control of Wave Energy Converters. *Ocean Eng.* **2024**, *301*, 117430. [CrossRef]
48. Guo, R.; Wu, Y.; Ma, X.; Aggidis, G.; Zhao, N. Grid Integration of a Novel Linear-Generator-Based Wave Power Conversion System. In Proceedings of the 2024 International Conference on Smart Energy Systems and Technologies (SEST), Torino, Italy, 10–12 September 2024; pp. 1–6.
49. Pelosi, D.; Trombetti, L.; Piccio, A.; Ottaviano, P.A.; Gallorini, F.; Barelli, L. Seawater Batteries as Sustainable Energy Storage Systems for the Future Renewables-Based Grids. In Proceedings of the 2024 IEEE International Humanitarian Technologies Conference (IHTC), Bari, Italy, 25–27 November 2024; pp. 1–6.
50. Zhu, X.; Huang, X.; Xiao, X. A Model Predictive Control Strategy for Smoothing Power Fluctuations of Wave Energy Converter Arrays Using Supercapacitors. In Proceedings of the IEEE 2024 UKACC 14th International Conference on Control (CONTROL), Southampton, UK, 10–12 April 2024; pp. 19–24.

51. Blanco, M.; Santos-Herran, M.; Navarro, G.; Torres, J.; Najera, J.; Villaba, I.; Lafoz, M.; Ramírez, D. Meta-heuristic Optimisation Approach for Wave Energy Converter Design by Means of a Stochastic Hydrodynamic Model. *IET Renew. Power Gener.* **2021**, *15*, 548–561. [CrossRef]
52. Blanco, M.; Navarro, G.; Lafoz, M.; Pérez, J.I. Study of the Impact of Wave Energy Generation in the Frequency of an Island Electric Grid. In Proceedings of the 12th European Wave and Tidal Energy Conference, Cork, Ireland, 27 August–2 September 2017; Volume 12, pp. 1–8.
53. Harris, M. Officials Sign Agreement for Canary Islands Wave Power Development. Available online: http://www.hydroworld.com/articles/2014/02/officials-sign-agreement-for-canary-islands-wave-power-development.html (accessed on 1 September 2025).
54. Veigas, M.; Iglesias, G. A Hybrid Wave-Wind Offshore Farm for an Island. *Int. J. Green Energy* **2015**, *12*, 570–576. [CrossRef]
55. Veigas, M.; Iglesias, G. Potentials of a Hybrid Offshore Farm for the Island of Fuerteventura. *Energy Convers. Manag.* **2014**, *86*, 300–308. [CrossRef]
56. Veigas, M.; Iglesias, G. Wave and Offshore Wind Potential for the Island of Tenerife. *Energy Convers. Manag.* **2013**, *76*, 738–745. [CrossRef]
57. Prediccion de Oleaje, Nivel Del Mar; Boyas y Mareografos—Puertos Del Estado. Available online: http://www.puertos.es/es-es/oceanografia/Paginas/portus.aspx (accessed on 26 December 2017).
58. École Centrale de Nantes; Kurnia, R.; Guillaume, D. NEMOH—Homepage. Available online: https://lheea.ec-nantes.fr/valorisation/logiciels-et-brevets/nemoh-presentation (accessed on 30 August 2024).
59. WEC-Sim (Wave Energy Converter SIMulator) [Computer Software]. Available online: https://github.com/WEC-Sim/WEC-Sim (accessed on 1 August 2025).
60. Eidsmoen, H. Hydrodynamic Parameters for a Two-Body Axisymmetric System. *Appl. Ocean Res.* **1995**, *17*, 103–115. [CrossRef]
61. Yu, Z.; Falnes, J. State-Space Modelling of a Vertical Cylinder in Heave. *Appl. Ocean Res.* **1995**, *17*, 265–275. [CrossRef]
62. Yu, Z.; Falnes, J. State-Space Modelling of Dynamic Systems in Ocean Engineering. *J. Hydrodyn.* **1998**, *10*, 1–17.
63. Blanco, M.; Santos-Herran, M.; Navarro, G.; Torres, J.J.; Najera, J.; García-Tabarés, L. Simplified Model of a Novel Direct-Drive PTO Based on an Azimuthal Linear Switched Reluctance Generator. In Proceedings of the 14th European Wave and Tidal Energy Conference (EWTEC), Plymouth, UK, 5–9 September 2021.
64. Hals, J.; Falnes, J.; Moan, T. A Comparison of Selected Strategies for Adaptive Control of Wave Energy Converters. *J. Offshore Mech. Arct. Eng.* **2011**, *133*, 31101. [CrossRef]
65. Lorenz, E.; Heinemann, D. Prediction of Solar Irradiance and Photovoltaic Power. In *Comprehensive Renewable Energy*; Elsevier: Amsterdam, The Netherlands, 2012; pp. 239–292.
66. Hoff, T.E.; Perez, R. Modeling PV Fleet Output Variability. *Sol. Energy* **2012**, *86*, 2177–2189. [CrossRef]
67. Boland, J. Time Series Modelling of Solar Radiation. In *Modeling Solar Radiation at the Earth's Surface*; Springer: Berlin/Heidelberg, Germany, 2008; pp. 283–312.
68. Boland, J. Time-Series Analysis of Climatic Variables. *Sol. Energy* **1995**, *55*, 377–388. [CrossRef]
69. Box, G. Box and Jenkins: Time Series Analysis, Forecasting and Control. In *A Very British Affair*; Palgrave Macmillan: London, UK, 2013; pp. 161–215.
70. Bishop, C.M. *Neural Networks for Pattern Recognition*; Oxford University Press: Oxford, UK, 1995; ISBN 9780198538493.
71. MacKay, D.J.C. A Practical Bayesian Framework for Backpropagation Networks. *Neural Comput.* **1992**, *4*, 448–472. [CrossRef]
72. Lauret, P.; Fock, E.; Randrianarivony, R.N.; Manicom-Ramsamy, J.-F. Bayesian Neural Network Approach to Short Time Load Forecasting. *Energy Convers. Manag.* **2008**, *49*, 1156–1166. [CrossRef]
73. MacKay, D.J.C. *Information Theory, Inference and Learning Algorithms*; Cambridge University Press: Cambridge, UK, 2003; ISBN 9780521642989.
74. Penny, W.D.; Roberts, S.J. Bayesian Neural Networks for Classification: How Useful Is the Evidence Framework? *Neural Netw.* **1999**, *12*, 877–892. [CrossRef] [PubMed]
75. Lafoz, M.; Blanco, M.; Ramirez, D. Grid Connection for Wave Power Farms. In Proceedings of the IEEE 2011-14th European Conference on Power Electronics and Applications (EPE 2011), Birmingham, UK, 30 August–1 September 2011; pp. 1–10.
76. Red Eléctrica de España, P.O. SENP 1: Funcionamiento de Los Sistemas Eléctricos Insulares y Extrapeninsulares 2019. Available online: https://www.ree.es/sites/default/files/2023-11/BOE-A-2019-18275_ministerio_para_la_transicion_ecologica.pdf (accessed on 1 September 2025).
77. Mazorra-Aguiar, L.; Díaz, F. Solar Radiation Forecasting with Statistical Models. In *Wind Field and Solar Radiation Characterization and Forecasting*; Springer: Cham, Switzerland, 2018; pp. 171–200.

applied sciences

MDPI

Article

Hybrid Wind–Redox Flow Battery System for Decarbonizing Off-Grid Mining Operations

Armel Robert, Baby-Jean Robert Mungyeko Bisulandu, Adrian Ilinca * and Daniel R. Rousse

Mechanical Engineering, T3E Research Group, École de Technologie Supérieure (ÉTS), Montréal, QC H3C 1K3, Canada; armel.robert.1@ens.etsmtl.ca (A.R.); baby-jeanrobert.mungyekobisulandu@etsmtl.ca (B.-J.R.M.B.); daniel.rousse@etsmtl.ca (D.R.R.)
* Correspondence: adrian.ilinca@etsmtl.ca

Abstract

Transitioning to sustainable energy systems is crucial for reducing greenhouse gas (GHG) emissions, especially in remote industrial operations where diesel generators remain the dominant power source. This study examines the feasibility of integrating a redox flow battery (RFB) storage system to optimize wind energy utilization at the Raglan mining site in northern Canada, with the goal of reducing diesel dependency, enhancing grid stability, and improving energy security. To evaluate the effectiveness of this hybrid system, a MATLAB R2024b-based simulation model was developed, incorporating wind energy forecasting, load demand analysis, and economic feasibility assessments across multiple storage and wind penetration scenarios. Results indicate that deploying 12 additional E-115 wind turbines combined with a 20 MW/160 MWh redox flow battery system could lead to diesel savings of up to 63.98%, reducing CO_2 emissions by 68,000 tonnes annually. However, the study also highlights a key economic challenge: the high Levelized Cost of Storage (LCOS) of CAD (Canadian dollars) 7831/MWh, which remains a barrier to large-scale implementation. For the scenario with high diesel economy, the LCOS was found to be CAD 6110/MWh, and the corresponding LCOE was CAD 590/MWh. While RFB integration improves system reliability, its economic viability depends on key factors, including reductions in electrolyte costs, advancements in operational efficiency, and supportive policy frameworks. This study presents a comprehensive methodology for evaluating energy storage in off-grid industrial sites and identifies key challenges in scaling up renewable energy adoption for remote mining operations.

Keywords: decarbonization; renewable energies; energy storage; redox flow batteries; wind energy; economic analysis

1. Introduction

In the past two decades, dependence on fossil fuels has steadily decreased due to resource depletion and growing concerns over greenhouse gas (GHG) emissions from their combustion. These challenges have prompted governments worldwide to implement strict energy policies to regulate and manage the utilization of both fossil and renewable energy sources. To mitigate GHG emissions, countries that have ratified the Paris Agreement have established regulatory frameworks that set maximum allowable emission thresholds for industries, with penalties for non-compliance. As a signatory to the Paris Agreement, Canada, which accounted for 0.72 Gt CO_2-equivalent emissions in 2013 (1.95% of global emissions at the time) [1], has committed to reducing national GHG emissions by 40–45% by

2030 and achieving net-zero emissions by 2050 [2–5]. In 2021, the Raglan mine's production process consumed 58.9 million liters of diesel, with electricity generation alone accounting for 73% of this consumption, resulting in 119,440 tons of CO_2-equivalent emissions. Self-propelled mining equipment accounted for the remaining 27%, emitting 44,400 tons of CO_2 equivalent. To meet these ambitious targets and minimize the environmental and economic risks associated with climate change, transitioning toward decarbonization through the adoption of renewable energy technologies and sustainable energy practices has become imperative [6].

According to Da Silva et al. [7], decarbonization is characterized by three fundamental challenges: (i) replacing fossil fuels with renewable energy sources, (ii) developing alternative solutions for sectors where reducing fossil fuel consumption is particularly difficult, and (iii) eliminating carbon dioxide emissions through advanced mitigation strategies. Du et al. [8] highlight the necessity of a comprehensive transformation in the global electricity sector to achieve effective decarbonization. Similarly, Balaban et al. [9] argue that decarbonizing an energy system requires significant structural changes across the electricity production chain, including generation, transmission, and consumption, as this sector alone accounts for nearly two-thirds of total GHG emissions. Dongsheng et al. [10] further emphasize that ensuring a resilient and sustainable energy future by 2050 is only feasible if power generation integrates cost-effective and scalable renewable solutions. Meanwhile, Obiora et al. [11] and Roshan Kumar et al. [12] stress that a transition to low-carbon energy infrastructures must precede decarbonization efforts, as this remains the most effective approach to reducing the severe environmental impacts of climate change.

Amid the ongoing energy transition, many industrial facilities are actively investing in renewable energy systems to reduce carbon emissions and enhance energy sustainability. In this context, the Raglan mining company in Nunavik, northern Quebec, has committed to decarbonizing its electricity generation network by expanding its wind farm. As a major producer of raw nickel, the site's operations encompass ore exploration, underground excavation, and rock processing, all of which require a reliable and continuous energy supply. Among the most widely adopted renewable alternatives, solar and wind energy are increasingly integrated into off-grid and industrial energy systems. However, due to their intermittent nature and weather dependence, these energy sources require advanced energy storage solutions to ensure grid stability and operational reliability. This study investigates the feasibility of storing surplus wind energy using redox flow batteries (RFBs) to regulate and optimize local electricity production.

For Asri et al. [13], it is, therefore, crucial to critically analyze the fundamental characteristics of energy storage systems (ESSs) in order to establish benchmarks for selecting the most suitable technology. Mahadevan et al. [14] claim that each ESS technology has unique strengths and limitations, influencing its applicability and suitability for specific applications. The choice of an ESS is influenced by factors such as energy density, cost, scalability, and the intended use case, whether for short-term grid balancing or long-term energy storage. It is crucial to consider the climatic conditions of a region before implementing any energy storage system [15]. Table 1 focuses on ESSs currently proficient in providing critical storage capacities of at least 20 MW.

Redox flow battery technology has gained significant attention as a viable solution for large-scale energy storage due to its high efficiency, long cycle life, and independent scalability of power and capacity [16–21]. These batteries are considered one of the most promising technologies for integrating renewable energy into electricity grids [22–24], offering low-cost and long-duration storage capabilities [25]. The large-scale deployment of variable renewable energy resources is heavily influenced by the availability of cost-effective, long-duration storage solutions [26]. Despite their advantages, flow batteries accounted for only 5% of the 16 GWh installed energy storage capacity worldwide in 2021, highlighting the need for further technological advancements and economic optimizations to expand their adoption.

Table 1. Comparison and characteristics of energy storage systems [13].

Storage System Type	Max Power Rating [MW]	Discharge Time	Max Cycles or Lifetime	Energy Density [Wh/L]	Efficiency
Pumped hydro	~3000+	4 h–16 h	30–60 years	0.2–2	70–85%
Compressed air	~1000	2 h–30 h	20–40 years	2–6	40–70%
Molten salt	100–150	Several hours	~30 years	70–210	80–90%
Li-ion battery	1–100+	1 min–8 h	1000–10,000 cycles	200–400	85–95%
Lead–acid	1–100	1 min–8 h	6–40 years	50–80	80–90%
Flow battery	10–100+	Several hours	12,000–14,000 cycles	20–70	60–85%
Hydrogen	~100+	Minutes–weeks	5–30 years	~600 (at 700 bar)	25–45%
Flywheel	1–20	Seconds–minutes	20,000–100,000+ cycles	20–80	70–95%

A flow battery is a reversible and rechargeable electrochemical system that stores energy in chemical form and converts it back into electrical energy when needed. The fundamental operating principle of these batteries involves the circulation of an electrolytic solution (electrolyte) from one or more storage reservoirs into electrochemical cells facilitated by pumps [16,27]. These cells, consisting of two half-cells (positive and negative), serve as the site for redox reactions, enabling the storage and release of energy. Flow batteries can be configured in series and/or parallel to achieve the desired voltage and current characteristics. They are often considered a hybrid technology between fuel cells and conventional batteries, where hydrogen (H_2) and oxygen (O_2) gases in fuel cells are replaced by electrolytes capable of storing electrical charges [28]. The schematic representation of a flow battery system is illustrated in Figure 1.

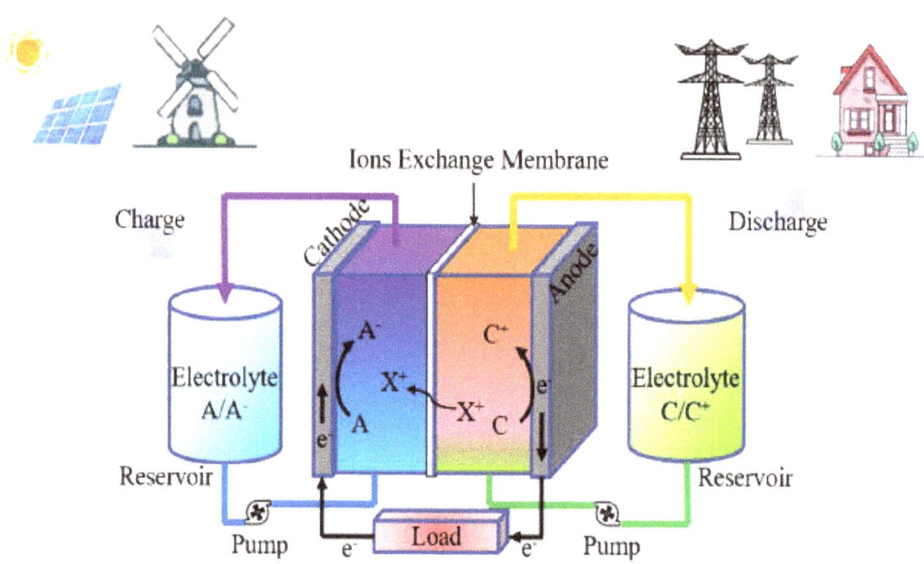

Figure 1. Schematics and working principle of redox flow batteries [16].

Several studies have explored the application of flow batteries for energy storage in different renewable energy systems. Ouyang et al. [25] investigated the performance of a vanadium redox flow battery (VRFB) within a microgrid integrating a biomass gasifier and a solid oxide fuel cell designed for electricity generation in rural environments (Figure 2). Their findings indicate that the electrolyte flow rate plays a crucial role in battery performance, with an efficiency of 84% achieved during peak shaving operations.

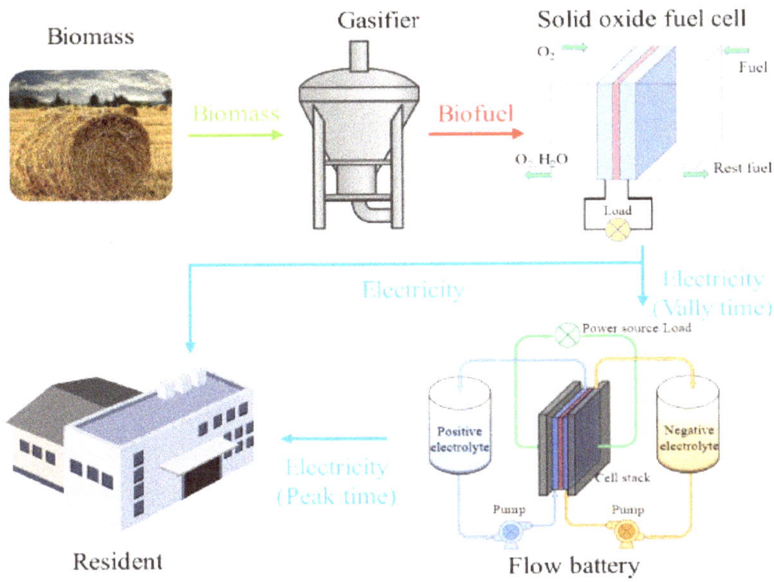

Figure 2. Flow battery in a renewable energy microgrid [25].

Zheng et al. [29] examined the integration of a flow battery system to address the intermittency of renewable energy sources while managing fluctuating energy demand. The proposed energy system featured a hybrid setup comprising a partially covered parabolic mirror thermal photovoltaic collector, a vanadium redox flow battery, thermal energy storage, and an absorption chiller/heat pump. Their comparative analysis between a VRFB-based system and a Power-to-Hydrogen-to-Power (P2H2P) system revealed that the VRFB system achieved an exergy efficiency of 78.52%, outperforming the P2H2P system, which had an efficiency of 41.73%.

Yesilyurt et al. [30] explored the feasibility of powering an air-source hybrid heat pump using a vanadium redox flow battery-based energy storage system. Their study selected two key performance indicators: exergy destruction and exergy efficiency. Using a MATLAB Simulink model, they analyzed a system where solar photovoltaic panels and wind turbines charged the VRFB. The results showed that integrating wind turbines with the VRFB was more effective, as the photovoltaic-based system exhibited significant exergy destruction and lower exergy efficiency.

In a different study, Oshnoei et al. [31] presented a load frequency control model for a wind power system and redox flow battery storage. Their findings demonstrated that frequency-based VRFB modeling yielded superior performance compared to conventional VRFB modeling techniques, highlighting the potential of flow batteries in enhancing grid stability.

Among the different types of flow batteries, all-vanadium redox flow batteries (VRFBs) stand out due to their flexibility, adaptability, and scalability. Reynard and Girault [32] investigated redox flow battery technology for energy storage and clean hydrogen production, integrating VRFBs with catalytic reactors to enhance efficiency. Huang et al. [33] estimated the overall efficiency of such systems to range between 70% and 85%, while Ulaganathan et al. [34] reported efficiencies exceeding 83%, a value often used as a benchmark for sizing battery systems.

One key advantage of all-vanadium flow batteries is their ability to independently scale storage capacity and power output. Hosseiny and Wessling [35] and Skyllas-Kazacos et al. [36] emphasize that increasing storage capacity is directly proportional to the electrolyte volume, necessitating larger electrolyte reservoirs for extended operation. Zhang et al. [36] further highlight that increasing the battery's storage capacity and cycle duration can effectively reduce capital expenditure (CAPEX) per kilowatt-hour, improving the economic feasibility of large-scale deployments.

The economic viability and resilience of hybrid energy system solutions depend on a comprehensive evaluation of economic and reliability factors during the design phase [37]. In the literature, numerous studies have investigated the investment costs and financial feasibility of redox flow battery (RFB) storage systems for large-scale energy applications. Cremoncini et al. [38] classify two primary approaches to the techno-economic analysis of RFBs: The first focuses on material and investment costs. In contrast, the second incorporates Levelized Cost of Storage (LCOS) models to assess long-term financial sustainability. Darling [26] conducted a detailed techno-economic study on redox flow batteries, demonstrating that capital and operational costs, combined with projected energy throughput, are key determinants of LCOS.

According to IRENA [39], the costs of vanadium and zinc/bromine flow batteries could decline by more than 60% over the next two decades, driven by technological advancements and growing market interest. Similarly, Schmidt et al. [40] predict a steady reduction in redox flow battery costs due to economies of scale and material innovations. Minke and Turek [41] emphasize that assessing the techno-economic feasibility of RFBs remains a significant challenge for large-scale projects, where system costs must be evaluated based on material composition, system design, and modeling methodologies.

Zheng et al. [42] emphasize the need for a deeper understanding of the technical and economic performance parameters of flow battery technologies to enhance their commercial viability. Zhang et al. [43] further highlight that cost challenges remain a significant barrier to the widespread adoption of RFBs. Noack et al. [44] advocate for optimization strategies in vanadium redox flow battery systems, particularly in reducing material costs, optimizing bipolar plate resistances, and improving energy storage media efficiency.

To support cost reductions, Viswanathan et al. [45] developed a cost model for redox flow batteries, revealing that the energy-to-power ratio of the storage system is a crucial factor in minimizing costs across various battery chemistries. Additionally, LCOE reductions are influenced by integrating advanced materials (e.g., novel alloys), improved manufacturing processes, and higher nominal power capacities [46]. Currently, the levelized cost of energy (LCOE) for redox flow batteries is estimated at CAD 0.0503/kWh [29]. Ouyang et al. [25] report a payback period of 7.33 years for a microgrid equipped with a vanadium redox flow battery, further reinforcing the economic feasibility of this technology. In the literature, LCOS models for redox flow batteries have been widely utilized to evaluate cost-effectiveness across different storage applications [38,40].

This study evaluates the feasibility of integrating a redox flow battery storage system with an expanded wind farm at the Raglan mining site. This research utilizes a MATLAB R2024b-based simulation model to evaluate the technical, economic, and environmental impacts, with a focus on diesel savings, CO_2 emissions reduction, and system reliability. Unlike previous studies focusing on standalone storage systems, this research integrates a comprehensive techno-economic analysis, comparing multiple wind–battery hybrid scenarios. Specifically, it examines how different battery capacities (80 MWh, 120 MWh, 160 MWh) and wind penetration levels (45–75%) affect the Levelized Cost of Energy (LCOE) and Levelized Cost of Storage (LCOS), key determinants of project feasibility [26,44].

The key research questions guiding this study are as follows:

1. What is the optimal configuration of wind energy and redox flow storage to maximize diesel savings at the Raglan mining site?
2. What are the economic implications of integrating large-scale redox flow batteries, particularly in Arctic conditions?
3. How does energy storage improve grid stability, and what policy incentives could enhance financial feasibility?

This article is part of a broader research initiative exploring various energy storage technologies to support the decarbonization of remote mining operations in northern Quebec. In addition to the redox flow battery system analyzed in this study, parallel investigations are underway on battery energy storage systems (BESS), hydrogen storage, and PHSP systems [47,48]. These complementary studies evaluate each option based on technical performance, economic viability, and environmental compatibility within off-grid Arctic contexts. However, due to the ongoing comparative analysis, this paper focuses solely on the techno-economic feasibility of implementing a redox flow battery system in the Raglan hybrid electricity network.

Building on this context, the present work investigates the potential for significantly reducing or even eliminating diesel dependency at the Raglan mine by integrating a wind farm coupled with a redox flow battery storage system. Specifically, it examines how this hybrid "Diesel Thermal Power Plants–Wind Farm–redox battery" configuration can address the intermittency of wind energy and maintain a stable electricity supply across the mine's 25 kV-G network. The study involves the overall dimensioning of the wind–redox battery system, assessing its ability to meet electricity and heating demands, and techno-economic feasibility analysis to support future investment and implementation decisions.

The findings of this study offer valuable insights for industrial energy planners, policymakers, and renewable energy developers, providing a scalable model for integrating storage systems in off-grid mining operations.

Given that this work focuses on decarbonizing the Raglan mine through the sizing and optimization of a hybrid energy system based on redox flow battery storage, the model aims to minimize the levelized cost of electricity (LCOE) while maximizing system performance. Accordingly, the outputs of the optimization process include

- Economic indicators: capital and operating expenditures, energy price, payback period, and net present value (NPV);
- Technical indicators: installed capacity and power, number of wind turbines, wind penetration rate, number of generators in operation or removed, redox battery capacity, and greenhouse gas (GHG) emission reduction.

The specific objective of this work is to study the feasibility of integrating a redox flow battery (RFB) storage system into the Raglan mine site's hybrid grid to reduce diesel dependence, improve grid stability, and enhance energy security (optimizing wind farm energy production and addressing wind farm intermittency).

Table 2 summarizes relevant work on RFB applications.

Table 2. Summary of works applying RFB technology.

References	Context-Application	Storage System	Results
Zheng et al. [29]	To address the intermittency of renewable energy sources To manage fluctuating energy demand	VRFB system	Exergy efficiency of 78.52%
Yesilyurt et al. [30]	Powering a hybrid air-source heat pump	VRFB system	Efficiency of wind turbine integration with VRFB
Oshnoei et al. [31]	Load frequency control model for a wind power system	VRFB system	Higher performance due to the use of frequency-based VRFB
Reynard and Girault [32]	Energy storage Clean hydrogen production, integrating VRFBs with catalytic reactors to enhance efficiency	VRFB system	Improving the efficiency of catalytic reactors
Ouyang et al. [25]	Performance of a VRFB within a microgrid integrating a biomass gasifier and a solid oxide fuel cell designed for electricity generation in rural environments	VRFB system	Battery performance is highly dependent on the flow of the electrolyte. An efficiency of 84% was achieved during peak shaving operations.

The rest of this paper is organized as follows: Section 2 briefly describes the energy system; Sections 3 and 4 detail the methodology, including the design and sizing of the wind energy storage system, as well as the assumptions and modeling framework used to assess the technical, economic, and environmental feasibility of integrating redox flow batteries (RFBs); Section 5 presents the validation of the model with global data at Raglan mine; Section 6 discusses the results, including economic feasibility and grid performance; and Section 7 outlines the conclusions and recommendations for future energy storage projects.

2. System Description

The Raglan mining site, located in northern Quebec, Canada, operates a localized energy network known as "Network 25 kV-G," which supplies both electricity and heat to the site. This network comprises twenty diesel generators and a wind farm with two operational wind turbines. Among these twenty generators, six EMD models (see Figure 3) are equipped with heat recovery loops, forming a cogeneration system that meets the site's thermal energy demands.

As part of its decarbonization strategy, the Raglan mining company is committed to reducing diesel consumption, a major contributor to the site's greenhouse gas (GHG) emissions. Given the high wind energy potential in the region, the company plans to expand the existing wind farm by adding additional wind turbines to replace the diesel generators scheduled for decommissioning. However, due to the intermittent nature of wind energy, integrating an energy storage solution is essential to ensure grid stability, reliability, and continuous energy supply. To address these challenges, the company is considering a redox flow battery (RFB) storage system to help mitigate wind energy fluctuations and regulate the 25 kV-G network during peak and low-demand periods.

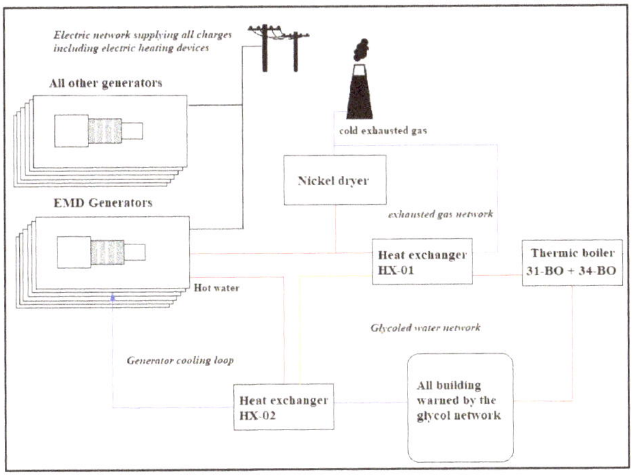

Figure 3. Scheme of the electrical and heating networks.

The VRB ENERGY all-vanadium redox flow battery, manufactured by a Vancouver-based Canadian company, has been selected for this application (see Figure 4). This battery technology offers high efficiency, long cycle life, and independent scalability of power and capacity, making it well-suited for large-scale renewable energy storage at the Raglan site.

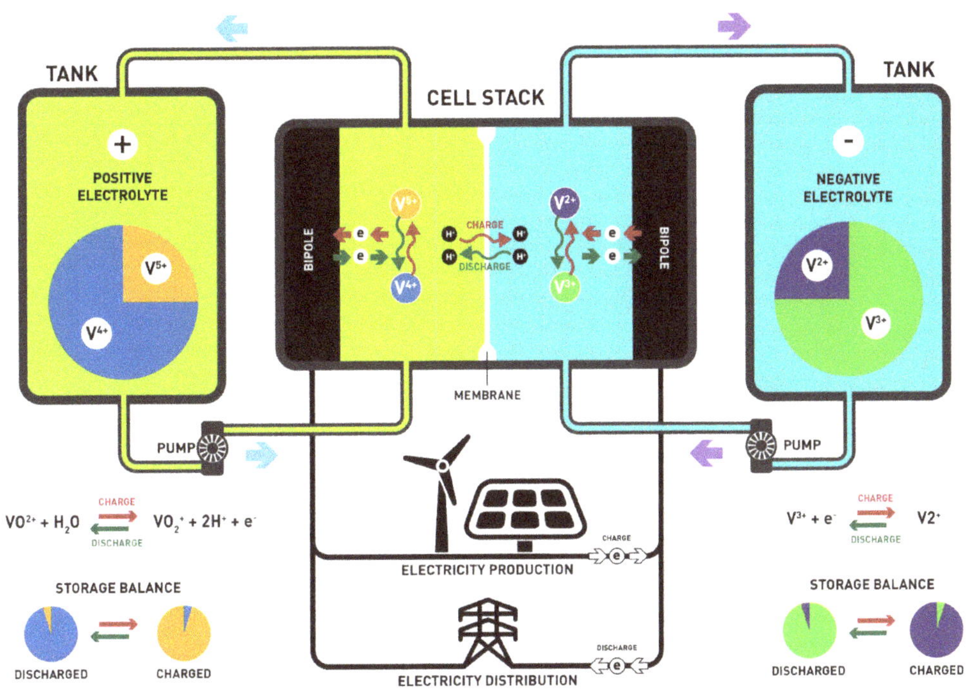

Figure 4. VRB ENERGY all-vanadium flow battery: battery model VRB-ESS, 1 MW, 500 MWh of energy storage capacity installed or in development [49].

The project's first phase focuses on optimizing wind farm production, ensuring a complete transition from diesel generators, meeting electrical and thermal energy demands, and appropriately sizing the redox flow battery required for the site's operations. The successful implementation of this hybrid wind–storage system plays a critical role in achieving the site's long-term energy transition and sustainability goals.

3. Methodology

This study aims to design and optimize a wind energy storage system utilizing redox flow batteries (RFBs) at the Raglan mining site, thereby reducing diesel consumption and decarbonizing electricity production. The MATLAB-based simulation model developed for this study integrates wind power generation, battery storage dynamics, load demand, and economic feasibility analysis. The main equations used in the model are presented in Appendix A.

3.1. Assumptions

The simulation is based on the 2020 operational data from the Raglan site, considering the following key assumptions:

- Energy demand: The mine's annual electricity consumption is 145,195 MWh, requiring 34.2 million liters of diesel. The daily average electricity consumption of the site is shown in Figure 5.

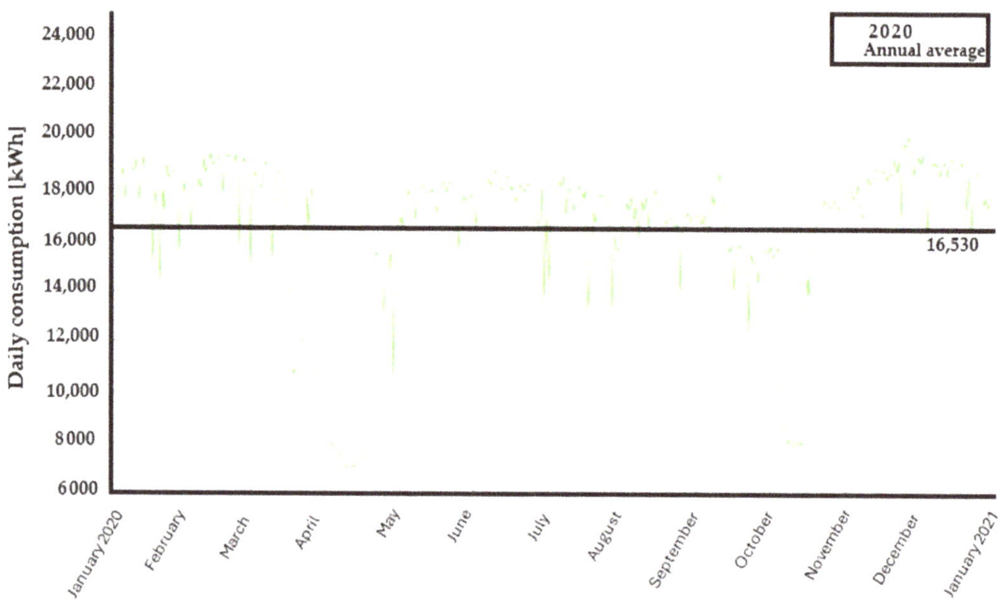

Figure 5. Daily electricity consumption of the 25 kV network in 2020.

- Wind resource: Wind speed data from 2020 indicates monthly average speeds between 6 and 9 m/s (see Figure 6) with an annual average of 8.5 m/s at the wind turbine hub height (92 m).
- Ambient temperature: The average temperature on site is −15 °C, which affects air density (1.367 kg/m^3 instead of the standard 1.225 kg/m^3).
- Wind farm expansion: The existing two wind turbines (Enercon E-82) are expected to be supplemented with newer E-115 EP3 E4 turbines, each with a rated power of 4260 kW (92 m hub height) and a power curve illustrated in Figure 7.
- Battery storage: The selected VRB ENERGY all-vanadium flow battery is installed in a heated facility to mitigate cold-related efficiency losses.

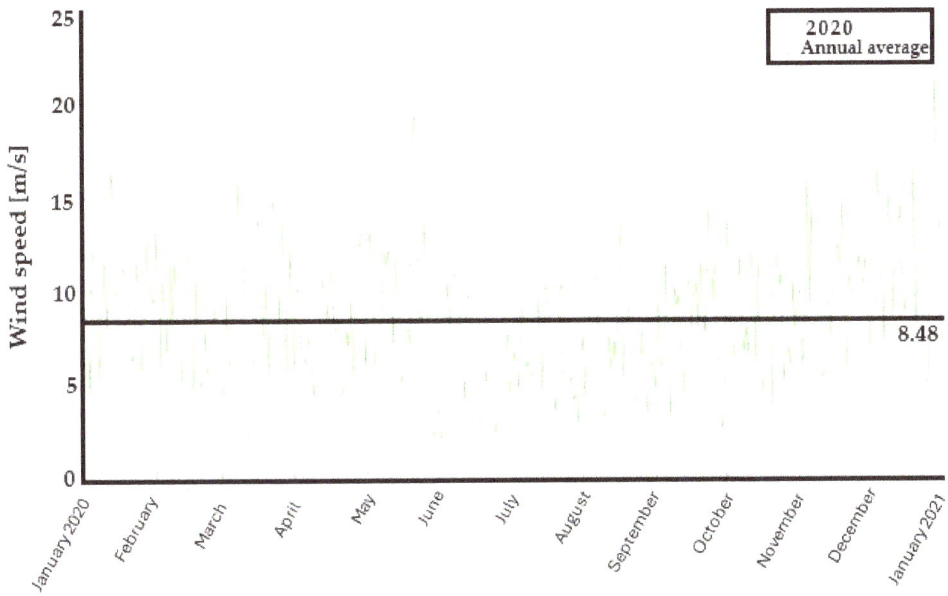

Figure 6. Wind speed in 2020 in Raglan.

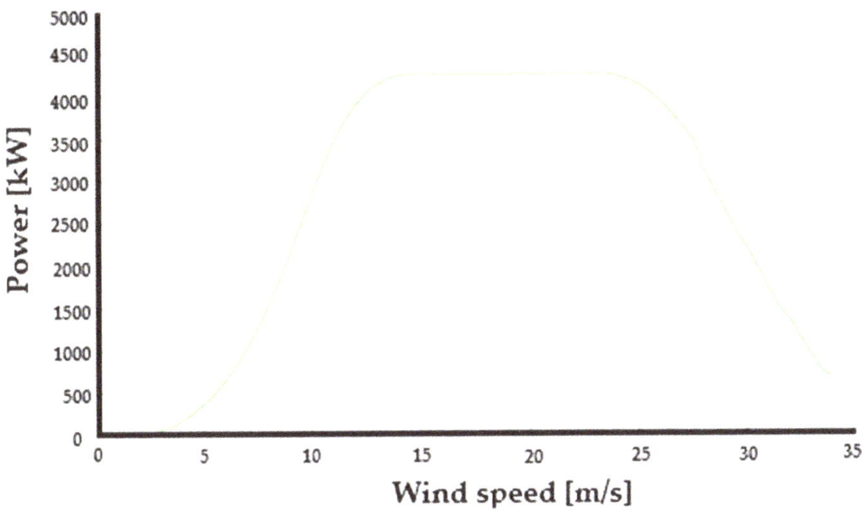

Figure 7. The power curve of the E-115 EP3 E4 wind turbine.

3.2. Wind Energy Production in Mining Sites: Forecasting, Optimization, and Utilization

Given that wind production is highly dependent on wind speed, this study used wind speed data from several years (2017–2018–2019–2020–2021). The year 2020 was chosen as it was the most representative compared with the other years. An average speed of approximately 8.5 m/s was therefore established over the year. This is homogeneous. Figure 8 shows the occurrence of wind speeds at quarter-hourly intervals over the year 2020.

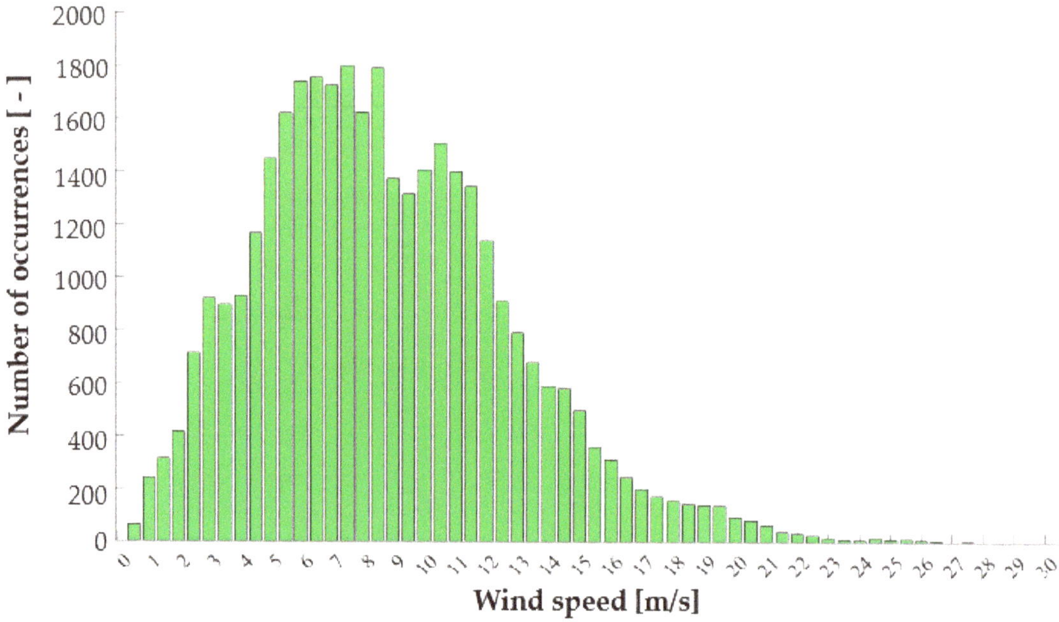

Figure 8. Frequency of occurrence of different wind speeds in 2020.

The distribution of Figure 8 is made possible using Equation (A5) in Appendix A.

From the wind data (wind potential of the site) of the Raglan mining site, the power available for each of the wind turbines is deduced using the information provided by the manufacturer. For example, the E-115 EP3 E4 wind turbine (Enercon, Aurich, Germany) produces 3472 MW, while the E-82 E4 wind turbine (Enercon, Aurich, Germany) produces 1880 MW when the wind speed reaches 11 m/s. As wind speeds are dependent on the seasons and the weather, energy production from wind turbines is also dependent on them. Except for violent winds (storms, etc.) with speeds of up to 26.5 m/s, wind production is higher when the wind is stronger. Thus, the wind distribution at the Raglan site follows a Weibull distribution, and its shape is characteristic of it (see Figure 8).

Considering that wind energy production is intermittent, the project plans to couple the wind farm with a redox flow battery storage system.

To study the periods without wind at Raglan, an assumption is made that considers a wind speed of less than 3 m/s to be a period without wind. This assumption is based on the fact that the E-115 wind turbine does not produce even 80 kW below 3 m/s, which is required for de-icing. In 2020, there was 643 h without wind over the whole year. The longest period without wind was 28 consecutive hours. During this 643 h, almost 100% of electricity production would have to rely on diesel generators (see Figure 9).

The periods without wind, representing more than 61% of cases, are shorter than eight hours. This indicator can be used to determine the battery size and its storage life.

3.3. *Sizing of the Energy System*

Raglan mine is set to expand its renewable energy capacity by 2030, with wind energy selected as the primary source due to the region's abundant wind resources and the success of its initial wind turbine project, which currently operates two turbines. The additional wind energy generated should be integrated into the 25 kV-G network, marking a significant milestone in the site's decarbonization efforts. This transition reduces reliance on diesel generators, enhances energy sustainability, and aligns with broader carbon reduction targets.

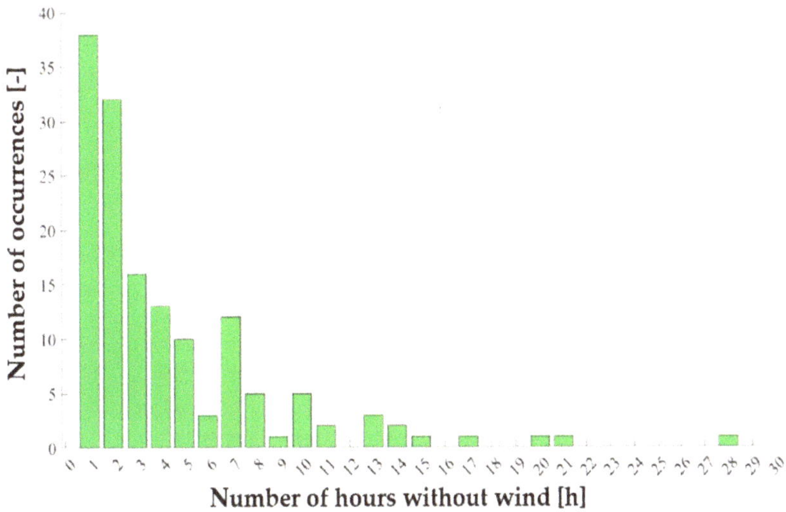

Figure 9. Number of occurrences of windless periods.

3.3.1. Sizing, Installation, and Cost of the Wind Farm

Proper sizing and installation of the wind farm are critical, as site-specific characteristics directly influence turbine performance. The Raglan mining site in northern Canada experiences extreme cold temperatures throughout the year, making wind energy more suitable than solar power due to limited sunlight availability. Among the various renewable energy sources, wind power is particularly well-adapted to the climatic conditions of northern Canada [48].

As mentioned, the Raglan site currently operates two Enercon E-82 E4 wind turbines, each with a nominal power output of 3 MW, integrated into the 25 kV grid network. To expand the site's renewable energy capacity and advance its decarbonization objectives, the current project proposes deploying additional wind turbines of the latest ENERCON model, the E-115 EP3 E4. This upgraded model, featuring a 92 m mast and a nominal power rating of 4.260 MW, represents a 42% increase in capacity compared to the existing turbines. Table 3 summarizes the key characteristics of the wind turbines at the Raglan mine site.

Table 3. Characteristics of the wind turbines at the Raglan mine site.

Types of Wind Turbine	Minimum Wind Speed for a Wind Turbine Operation	Wind Speed to Reach Maximum Power	Maximum Operating Wind Speed of the Wind Turbine
E-82	3 m/s	17 m/s	26 m/s
E-115	2 m/s	14.5 m/s	34 m/s

The E-115 wind turbine is significantly larger than its predecessor, featuring a rotor diameter of 115.7 m and a sweep area of approximately 10,516 m^2. One of its key advantages is the ability to generate higher energy output while reducing the number of turbines required, thereby minimizing land use and environmental impact. Additionally, this model is specifically designed to withstand Arctic conditions, including frost, extreme cold, and severe storms. Given the site's environmental challenges, the estimated operational lifespan of the wind turbines is conservatively set at 15 years for design purposes. However, the manufacturer specifies a lifespan of 25 years under standard operating conditions.

Due to the extensive size of the Raglan mining site, careful selection of the wind farm's location is essential to ensure optimal connectivity to the 25 kV-G electrical network while facilitating access for maintenance teams. The area surrounding Katinniq has been identified as the most suitable location for the additional wind turbines. LiDAR (Light Detection and Ranging) studies are planned to further refine site selection and optimize turbine placement, enabling the accurate assessment of wind characteristics, including speed, intensity, and directional patterns.

A thorough analysis of investment costs is crucial in large-scale projects, as it informs key decisions regarding equipment selection, financial feasibility, and trade-offs between technical performance and site-specific constraints. Understanding these costs is essential for optimizing project design while balancing economic and environmental considerations.

In recent years, the cost of wind energy projects has significantly declined and is expected to continue decreasing in the coming decades. According to Stehly and Duffy [50], this trend is driven by improvements in turbine performance, reduced energy losses and operational risks, enhanced cost management, and advances in materials and construction techniques.

To evaluate the financial feasibility of the wind energy expansion at Raglan, this study employs the Levelized Cost of Energy (LCOE) metric. LCOE is a key economic indicator that provides a comparative assessment of different energy technologies. It consists of two main components: capital expenditure (CAPEX), which represents upfront investment costs, and operating expenditure (OPEX), which includes maintenance and operational expenses. LCOE is typically expressed in CAD/kWh or CAD/MWh and is influenced by five primary factors: initial capital investment, ongoing operational costs, system performance, financial and tax assumptions, and project lifespan.

The study also incorporates data from the IRENA report [46], which provides relevant insights specific to Canada, including the Raglan mining site. Based on these indicators and the referenced report, the cost analysis yielded the results in Table 4.

Table 4. Wind CAPEX, OPEX, and LCOE in Canada and Raglan.

Localization	OPEX (CAD/kW·yr)	OPEX (CAD/MWh)	CAPEX (CAD/kW)	CAPEX (CAD/MWh)	LCOE (CAD/MWh)
Canada	46.9	11.84	1833	23.15	35.0
Raglan	117.6	30.76	4857	84.69	115.5

The results indicate a substantial increase in both OPEX (150.8%) and CAPEX (164.9%) for the Raglan site compared to the national average, primarily due to the site's extreme environmental conditions, including harsh winters and remote location.

The price of an ENERCON E-115 EP3 E4 wind turbine (4.260 MW) for the Raglan site is calculated at approximately CAD 20,690,000 using the cost estimates from Table 4. A simplifying assumption was adopted to determine the total wind farm investment cost, wherein the cost of a single wind turbine is multiplied by the total number of turbines planned for deployment.

3.3.2. Sizing, Implementation, and Cost of the Storage System

Despite the abundant wind energy potential at the Raglan site, an energy storage solution is essential to mitigate wind power's intermittency and ensure grid stability.

This study assesses the feasibility of integrating redox flow batteries (RFBs) as the primary storage technology, given their proven advantages, pre-installation readiness, and technological maturity.

To accurately size the storage system for the Raglan mining site, this study relies on agency reports, laboratory data, and direct manufacturer specifications. Given the region's extreme climatic conditions, a critical design consideration is protecting batteries from subzero temperatures. While redox flow battery systems are typically delivered in insulated containers, prolonged exposure to temperatures below −5 °C can cause performance degradation. Specifically, low temperatures can impair electrolyte flow, reduce electrochemical efficiency, and shorten system lifespan. To mitigate these risks, the storage units are housed in a heated shed, ensuring protection against snow, wind, and frost while maintaining optimal operating conditions.

Two key parameters determine the size of the shed:

- The footprint of the redox battery system;
- The required storage capacity.

This study directly consulted battery manufacturers since literature sources do not provide standard values for redox battery footprints. The responses varied between 30 m^2/MWh and 80 m^2/MWh, with an average value of 50 m^2/MWh selected for this project. Given the large footprint of the mining site, the battery storage facility is strategically located near the main complex in Katinniq, ensuring efficient integration into the mine's energy circuit.

As with the wind farm investment analysis, the cost estimation for redox batteries is a critical aspect of the feasibility assessment. Due to the harsh environmental conditions at Raglan, cost adaptations are necessary to account for additional infrastructure requirements. This study uses the Levelized Cost of Storage (LCOS) metric to quantify the financial viability of storage deployment.

LCOS represents the cost per kWh of stored electricity over the system's lifetime, incorporating capital (CAPEX) and operational (OPEX) expenses. This metric provides a long-term perspective on the economic feasibility of storage solutions, where a lower LCOS indicates greater profitability [51]. Table 5 presents the estimated LCOS for the Raglan site based on different storage durations.

Table 5. LCOS of redox systems at Raglan (CAD/MWh).

Parameters	10 MW, Raglan 4 h	6 h	8 h
Storage system			
Storage unit and electrolyte	118.1	105.2	98.3
Storage balance	23.6	21.0	19.8
Energy system			
Power supply equipment	14.3	9.5	7.1
Control and communication	0.9	0.6	0.4
System Integration	21.9	19.3	18.0
System balance			
Engineering, transportation, and construction	83.0	73.0	68.7
Project development	41.8	37.2	34.9

Table 5. Cont.

Parameters	10 MW, Raglan 4 h	6 h	8 h
Network Integration	9.0	6.0	4.5
Total CAPEX	312.5	271.8	251.8
Total OPEX	35.3	33.9	33.3
Total LCOS	347.8	305.8	285.0

The estimated LCOS values of 347.8, 305.8, and 285.0 CAD/MWh correspond to 4 h, 6 h, and 8 h storage durations, respectively. These values align with literature estimates and manufacturer-provided data, confirming their validity as benchmarks for cost assessment. Note here that LCOS also represents the total lifetime cost of the storage system (including operating and maintenance costs), divided by the total amount of electricity that the system is expected to store during its lifetime [51]. Based on the above cost assessments, Table 6 presents the final investment costs for the redox battery system and associated infrastructure, specifically for a 20 MW system at Raglan.

Table 6. Final costs of redox batteries.

Parameters	20 MW, Raglan 4 h	6 h	8 h
Electrochemical system (M CAD)	117.0	152.7	188.6
Area required (m^2)	4000	6000	8000
Shed price (M CAD)	32.5	48.8	65.0
Total cost (M CAD)	149.5	201.4	253.6

With the energy system components and financial parameters established, we develop a simulation model that integrates these factors into a comprehensive analysis. This model enables scenario testing and the evaluation of different configurations of wind–storage integration, optimizing the economic and technical feasibility of the renewable energy transition at Raglan.

4. Design, Sizing, and Modeling of the Wind Energy and Storage System

A MATLAB-based simulation model was developed to evaluate the performance of the Raglan mine energy network following the integration of additional wind turbines and energy storage. This model enables the simulation of various operational scenarios by incorporating multiple input parameters, including

- Wind energy production;
- Electricity and heat generation based on generator operation;
- Diesel fuel consumption;
- Energy storage dynamics (charging and discharging cycles).

The model operates on an hourly time scale over an entire year, allowing for a detailed analysis of key performance indicators such as

- Renewable energy penetration rate;
- Wind energy curtailment losses;
- Diesel fuel savings;
- Greenhouse gas (GHG) emissions reduction;
- Economic performance metrics (including financial feasibility analysis).

The MATLAB-based dynamic model provides a crucial tool for predicting system behavior under various configurations, eliminating the need for physical testing. It enables the identification of optimal energy management strategies while ensuring grid stability and cost-effectiveness.

The simulation framework is structured into four interdependent sub-models, which operate in an iterative process to capture the complex interactions between energy production, storage, and demand, as shown in the flowchart in Figure 10.

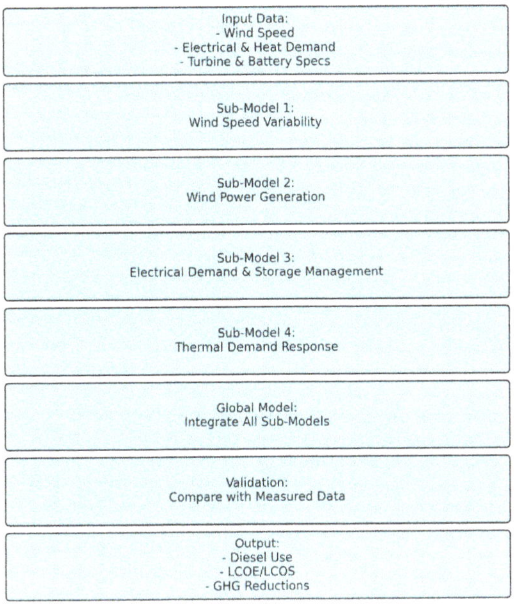

Figure 10. Modeling framework flowchart.

4.1. Model Input Data

The MATLAB-based simulation program developed for this study is fully configurable, enabling flexible parameter adjustments tailored to specific simulation objectives. The operating parameters are structured within a "structure array" to enhance readability and clarity, facilitating efficient data organization and processing within the MATLAB workspace (see Figure 11). Key configurable parameters include

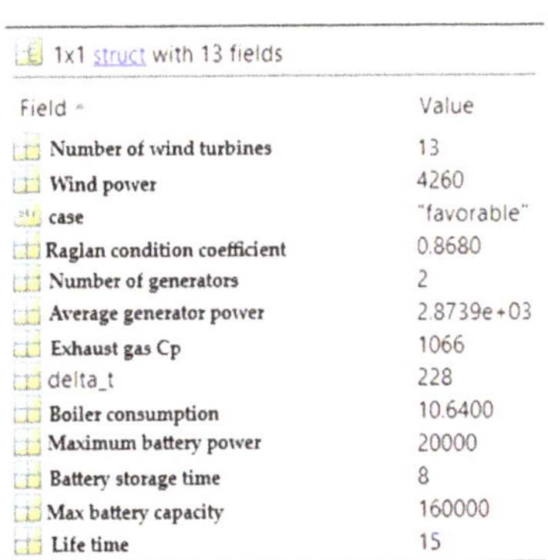

1x1 struct with 13 fields

Field	Value
Number of wind turbines	13
Wind power	4260
case	"favorable"
Raglan condition coefficient	0.8680
Number of generators	2
Average generator power	2.8739e+03
Exhaust gas Cp	1066
delta_t	228
Boiler consumption	10.6400
Maximum battery power	20000
Battery storage time	8
Max battery capacity	160000
Life time	15

Figure 11. Structure array of model input parameters.

- Number of wind turbines;
- Number of diesel generators in operation;
- Maximum battery power (maximum_battery_power);
- Battery storage duration (battery_storage_time);
- Total battery storage capacity (max_battery_capacity).

The structure array is linked to an internal MATLAB folder containing the essential datasets from the Raglan mining site, which are required for accurate simulation. These include

- Hourly heat demand data (2021);
- Wind speed data at 15 min intervals (2020);
- Electricity consumption data from the 25 kV network (hourly, 2020);
- The power curve of the ENERCON E-115 wind turbine.

By integrating these site-specific datasets, the MATLAB model enables a realistic and data-driven simulation, allowing for accurate assessments of wind energy integration, energy storage performance, and diesel displacement strategies at the Raglan mining site.

4.2. Sub-Model 1: Wind Speed Model

This sub-model utilizes the "2020 Wind Data, 15 Minutes" dataset to simulate variations in wind speed under different conditions. For simulation purposes, three distinct wind scenarios were defined:

- Unfavorable: Wind speed decreases between −10% and −5%;
- Random: Wind speed fluctuates within a range of −5% to +5%;
- Favorable: Wind speed increases between +5% and +10%.

These conditions enhance the model's flexibility, reflecting the natural variability of wind speeds, which fluctuate continuously. The simulation of wind speed variations is controlled by the following formula (Equation (1) and Equation (A4) in Appendix A):

$$r = a + (b - a) * rand(N, 1) \tag{1}$$

where

- a and b are the lower and upper bounds of the wind speed variation interval;
- N represents the random numbers within the interval (a,b);
- $rand(N,1)$ is the MATLAB random function that generates values within the specified range.

A 15 min time step is used in the model, ensuring high-resolution variability in wind speed. Equation (1) serves as the MATLAB function for wind speed randomization, allowing for dynamic simulation of the 25 kV grid response, particularly regarding

- Wind turbine performance fluctuations;
- Energy storage system behavior under variable wind conditions;
- Grid stability under different wind penetration scenarios.

Additionally, this sub-model enables further analysis of critical wind conditions, such as

- The number of hours per year with no wind;
- Consecutive hours with zero wind generation;
- Consecutive hours without peak wind speeds.

By incorporating these stochastic wind variations, the model ensures a realistic assessment of wind energy intermittency and its impact on grid stability and energy storage requirements at the Raglan mining site.

4.3. Sub-Model 2: Wind Power Generation Model

The wind power generation model is based on Sub-Model 1, utilizing the same 15 min time step for consistency. This sub-model incorporates the wind turbine's power curve (see Figure 7) and simulates energy production under varying wind conditions.

The model utilizes a MATLAB function, such as the 'SEARCHV' function in Microsoft Office, to accurately map wind speed values to corresponding power outputs, ensuring precise alignment between input wind speeds and turbine-generated power.

Given the harsh environmental conditions at Raglan (including cold temperatures, frost, and storms), a correction factor is applied to account for production inefficiencies. This factor was determined using real-world production data from the existing E-82 wind turbines, reflecting the actual site conditions. The correction factor is calculated as follows (Equation (2) and Equation (A8) in Appendix A):

$$Corrective\ factor = \frac{Real\ production\ of\ the\ E-82\ wind\ turbines}{Production\ obtained\ using\ its\ power\ curve\ from\ the\ supplier} \tag{2}$$

Data analysis from 2019 to 2021 revealed that the E-82 wind turbines at Raglan produced only 86.8% of their theoretical capacity due to site-specific constraints. The same correction factor (0.868) was applied to the E-115 wind turbines to ensure realistic projections.

The time step is increased from 15 min to 1 h to facilitate further analysis, allowing for the following:

- Estimation of the wind farm's load factor;
- Calculation of total wind energy production (based on the number of new turbines installed);
- Computation of average hourly energy output per turbine.

Following multiple simulation iterations, the average load factor of the E-115 wind turbines was determined to be 43.6%. Subsequent calculations use this value to refine energy output estimations and assess the wind farm's overall performance under Raglan's environmental conditions.

4.4. Sub-Model 3: Model of the Response to Electrical Demand

The electrical demand response model simulates the behavior of the 25 kV network under different load conditions. This sub-model is essential for monitoring and controlling:

- Battery management (state of charged and discharged energy);
- Unused surplus energy;
- Generator consumption.

To introduce flexibility, a randomization factor is applied to simulate variations in electrical demand. The model uses Equation (1) (previously introduced) with a random fluctuation range of −5% to +10%, ensuring a realistic simulation of demand changes.

A key objective of this sub-model is to ensure that the 25 kV network maintains a surplus of energy, enabling efficient battery storage and load management. Additionally, it anticipates future increases in electrical demand, given that the Raglan mine plans to expand nickel production.

Complete decarbonization is not immediately feasible since the ore-drying process relies entirely on diesel generators. Thus, the model accounts for the continued operation of some generators to meet heat demand, ensuring a pragmatic transition to renewable energy.

Two scenarios were tested to analyze the behavior of the 25 kV network under different supply and demand conditions:

1. Electricity demand is satisfied:
    - The energy supply exceeds demand, generating excess power.
    - The surplus is directed toward charging storage batteries.
    - If batteries are fully charged, the excess energy is curtailed (rejected).
    - This is the ideal case for the Raglan mine, as it maximizes the utilization of renewable energy.
2. Electricity demand is not satisfied:
    - The 25 kV network draws from battery storage to cover the deficit.
    - If battery reserves are insufficient, additional power must be generated.
    - In this case, an emergency recourse to diesel generators is triggered.
    - Each generator has an estimated power output of 2.88 MW.

Figure 12 illustrates Sub-Model 3's operating structure and details how energy demand, supply, and storage interact within the MATLAB simulation framework.

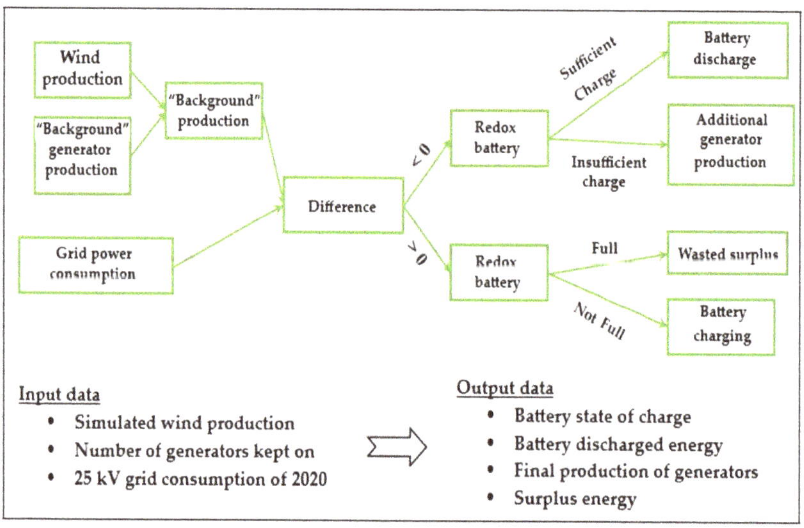

Figure 12. Management of the electrical part of the MATLAB model.

4.5. Sub-Model 4: Thermal Demand Response Model

At the Raglan mine site, heat is generated through two primary sources:

- Heat recovery from EMD engine blocks;
- Diesel-fired boilers.

The thermal energy sub-model is based on operational data from 2021, during which 87% of the energy produced by diesel generators was recovered and utilized for heating through the cooling circuit—a particularly critical process during the cold months at Raglan.

The model integrates specific heat recovery values from two key sources:

- Exhaust gas recovery: 5.12 MWh;
- 31-HX-02 heat exchanger recovery: 12.35 MWh.

Additionally, the model predicts the contribution of wind energy to the heating circuit. When the diesel generator output is insufficient, surplus wind energy is directed to compensate for the heat demand.

Figure 13 illustrates a simplified schematic of the thermal management sub-model, depicting how different energy sources interact to maintain the site's heating requirements.

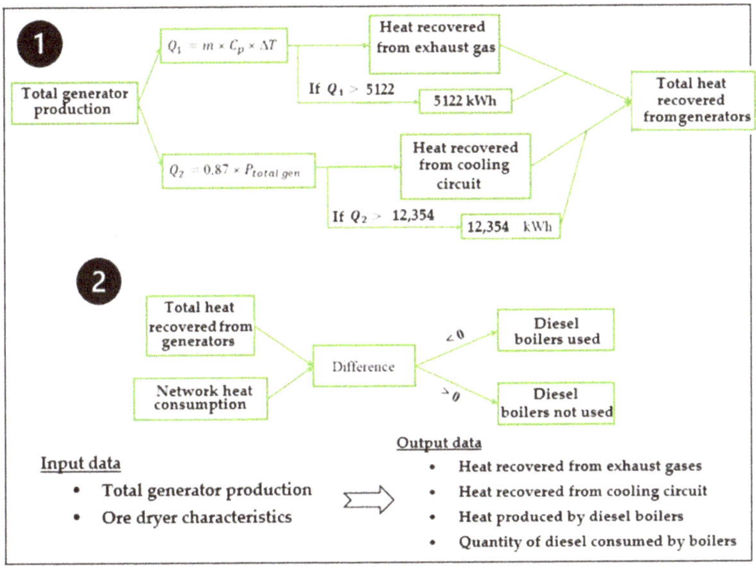

Figure 13. Thermal part management.

5. Validation of the Global Model

The global model integrates the four sub-models presented earlier, providing a comprehensive simulation of the Raglan energy system. This model accounts for both existing infrastructure and newly proposed installations, ensuring a realistic representation of energy production, storage, and consumption at the site.

To guarantee accuracy and computational efficiency, the MATLAB-based program has been thoroughly tested for

- Error-free execution;
- Optimized calculation times;
- Seamless operation with a single command.

The MATLAB simulation results are automatically exported to an Excel file using the "writetable" function to facilitate further analysis and generate structured data tables.

Its predictions were compared against actual operating data from the Raglan mining site to validate the model's accuracy. The results are presented in Table 7.

Table 7. Comparison between actual plant data and the present work.

Characteristics	Modeled Value	Real Value	Relative Error (%)
Diesel consumed by generators (L)	37,979,558	37,141,356	2.26
Diesel consumed by boilers (L)	111,815	111,182	0.57
Total diesel consumed by the 25 kV network (L)	38,091,374	37,252,539	2.25
Wind power production (MWh)	16,286	16,871	3.59

The low relative errors observed in all key parameters confirm the accuracy and reliability of the simulation. The results demonstrate that the model provides a realistic approximation of the Raglan mine's energy system, making it a valuable tool for assessing new renewable energy integration and optimizing system performance.

6. Results and Discussion

6.1. Data

Table 8 presents the data and operating parameters used for modeling and simulation. These values are essential for sizing the wind farm's energy storage system and estimating the electricity and heat demand at the Raglan site.

Table 8. Model input and output data.

Data at the Entrance	Value
Exchange rate as of 21 September 2022	1.34 CAD/USD
Load factor (%)	32%
Lifetime of project (yr)	15.00
IRENA Canada Wind OPEX (CAD/kW.yr)	35.00
IRENA Canada Wind CAPEX (CAD/kW.yr)	1368
Exits	**Value**
CAPEX (CAD/MWh)	49.24
OPEX (CAD/MWh)	24.15
LCOE (CAD/MWh)	73.40

Several additional site-specific parameters were used in the modeling process:

- Wind speed at Raglan: Monthly average wind speed is typically between 6 and 9 m/s.
- The overall efficiency of electricity production by generators is 3.78 kWh/L, corresponding to an electricity generation efficiency of 35.5%.
- Air density: Approximately 1.37 kg/m^3, based on an average atmospheric temperature of −15 °C, a common condition at Raglan.

The following assumptions were made regarding energy production and consumption data:

- Wind energy production data from 2020 was used as the baseline for modeling wind power generation.
- Electricity consumption data from 2020 was used to model the electrical demand of the mine.
- Thermal energy consumption data from 2021 was used to represent the mine's heating requirements.

By integrating these parameters and historical datasets, the model accurately represents the energy dynamics at the Raglan site, supporting the optimization of wind energy integration, energy storage, and system performance.

6.2. Model Results

This section presents the simulation results of the proposed model, which aims to decarbonize the Raglan mining site by integrating a vanadium flow battery storage system coupled with a wind farm.

The primary objective of this study is to develop optimized solutions that

- Reduce diesel consumption, lowering operational costs and reliance on fossil fuels;

- Decrease greenhouse gas (GHG) emissions, mitigating the environmental impact of mining activities.

The following subsections provide an in-depth analysis of the model's performance, including energy production, storage efficiency, and the potential for diesel displacement through the integration of renewable energy.

6.2.1. General Trend Results

To significantly reduce diesel consumption, a large penetration of wind energy in the energy mix is necessary, which implies the installation of additional wind turbines. This is the only way to reduce diesel consumption and the number of installed generators. The different simulations focus on three main variables: the number of generators to be left running permanently throughout the year, which varies from zero to three; the battery capacity, which can take values of 0 MWh, 80 MWh, 120 MWh, and 160 MWh; and the additional number of E-115 wind turbines to be installed, which varies from 0 to 12.

Tables 9 and 10 provide a visual representation of the different configurations, enabling easy identification of diesel savings. It presents the diesel consumption reduction (as a percentage of diesel saved compared to the status quo) for the different configurations, considering all combinations of the number of new wind turbines, the battery capacity, and the number of generators to be kept running. Table 9 shows that significant diesel savings of up to 63.98% can be achieved. This is only possible by stopping all generators while increasing the number of wind turbines and the storage capacity of the flow battery.

Table 9. Amount of diesel saved, in the cases of 120 MWh and 160 MWh redox battery.

Number of Wind Turbines	160 MWh Redox Battery				120 MWh Redox Battery			
	0 gen	1 gen	2 gen	3 gen	0 gen	1 gen	2 gen	3 gen
12	−63.98%	−58.66%	−51.98%	−41.84%	−62.72%	−57.43%	−50.82%	−40.88%
11	−62.83%	−57.69%	−51.27%	−41.28%	−61.56%	−56.44%	−50.08%	−40.37%
10	−61.45%	−56.59%	−50.35%	−40.79%	−60.20%	−55.34%	−49.12%	−39.76%
9	−59.82%	−55.34%	−49.33%	−40.13%	−58.63%	−54.10%	−48.15%	−39.08%
8	−57.85%	−53.71%	−48.14%	−39.32%	−56.71%	−52.58%	−46.95%	−38.23%
7	−55.19%	−51.75%	−46.75%	−38.29%	−54.20%	−50.68%	−45.61%	−37.17%
6	−51.24%	−49.12%	−44.79%	−37.13%	−50.51%	−47.99%	−43.78%	−36.02%
5	−45.15%	−44.70%	−42.17%	−35.43%	−44.82%	−44.12%	−41.11%	−34.44%
4	−37.87%	−37.96%	−37.46%	−32.95%	−37.79%	−37.70%	−36.86%	−32.18%
3	−30.06%	−29.83%	−29.52%	−27.95%	−29.96%	−29.86%	−29.46%	−27.49%
2	−20.62%	−20.62%	−20.34%	−19.21%	−20.65%	−20.59%	−20.29%	−19.20%
1	−10.14%	−10.20%	−10.07%	−9.23%	−10.19%	−10.17%	−10.17%	−9.21%
0	0.73%	0.74%	0.61%	0.96%	0.71%	0.66%	0.68%	1.09%

Legend: Dark green = very good economy; Light green = good economy; Yellow = fairly good economy; Red = bad economy; gen = generator.

The reduction in diesel consumption is directly linked to the penetration of wind energy into the Raglan mine's energy mix; logically, the greater the number of wind turbines built, the more significant the proportion of wind energy in Raglan's energy mix. As a result, diesel consumption and the number of operating generators decrease. As such, the diesel savings increase by reducing the number of generators in operation, as well as by increasing battery capacity and wind turbine capacity. By leaving three generators in operation, we can achieve diesel savings of 35.44%, 39.66%, 40.88%, and 41.84% for the 0 MWh (without battery), 80 MWh, 120 MWh, and 160 MWh battery sizes. Besides this, the best diesel savings options, in the case of removing all generators and integrating 12 additional wind turbines into the grid, amount to 56.68%, 61.27%, 62.72%, and 63.98%, respectively, for the 0 MWh (without battery), 80 MWh (see Table 10), 120 MWh, and

160 MWh batteries (see Table 9). In contrast, with four wind turbines added to the grid and one generator in continuous operation, the configuration with the largest storage capacity (160 MWh battery) remains the best option, with an estimated diesel savings rate of 37.96%.

Table 10. Amount of diesel saved, in the cases of without redox battery and 80 MWh redox battery.

Number of Wind Turbines	80 MWh Redox Battery				Without Redox Battery			
	0 gen	1 gen	2 gen	3 gen	0 gen	1 gen	2 gen	3 gen
12	−61.27%	−56.06%	−49.49%	−39.66%	−56.68%	−51.31%	−44.79%	−35.44%
11	−60.10%	−55.05%	−48.74%	−39.16%	−55.43%	−50.34%	−43.93%	−34.76%
10	−58.79%	−53.93%	−47.82%	−38.49%	−54.17%	−49.16%	−43.00%	−34.10%
9	−57.22%	−52.62%	−46.75%	−37.79%	−52.51%	−47.80%	−41.88%	−33.29%
8	−55.29%	−51.14%	−45.56%	−36.92%	−50.60%	−46.15%	−40.63%	−32.23%
7	−52.92%	−49.33%	−44.11%	−35.91%	−48.45%	−44.40%	−39.07%	−31.19%
6	−49.56%	−46.73%	−42.33%	−34.69%	−45.82%	−42.09%	−37.31%	−29.84%
5	−44.39%	−43.23%	−39.96%	−33.20%	−42.28%	−39.37%	−35.05%	−28.21%
4	−37.69%	−37.37%	−36.13%	−30.95%	−37.01%	−35.46%	−32.12%	−26.10%
3	−29.94%	−29.72%	−29.21%	−26.90%	−29.62%	−29.14%	−27.56%	−23.15%
2	−20.61%	−20.52%	−20.23%	−19.08%	−20.51%	−20.27%	−19.61%	−17.86%
1	−10.14%	−10.08%	−10.09%	−9.19%	−10.14%	−10.04%	−9.71%	−8.76%
0	0.71%	0.63%	0.51%	1.11%	0.64%	0.66%	0.72%	1.32%

Legend: Dark green = very good economy; Light green = good economy; Yellow = fairly good economy; Red = bad economy; gen = generator.

Scenario of Two Generators in Operation and 120 MWh Battery: Impact of Adding Wind Turbines to the 25 kV-G Network

This scenario evaluates the performance of a hybrid system with two continuously operating diesel generators and a 120 MWh redox flow battery. The number of additional E-115 wind turbines varies from 0 to 12 to assess their impact on system costs and operation. The results are shown in Figure 14:

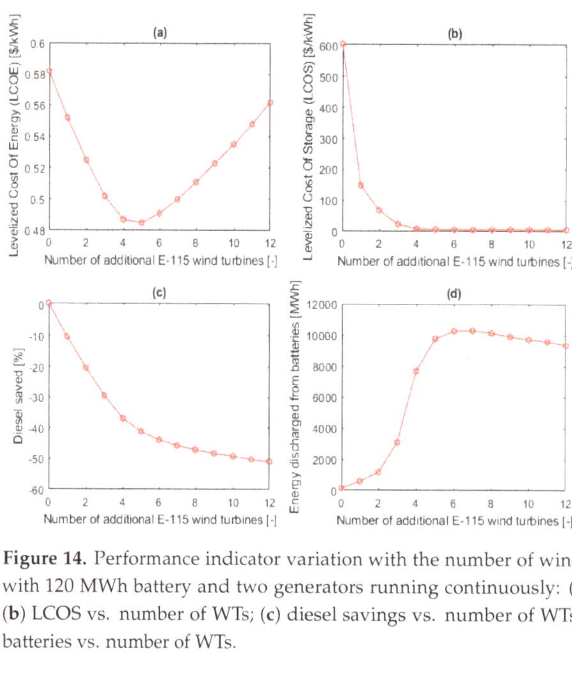

Figure 14. Performance indicator variation with the number of wind turbines (WTs) for a system with 120 MWh battery and two generators running continuously: (**a**) LCOE vs. number of WTs; (**b**) LCOS vs. number of WTs; (**c**) diesel savings vs. number of WTs; (**d**) energy discharged from batteries vs. number of WTs.

(a) Levelized Cost of Energy (LCOE): The LCOE initially decreases as wind turbines are added, reaching a minimum of around four turbines, beyond which further additions lead to a gradual increase. This reflects that the optimal integration point of wind energy before overcapacity leads to diminishing returns (see Equation (A1) in Appendix A).

(b) Levelized Cost of Storage (LCOS): LCOS sharply decreases with the addition of wind turbines and approaches zero after four turbines, indicating maximal economic utilization of the battery system beyond that point (see Equation (A2) in Appendix A).

(c) Diesel Savings: Increasing wind capacity consistently reduces diesel usage. However, the rate of diesel savings declines with each additional turbine, illustrating a saturation effect.

(d) Battery Discharge Energy: The energy discharged from the battery increases with added wind power, peaking at around six turbines. Beyond this point, battery usage stabilizes or slightly declines, suggesting that excess wind generation reduces reliance on storage.

This analysis highlights the trade-offs between wind capacity, storage utilization, and diesel consumption. An optimal configuration emerges with four to six additional turbines, effectively balancing cost and performance.

In addition to the analysis of the number of installed wind turbines, we explore the variation in the previous parameters with the wind penetration rate related to both installed capacity and the specific site operating conditions:

$$Wind\ Penetration(\%) = WP(\%) = \frac{E_{wind}}{E_{load}} \times 100 \quad (3)$$

As such, in Figure 15, we account for the actual wind energy contribution over a year, not just the installed capacity. Although the overall trends in Figure 15 resemble those in Figure 14, key distinctions emerge in subfigures (b) and (c):

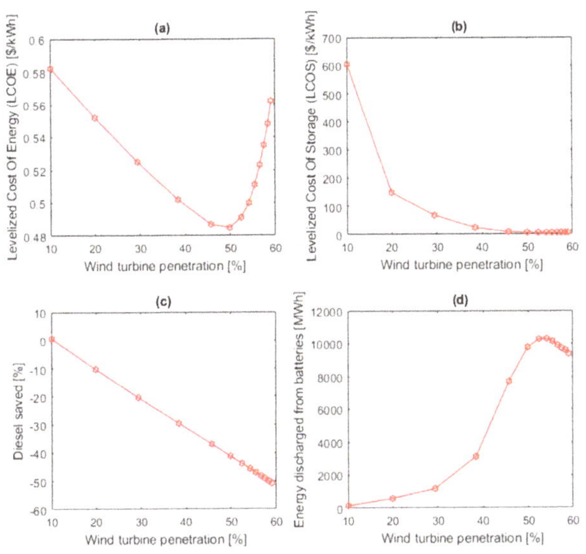

Figure 15. Performance indicator variation with the wind power penetration rate (WPP, %) for a system with 120 MWh battery and two generators running continuously: (**a**) LCOE vs. WPP; (**b**) LCOS vs. WPP; (**c**) diesel savings vs. WPP; (**d**) energy discharged from batteries vs. WPP.

(a) The Levelized Cost of Energy (LCOE) steadily decreases as wind penetration increases from 10% to 50%, reaching a minimum of approximately CAD 490/MWh at 50%. Beyond this point, the LCOE rises, indicating an optimal penetration level of around 50%.

(b) The Levelized Cost of Storage (LCOS) experiences a sharp decline between 10% and 20% wind penetration, after which the reduction becomes more gradual. LCOS reaches zero at 50% penetration, confirming this as the most economically efficient point for storage operation.

(c) Diesel consumption shows an almost linear decrease throughout the entire range of wind penetration, reinforcing a strong negative correlation between wind contribution and fossil fuel reliance.

(d) Battery discharge increases with wind penetration, peaking around 55% and then slightly decreasing. This suggests a saturation point in the contribution of storage to system balancing.

Together, these results confirm that a wind penetration rate of around 50% offers the best trade-off between cost reduction, storage utilization, and diesel offset in this configuration.

Impact of the Number of Operating Generators and Additional Wind Turbines on the Diesel Savings

For a system with a 120 MWh battery, Figure 16 illustrates the impact of wind turbine additions on diesel savings under different generator operation scenarios (0 to 3 generators running). The most significant diesel reduction occurs when no generators operate permanently, resulting in over 150% savings compared to the baseline. The >100% diesel savings reflect not only avoided electricity production from diesel but also the replacement of diesel used for thermal needs (e.g., building or process heating) through excess wind energy and heat recovery. This indicates net-negative diesel consumption relative to the baseline for electricity generation only when combined power and heat offset benefits are considered.

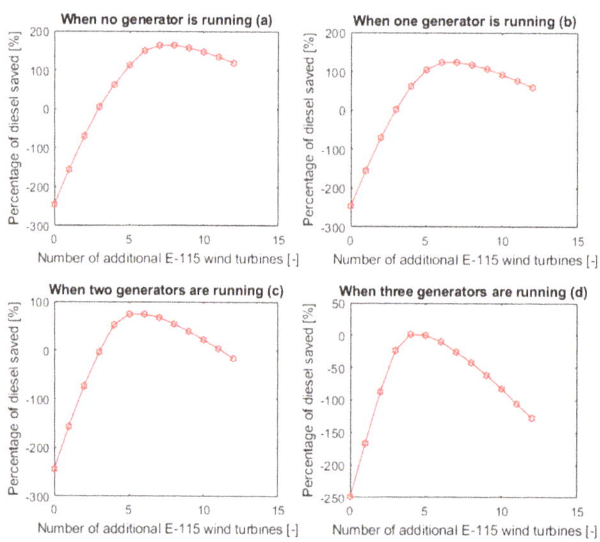

Figure 16. Share of diesel saved for different numbers of continuously running generators, depending on the number of wind turbines, with 160 MWh redox battery storage.

The relative diesel savings decrease as more generators remain permanently online (Figure 16b–d). This demonstrates that continuously operating diesel generation limits the economic and environmental benefits of wind integration.

6.2.2. Results for the Most Economical Scenario

Economic performance is a primary driver for technology adoption in any industrial context, especially for mining operations. This section presents the results for the most cost-effective scenario over a 15-year project horizon.

As it is currently not feasible to fully disconnect all diesel generators due to ore processing and operational constraints, the most economical scenario involves maintaining two generators running continuously, integrating five E-115 wind turbines, and including a redox flow battery system rated at 20 MW/80 MWh. Under these conditions, the configuration with continuous operation of two generators outperforms the three-generator alternative in terms of both cost and environmental performance.

Table 11 presents the main characteristics of the most economical scenario. This scenario reduces 42,429.6 tCO_2 emissions, equivalent to 39.96% of the diesel consumed in Raglan annually. It also allows for savings of CAD 155.8 million. The return on investment is estimated at ten years.

Table 11. Results of the most economical scenario.

Data Type	Value	Unit
Wind CAPEX (5 wind turbines)	103.4	M CAD
CAPEX redox (battery and hangar)	149.5	M CAD
TOTAL CAPEX	253.0	M CAD
CAPEX additional cost	236.3	M CAD
TOTAL Economy	155.8	M CAD
LCOE complete system	460	CAD/MWh
LCOS complete system	7830	CAD/MWh
Wind integration	80.48	%
Wind penetration	48.71	%
Diesel saved	39.96	%
Quantity of GHG avoided	42,429	tCO2 eq.

"CAPEX Additional Cost" refers to the incremental investment required to deploy the hybrid system—batteries, wind turbines, and supporting infrastructure—compared to a business-as-usual scenario (diesel-only operation). CAPEX is calculated using Equation (A7) in Appendix A. Equation (A6) allows for the estimation of OPEX.

The high LCOS (CAD 7831/MWh) results from including building infrastructure costs (~25% of total) and the high capital intensity of redox flow storage, especially in remote, harsh-climate environments like Raglan.

6.2.3. Scenarios with the Highest Diesel Savings

As observed in Tables 9 and 10, the scenario with the highest diesel savings involves a large-scale hybrid configuration comprising a wind farm of 14 turbines, of which 12 are additional E-115 units, coupled with a 160 MWh redox flow battery. In its initial form, no diesel generators are running continuously. Under these conditions, the system achieves substantial environmental benefits, with an estimated reduction of 68,000 tCO_2 equivalent, corresponding to a diesel savings rate of 63.98%. However, this scenario is infeasible at Raglan, primarily due to operational constraints related to ore drying.

Considering the actual operation conditions, the ore drying process at Raglan relies entirely on heat recovered from the exhaust gases of diesel engine blocks. Since the process requires continuous thermal input, a realistic scenario requires keeping two generators running in the implemented scenario. Thus, the finalized configuration includes 12 additional E-115 wind turbines, a 160 MWh redox flow battery system, and two diesel generators operating continuously. The technical and economic results for this high diesel-reduction scenario are summarized in Table 12. It is noted that the LCOE in the case of the most economical scenario is lower (Table 11), compared to that of the scenario with high diesel savings. This is explained by the fact that the total cost of energy production with high diesel savings uses a large share of energy from wind turbines, the production cost of which, at the Raglan site, is high, associated with this, the limited lifespan (15 years for wind turbines at Raglan, while the LCOE is often calculated for lifespans of 20 or 25 years) of renewable energy equipment (due to the cold climate, difficult maintenance). In the case of LCOS, the scenario with the greatest diesel savings presents a low value, i.e., CAD 6110/MWh (see Table 12).

Table 12. Results of the scenario with high diesel economy.

Data Type	Value	Unit
Wind CAPEX (12 wind turbines)	248.3	M CAD
CAPEX redox (battery and hangar)	253.6	M CAD
TOTAL CAPEX	501.9	M CAD
CAPEX additional cost	485.2	M CAD
TOTAL Economy	−16.3	M CAD
LCOE complete system	590	CAD/MWh
LCOS complete system	6110	CAD/MWh
Wind integration	45.81	%
Wind penetration	60.09	%
Diesel saved	51.98	%
Quantity of GHG avoided	55,203	tCO2 eq.

6.3. Discussion

The results presented in this study are based on a model built with several conservative assumptions. Consequently, real-world outcomes may be even more favorable. For instance, actual wind power generation could exceed projections, battery utilization might be higher, diesel prices could rise, and the cost of redox flow technology may decrease over time.

One of the most significant cost drivers in the current analysis is the purchase and installation of redox flow batteries. As this technology has not yet reached full industrial maturity, large-scale systems like those required at Raglan remain rare and costly. Additionally, the need to construct a dedicated facility to house the tanks and electrochemical stacks adds substantial capital costs, unlike storage systems that do not require such infrastructure. Despite these limitations, most configurations analyzed yield a positive return on investment.

Soon, Raglan may be eligible for government subsidies to support decarbonization projects. If such funding is secured, it would reduce total project costs and significantly improve profitability. However, since the amount and timeline of potential subsidies remain uncertain, they have not been included in the current financial assessment.

Another potential but unquantified benefit is the commercial value of decarbonized or "green" nickel. Once Raglan completes its decarbonization process, it could position its nickel as a low-carbon product, an attractive feature for the growing electric vehicle (EV) industry. This could give Raglan a strategic commercial advantage in emerging clean technology markets.

The study also highlights that removing diesel generators from the 25 kV-G network is more effective in reducing emissions than simply increasing storage capacity. This aligns with Raglan's plans for the second phase of its decarbonization strategy, which involves retiring the remaining EMD diesel generators. To enable this, an alternative ore-drying process must be identified that does not rely on engine exhaust heat.

Once diesel generators are fully phased out, wind energy penetration is expected to increase significantly. At that point, a new challenge will emerge: ensuring the quality and stability of electricity integrated into the network. This will likely require careful management of power electronics, grid-forming inverters, and control systems to maintain voltage and frequency standards in a diesel-free environment.

7. Conclusions and Perspectives

This study explores the decarbonization of the Raglan mining site by expanding its wind farm and integrating a redox flow battery energy storage system. To set the record straight, a brief comparison of different energy storage technologies is also presented and discussed.

In the first phase, we analyzed wind potential data at the Raglan site to assess the feasibility of installing additional wind turbines that could replace the site's diesel generators, which currently meet its electrical needs. The second phase focused on identifying the technical requirements for sizing and implementing the storage system, particularly by analyzing the site's electrical and thermal demand. These factors are essential to ensure optimal operation of the combined wind–battery system.

A hybrid simulation model, consisting of four sub-models, was developed using MATLAB and Excel. The model evaluated various scenarios, including

* Full or partial decommissioning of diesel generators;
* The number of additional E-115 wind turbines to be added;
* The ideal energy storage capacity of the redox flow battery.

Each scenario was assessed using two key economic indicators: the levelized cost of Energy (LCOE) and the Levelized Cost of Storage (LCOS). The results indicate that completely removing diesel generators is currently not feasible, primarily because some processes, such as ore drying, depend on the residual heat from diesel engines. Consequently, maintaining at least two generators in continuous operation is necessary.

The study also highlights the significant cost of implementing redox flow batteries under current market conditions. The LCOS, estimated at 7831 CAD/MWh, is high due to the price and infrastructure requirements of the storage system. As a result, battery utilization remains low, making the widespread integration of redox storage at Raglan technically possible but economically challenging under current conditions.

While this work demonstrates the techno-economic feasibility of integrating wind energy and redox storage for decarbonization, it does not address the grid stability or power quality issues, such as voltage fluctuations, frequency regulation, grid inertia, and stability during high renewable penetration. These aspects are crucial for ensuring reliable and safe operation, particularly in remote and weather-exposed environments such as Raglan. They will be investigated in future work using dynamic power system simulation tools.

Additionally, model results were compared with literature benchmarks for validation. While the current model is robust, further work is needed to refine the operating parameters and reduce reliance on assumptions. This will ensure more accurate and scalable projections for wind integration.

Future studies will focus on the following aspects:

- Detailed sensitivity and uncertainty analysis of key parameters (wind variability, energy demand, climate trends);
- Probabilistic modeling to assess long-term performance and reliability under varying environmental and operational conditions;
- Power system simulations to evaluate voltage and frequency stability under high-renewable-integration scenarios.

These next steps will provide a more comprehensive understanding of the technical and economic dynamics of large-scale decarbonization of remote industrial sites.

Author Contributions: Conceptualization, A.R. and D.R.R.; methodology, A.R., D.R.R., B.-J.R.M.B. and A.I.; software, A.R.; validation, D.R.R. and B.-J.R.M.B.; formal analysis, A.R., D.R.R., B.-J.R.M.B. and A.I.; investigation, A.R. and D.R.R.; resources, D.R.R.; data curation, A.R.; writing—original draft preparation, A.R., D.R.R. and B.-J.R.M.B.; writing—review and editing, D.R.R. and A.I.; visualization, A.R. and B.-J.R.M.B.; supervision, D.R.R.; project administration, D.R.R.; funding acquisition, D.R.R. All authors have read and agreed to the published version of the manuscript.

Funding: The authors thank Glencore and Hatch industrial partners for their support during this project. They also acknowledge the FRQ-NT's financial support via grants 322720, 340327, and 364164, as well as NSERC's grant RGPIN-2025-04831 and the Michel Trottier private donation to the T3E research group.

Institutional Review Board Statement: Not applicable.

Informed Consent Statement: Not applicable.

Data Availability Statement: The original contributions presented in this study are included in the article. Further inquiries can be directed to the corresponding author.

Acknowledgments: The authors acknowledge the data and information from the industrial partners Glencore and Hatch. The authors did not use generative AI technologies while preparing this work. The authors utilized AI-assisted technologies, including Grammarly (www.grammarly.com) and Antidote (www.antidote.info), to enhance formulation and eliminate grammatical errors. After using this tool/service, the author(s) reviewed and edited the content as needed and take(s) full responsibility for the publication's content.

Conflicts of Interest: The authors declare no conflicts of interest.

Appendix A

Table A1 gives a summary of the main equations used in this model.

Table A1. Main equations used in this model.

Equation	Equation Number	Variable	Observation	Meaning of Terms
$LCOE\left[\frac{\$}{kWh}\right] = \frac{\sum_{t=1}^{T} \frac{CAPEX_t[\$] + OPEX_t[\$]}{(1+i[\%])^t}}{\sum_{t=1}^{T} \frac{Q_t[kWh]}{(1+i[\%])^t}}$	(A1)	$LCOE$	Equation taken from the IRENA [43] report "Renewable power generation costs 2021"	$CAPEX_t$: investment expenditure for year t $OPEX_t$: operating and maintenance and fuel expenses during year t Q_t : electricity production during year t i : discount rate T : system lifetime
$LCOS\left[\frac{\$}{kWh}\right] = \frac{CAPEX_t[\$] + \sum_{t=1}^{T} \frac{OPEX_t[\$]}{(1+i[\%])^t} + \sum_{t=1}^{T} \frac{Charging\ cost[\$]}{(1+i[\%])^t} + \frac{End-of-life\ cost[\$]}{(1+i[\%])^{T+1}}}{\sum_{t=1}^{T} \frac{Elec_{discharged}[kWh]}{(1+i[\%])^t}}$	(A2)	LCOS	The LCOS equation is taken from the work of Schmidt et al. [38].	$Charging\ cost$: cost of electricity (or more broadly of energy) needed to power the system. $End-of-life\ cost$: either the dismantling cost or the value of the installation at the end of the system's life (salvage value) $Elec_{discharged}$: annual amount of electricity discharged by the system
$Q = m \times Cp \times \Delta T$	(A3)	Amount of heat	The specific heat capacity of the exhaust gases is taken as 1066 J/kg.K	m : mass Cp : The specific heat capacity of the exhaust gases ΔT : Temperature variation
$r = a + (b-a) * rand(N,1)$	(A4)	Wind speed	The simulation of wind speed variations	a and b are the lower and upper bounds of the wind speed variation interval N represents the random numbers within the interval (a,b) rand(N,1) is the MATLAB random function that generates values within the specified range.
$f_\lambda(x;\lambda,k) = \begin{cases} \frac{k}{\lambda}\left(\frac{x}{k}\right)^{k-1} e^{-\left(\frac{x}{k}\right)^k} ; if\ x \geq 0 \\ 0 \quad ; if\ x < 0 \end{cases}$	(A5)	The Weibull distribution	The Weibull distribution is used to estimate the annual mean wind speed at Raglan.	λ the scale factor, k the shape factor and; x the wind speed.
$OPEX\left[\frac{\$}{MWh}\right] = \frac{1000}{Fc \times 8760} \times OPEX\left[\frac{\$}{kW.an}\right]$	(A6)	OPEX	LCOE corresponds to the addition of CAPEX and OPEX.	Fc: load factor
$CAPEX\left[\frac{\$}{MWh}\right] = \frac{1000}{Fc \times 8760 \times D} \times CAPEX\left[\frac{\$}{kW}\right]$	(A7)	CAPEX	LCOE corresponds to the addition of CAPEX and OPEX.	Fc: load factor D: project lifespan
$Corrective\ factor = \frac{Real\ production\ of\ E-82\ wind\ turbines}{production\ obtained\ using\ its\ power\ curve\ from\ the\ supplier}$	(A8)	correction factor		The correction factor is applied to account for production inefficiencies. It takes into account the harsh environmental conditions in Raglan (cold, frost, storms, etc.).

References

1. United Nations. Framework convention on climate change. In Proceedings of the Report of the Conference of the Parties on Its Twenty-First Session, Paris, France, 30 November–11 December 2015; Part One: Proceedings. p. 42.
2. Jahangiri, Z.; Hendriks, R.; McPherson, M. A Machine Learning Approach to Analysis of Canadian Provincial Power System Decarbonization. *Energy Rep.* **2024**, *11*, 4849–4861. [CrossRef]
3. Government of Canada. Canada's Official Greenhouse gas Inventory. *Annex 13—Electricity in Canada.* Available online: https://open.canada.ca/data/en/dataset/779c7bcf-4982-47eb-af1b-a33618a05e5b (accessed on 9 January 2025).
4. Government of Canada. Canadian Net-Zero Emissions Accountability Act. Available online: https://www.canada.ca/en/services/environment/weather/climatechange/climate-plan/net-zero-emissions-2050/canadian-net-zero-emissions-accountability-act.html (accessed on 9 January 2025).
5. Saffari, M.; McPherson, M. Assessment of Canada's Electricity System Potential for Variable Renewable Energy Integration. *Energy* **2022**, *250*, 123757. [CrossRef]
6. Nielsen, A.S.; Del Alamo Serrano, G.; Schanche, T.L.; Burheim, O.S. Reducing CO_2 Emissions, Energy Consumption, and Decarbonization Costs in Manganese Production by Integrating Fuel-Assisted Solid Oxide Electrolysis Cells in Two-Stage Oxide Reduction. *Appl. Energy* **2025**, *377*, 124572. [CrossRef]
7. Da Silva, F.T.F.; Garcez Lopes, M.S.; Asano, L.M.; Angelkorte, G.; Brambilla Costa, A.K.; Szklo, A.; Schaeffer, R.; Coutinho, P. Integrated Systems for the Production of Food, Energy and Materials as a Sustainable Strategy for Decarbonization and Land Use: The Case of Sugarcane in Brazil. *Biomass Bioenergy* **2024**, *190*, 107387. [CrossRef]
8. Du, Y.; Shen, X.; Kammen, D.M.; Hong, C.; Nie, J.; Zheng, B.; Yao, S. A Generation and Transmission Expansion Planning Model for the Electricity Market with Decarbonization Policies. *Adv. Appl. Energy* **2024**, *13*, 100162. [CrossRef]
9. Balaban, G.; Dumbrava, V.; Lazaroiu, A.C.; Kalogirou, S. Analysis of Urban Network Operation in Presence of Renewable Sources for Decarbonization of Energy System. *Renew. Energy* **2024**, *230*, 120870. [CrossRef]
10. Dongsheng, C.; Ndifor, E.Z.; Temidayo Olayinka, A.-O.; Ukwuoma, C.C.; Shefik, A.; Hu, Y.; Bamisile, O.; Dagbasi, M.; Uzun Ozsahin, D.; Adun, H. An EnergyPlan Analysis of Electricity Decarbonization in the CEMAC Region. *Energy Strategy Rev.* **2024**, *56*, 101548. [CrossRef]
11. Obiora, S.C.; Bamisile, O.; Hu, Y.; Ozsahin, D.U.; Adun, H. Assessing the Decarbonization of Electricity Generation in Major Emitting Countries by 2030 and 2050: Transition to a High Share Renewable Energy Mix. *Heliyon* **2024**, *10*, e28770. [CrossRef]
12. Roshan Kumar, T.; Beiron, J.; Marthala, V.R.R.; Pettersson, L.; Harvey, S.; Thunman, H. Combining Exergy-Pinch and Techno-Economic Analyses for Identifying Feasible Decarbonization Opportunities in Carbon-Intensive Process Industry: Case Study of a Propylene Production Technology. *Energy Convers. Manag. X* **2025**, *25*, 100853. [CrossRef]
13. Asri, L.I.M.; Ariffin, W.N.S.F.W.; Zain, A.S.M.; Nordin, J.; Saad, N.S. Comparative Study of Energy Storage Systems (ESSs). *J. Phys. Conf. Ser.* **2021**, *1962*, 012035. [CrossRef]
14. Mahadevan, V.; S, S.R.; Rusho, M.A.; Yishak, S. Critical Review of Energy Storage Systems: A Comparative Assessment of Mechanisms, Advantages, Challenges, and Integration with Renewable Energy. *Results Eng.* **2025**, *26*, 105589. [CrossRef]
15. Babaharra, O.; Choukairy, K.; Faraji, H.; Khallaki, K.; Hamdaoui, S.; Bahammou, Y. Thermal Performance Analysis of Hollow Bricks Integrated Phase Change Materials for Various Climate Zones. *Heat Transf.* **2024**, *53*, 2148–2172. [CrossRef]
16. Liu, X.; Pan, L.; Rao, H.; Wang, Y. A Review of Transport Properties of Electrolytes in Redox Flow Batteries. *Future Batter.* **2025**, *5*, 100019. [CrossRef]
17. Thamizhselvan, R.; Naresh, R.; Ulaganathan, M.; Pol, V.G.; Ragupathy, P. Achieving Exceptional Cell Voltage (2.34 V) through Tailoring pH of Aqueous Zn-Br_2 Redox Flow Battery for Potential Large-Scale Energy Storage. *Electrochim. Acta* **2023**, *441*, 141799. [CrossRef]
18. Chen, H.; Zhang, X.; Zhang, S.; Wu, S.; Chen, F.; Xu, J. A Comparative Study of Iron-Vanadium and All-Vanadium Flow Battery for Large Scale Energy Storage. *Chem. Eng. J.* **2022**, *429*, 132403. [CrossRef]
19. Sun, J.; Wu, M.; Jiang, H.; Fan, X.; Zhao, T. Advances in the Design and Fabrication of High-Performance Flow Battery Electrodes for Renewable Energy Storage. *Adv. Appl. Energy* **2021**, *2*, 100016. [CrossRef]
20. *Standard 34-2010*; Designation and Safety Classification of Refrigerants. ANSI/ASHRAE: Atlanta, GA, USA, 2011.
21. Huang, Z.; Mu, A.; Wu, L.; Wang, H.; Zhang, Y. Electrolyte Flow Optimization and Performance Metrics Analysis of Vanadium Redox Flow Battery for Large-Scale Stationary Energy Storage. *Int. J. Hydrog. Energy* **2021**, *46*, 31952–31962. [CrossRef]
22. Loktionov, P.; Pichugov, R.; Konev, D. Neutralization Flow Batteries in Energy Harvesting and Storage. *J. Energy Storage* **2023**, *72*, 108467. [CrossRef]
23. Ren, J.; Li, Y.; Wang, Z.; Sun, J.; Yue, Q.; Fan, X.; Zhao, T. Thermal Issues of Vanadium Redox Flow Batteries. *Int. J. Heat Mass Transf.* **2023**, *203*, 123818. [CrossRef]
24. Zeng, D.; Mao, T.; Zhang, Z.; Dai, J.; Ouyang, J.; Xie, Z. A High-Performance Aqueous Eu/Ce Redox Flow Battery for Large-Scale Energy Storage Application. *Int. J. Heat Mass Transf.* **2024**, *233*, 125978. [CrossRef]

25. Ouyang, T.; Zhang, M.; Qin, P.; Tan, X. Flow Battery Energy Storage System for Microgrid Peak Shaving Based on Predictive Control Algorithm. *Appl. Energy* **2024**, *356*, 122448. [CrossRef]
26. Darling, R.M. Techno-Economic Analyses of Several Redox Flow Batteries Using Levelized Cost of Energy Storage. *Curr. Opin. Chem. Eng.* **2022**, *37*, 100855. [CrossRef]
27. Lourenssen, K.; Williams, J.; Ahmadpour, F.; Clemmer, R.; Tasnim, S. Vanadium Redox Flow Batteries: A Comprehensive Review. *J. Energy Storage* **2019**, *25*, 100844. [CrossRef]
28. CEA. RMN: Tout Savoir sur L'électrolyte Organique D'une Batterie à Flux Redox. Available online: https://www.cea.fr/drf/Pages/Actualites/En-direct-des-labos/2023/rmn--tout-savoir-sur-lelectrolyte-organique-dune-batterie-a-flux-redox.aspx#:~:text=Les%20batteries%20%C3%A0%20flux%20redox,de%20stocker%20des%20charges%20%C3%A9lectriques (accessed on 21 January 2025).
29. Zheng, N.; Wang, Q.; Ding, X.; Zhang, H.; Duan, L.; Wang, X.; Zhou, Y.; Sun, M.; Desideri, U. Techno-Economic Analysis of a Novel Solar-Based Polygeneration System Integrated with Vanadium Redox Flow Battery and Thermal Energy Storage Considering Robust Source-Load Response. *Appl. Energy* **2024**, *376*, 124288. [CrossRef]
30. Yesilyurt, M.S.; Ozcan, H.G.; Yavasoglu, H.A. Co-Simulation-Based Conventional Exergy Evaluation of a Hybrid Energy Generation-Vanadium Redox Flow Battery-Air Source Heat Pump System. *Energy* **2023**, *281*, 128301. [CrossRef]
31. Oshnoei, S.; Oshnoei, A.; Mosallanejad, A.; Haghjoo, F. Novel Load Frequency Control Scheme for an Interconnected Two-Area Power System Including Wind Turbine Generation and Redox Flow Battery. *Int. J. Electr. Power Energy Syst.* **2021**, *130*, 107033. [CrossRef]
32. Reynard, D.; Girault, H. Combined Hydrogen Production and Electricity Storage Using a Vanadium-Manganese Redox Dual-Flow Battery. *Cell Rep. Phys. Sci.* **2021**, *2*, 100556. [CrossRef]
33. Huang, Z.; Mu, A.; Wu, L.; Wang, H. Vanadium Redox Flow Batteries: Flow Field Design and Flow Rate Optimization. *J. Energy Storage* **2022**, *45*, 103526. [CrossRef]
34. Ulaganathan, M.; Aravindan, V.; Yan, Q.; Madhavi, S.; Skyllas-Kazacos, M.; Lim, T.M. Recent Advancements in All-Vanadium Redox Flow Batteries. *Adv. Mater. Interfaces* **2016**, *3*, 1500309. [CrossRef]
35. Hosseiny, S.S.; Wessling, M. Ion Exchange Membranes for Vanadium Redox Flow Batteries. In *Advanced Membrane Science and Technology for Sustainable Energy and Environmental Applications*; Elsevier: Amsterdam, The Netherlands, 2011; pp. 413–434. ISBN 978-1-84569-969-7.
36. Skyllas-Kazacos, M.; Menictas, C.; Lim, T. Redox Flow Batteries for Medium—to Large-Scale Energy Storage. In *Electricity Transmission, Distribution and Storage Systems*; Elsevier: Amsterdam, The Netherlands, 2013; pp. 398–441. ISBN 978-1-84569-784-6.
37. Emrani, A.; Achour, Y.; Sanjari, M.J.; Berrada, A. Adaptive Energy Management Strategy for Optimal Integration of Wind/PV System with Hybrid Gravity/Battery Energy Storage Using Forecast Models. *J. Energy Storage* **2024**, *96*, 112613. [CrossRef]
38. Cremoncini, D.; Di Lorenzo, G.; Frate, G.F.; Bischi, A.; Baccioli, A.; Ferrari, L. Techno-Economic Analysis of Aqueous Organic Redox Flow Batteries: Stochastic Investigation of Capital Cost and Levelized Cost of Storage. *Appl. Energy* **2024**, *360*, 122738. [CrossRef]
39. IRENA. *Electricity Storage and Renewables: Costs and Markets to 2030*; International Renewable Energy Agency: Abu Dhabi, United Arab Emirates, 2017; p. 132.
40. Schmidt, O.; Melchior, S.; Hawkes, A.; Staffell, I. Projecting the Future Levelized Cost of Electricity Storage Technologies. *Joule* **2019**, *3*, 81–100. [CrossRef]
41. Minke, C.; Turek, T. Materials, System Designs and Modelling Approaches in Techno-Economic Assessment of All-Vanadium Redox Flow Batteries—A Review. *J. Power Sources* **2018**, *376*, 66–81. [CrossRef]
42. Zheng, M.; Sun, J.; Meinrenken, C.J.; Wang, T. Pathways Toward Enhanced Techno-Economic Performance of Flow Battery Systems in Energy System Applications. *J. Electrochem. Energy Convers. Storage* **2019**, *16*, 021001. [CrossRef]
43. Zhang, M.; Moore, M.; Watson, J.S.; Zawodzinski, T.A.; Counce, R.M. Capital Cost Sensitivity Analysis of an All-Vanadium Redox-Flow Battery. *J. Electrochem. Soc.* **2012**, *159*, A1183–A1188. [CrossRef]
44. Noack, J.; Wietschel, L.; Roznyatovskaya, N.; Pinkwart, K.; Tübke, J. Techno-Economic Modeling and Analysis of Redox Flow Battery Systems. *Energies* **2016**, *9*, 627. [CrossRef]
45. Viswanathan, V.; Crawford, A.; Stephenson, D.; Kim, S.; Wang, W.; Li, B.; Coffey, G.; Thomsen, E.; Graff, G.; Balducci, P.; et al. Cost and Performance Model for Redox Flow Batteries. *J. Power Sources* **2014**, *247*, 1040–1051. [CrossRef]
46. IRENA. *Renewable Power Generation Costs in 2021*; International Renewable Energy Agency: Abu Dhabi, The Emirates, 2022; p. 204.
47. Tardy, A.; Rousse, D.R.; Mungyeko Bisulandu, B.-J.R.; Ilinca, A. Enhancing Energy Sustainability in Remote Mining Operations Through Wind and Pumped-Hydro Storage; Application to Raglan Mine, Canada. *Energies* **2025**, *18*, 2184. [CrossRef]
48. Durand, S.; Turcotte, P.; Haillot, D.; Rousse, D.R.; Arasteh, H.; Kalivogui, S.; Zhang, K.; Izquierdo, R.; Merabtine, A.; Maref, W.; et al. Decarbonization Strategies for Northern Canada: A Review of Renewable Energy and Energy Storage in off-Grid Communities. *Energy Convers. Manag. X* **2025**, *27*, 101055. [CrossRef]

49. VRB Energy. VRB Energy Announces UL1973 Certification for 1MW VRB-ESS. Available online: https://vrbenergy.com/vrb-energy-announces-ul1973-certification-for-1mw-vrb-ess/ (accessed on 12 June 2025).
50. Stehly, T.; Duffy, P. *2020 Cost of Wind Energy Review*; Renewable Energy: Golden, CO, USA, 2022; p. 77.
51. Solar Huawei. Coût du Stockage D'énergie Solaire: Guide Pour les Propriétaires. Available online: https://solar.huawei.com/tn/blog/tn/2024/solar-energy-storage-cost (accessed on 22 January 2025).

applied sciences

MDPI

Article

Estimation of Fuel Cell Power Demand on Commercial Vehicles Based on Improved Multiple Grey Prediction Method Considering Dynamic Time Window

Yuan Wang, Yingjia Li, Jianshan Lu * and Hongbo Zhou

College of Mechanical Engineering, Zhejiang University of Technology, Hangzhou 310023, China; wangyuan@zjut.edu.cn (Y.W.); lyj1184373494@163.com (Y.L.); zhouhb@zjut.edu.cn (H.Z.)
* Correspondence: lujianshan@zjut.edu.cn

Abstract: A frequently large variation in load conditions is of great impact on the service life of fuel cells during the operation of fuel cell vehicles and further increases the maintenance cost of the system. This study proposed a method of power demand prediction of the proton exchange membrane fuel cell (PEMFC) on vehicles in the actual traffic environment. The hybrid power system topology of fuel cells and power battery on commercial vehicles is selected to build a fuel cell model, and the accuracy of the fuel cell model is verified. An improved multiple grey prediction method is then proposed to predict the power demand during the sampling period of the fuel cell while considering a dynamic time window in the prediction period. Comparisons were made between this proposed model and the other two prediction models as a single-step prediction and multi-step prediction. Data of CHTC-HT and field testing working conditions were used to evaluate these three prediction models in fuel cell power demand. Results showed that the multiple grey method showed a better prediction performance than the other models, indicated by the lowest error value of 16.944% under the CHTC-HT condition, the lowest error value of 2.169% under stable conditions with less variable load and 1.930% under dynamic conditions with frequent load changes in field testing. This study of the demand power prediction can be devoted to pre-tuning the fuel cell system to avoid performance degradation caused by unanticipated power fluctuation.

Keywords: variable load condition; proton exchange membrane fuel cell; power demand prediction; grey prediction

1. Introduction

A fuel cell is an electrochemical power generation device that converts chemical energy directly into electrical energy [1]. As the main application scenario of hydrogen energy, the emergence of fuel cell technology provides a powerful solution for energy consumption and environmental degradation [2]. The research and commercialization of hydrogen energy and fuel cells are developing rapidly in China, the United States, Japan, and Europe [3]. Proton exchange membrane (PEM) fuel cell vehicles have thus become a major interest in academia and the automotive industry [4]. Presently, the main factors restricting the commercialization of fuel cells include durability and cost [5]. The durability problem of fuel cells is mainly reflected in its components, especially the proton exchange membrane and catalyst, which are easy to age and degrade during long-term operation. This directly affects the service life and maintenance frequency of the battery, increasing the cost of use.

Thus, it is necessary to enhance the durability of fuel cells through technological innovation to promote the commercialization of fuel cells.

The degradation of fuel cell performance is a complex problem with multiple coupling, and the harsh working conditions will lead to changes in its electrochemical active area and catalyst structural layer, which will further cause the degradation of fuel cell performance [6]. Conditions that affect the durability of fuel cells mainly include variable load, idle speed, high load, start-stop and other conditions [7]. Among them, variable load conditions are the most serious conditions that affect the durability of fuel cells and account for the highest proportion of performance degradation factors [8]. Therefore, predicting the fuel cell demand power in the next period and then adjusting the fuel cell system in advance when the vehicle demand power changes rapidly can avoid the melting of the catalyst and the polymerization of the catalyst particles caused by an excessive power adjustment rate, thus decreasing the fuel cell performance attenuation.

Accurate parameter prediction plays an important role in the lifetime of the PEMFC. Prediction methods used in the study of PEMFC performance can be divided into three categories: model-driven, data-driven and hybrid methods [9]. Model-driven methods usually use empirical or semi-empirical degradation models, combined with methods such as the extended Kalman filter (EKF) or particle filter to describe the behaviour of the system [10,11]. These methods use a small amount of data, but no accurate model has been established because of an extremely complex system structure of the PEMFC with an indistinct degradation mechanism [12]. Data-driven approaches such as machine learning [13] and data packet processing [14] analyze the currently recorded data to explore the declining trend in PEMFC performance. Based on the short-term change trend of time series data, researchers [15] use the linear regression relationship to extract input and output variables and establish a regression prediction model based on a machine learning algorithm. The hybrid method combines the advantages of model-driven and data-driven methods and makes use of the theoretical basis provided by the model-driven method and the prediction ability of the data-driven method to improve the accuracy and reliability of prediction. This approach is achieved by combining physical models with data-driven models or by integrating different predictive models, but the approach is complex and requires more time and resources to build and debug the model.

Due to its strong ability to deal with complex and nonlinear problems, data-driven methods especially perform well in the face of high-dimensional data and large amounts of historical data. Compared to model-driven approaches, data-driven approaches do not rely on deep understanding and assumptions about the internal mechanisms of the system and are therefore more flexible and adaptable when dealing with unknown or complex systems. At the same time, compared with the hybrid method, the data-driven method is more concise, avoiding the complexity of model combination and high computation. It is more suitable for vehicle application scenarios with high dynamic conditions. Therefore, grey system theory is used for data-driven prediction with the advantages of no requirements of large amounts of data. As a new interdisciplinary research field, grey system theory has achieved remarkable developments [16]. The GM(1,1) model is an important model in grey system theory and is a specific form of grey prediction model. It is an Accumulated Generating Operation (AGO), which can build an approximate linear difference equation to predict data trends. The GM can provide an effective prediction with a small amount of data and can adapt to different operating conditions, so the model can be applied to the research of power prediction. An improved GM (1,1) model has been proposed, which uses cubic spline interpolation to optimize background values and has better prediction accuracy in power investment prediction [17]. Based on the improved failure mode recognition algorithm and GM (1,1) model, a fatigue life prediction method for hyperstatic truss structures was established in one previous study, which proved the effectiveness of the fatigue life prediction method [18]. Researchers also used GM (1,1) to establish a prediction model to predict the future PM2.5 content in Changchun and found that it can be used as an effective method to predict the fog and haze weather [19]. Additionally, the GM (1,1) model has been proven in others' research to effectively predict the number of regional tourists [20] and predict the freight volume and freight turnover of Luoyang City to provide references for the decision making of logistic industry schemes [21].

A novel grey prediction model, which predicted four typical interval grey sequences, respectively, has effectively extended function transformation technology to the field of interval grey numbers for grey prediction [22]. Based on the Grey Verhulst model, an algorithm was proposed for predicting Internet topic trends with high accuracy [23]. Studies in vehicles found a new and effective method to analyze and predict the sound quality of vehicle internal noise by using grey system theory, and the method can be a useful tool for analyzing and predicting vehicle internal noise quality [24]. An enhanced grey predictive fuzzy controller (EGPFC) was also developed, which has better control performance in enhancing vehicle grip [25]. Especially in the field of fuel cell research, the grey model has been applied. Researchers propose a Grey–Markov chain power prediction energy management strategy for the fuel cell power generation system, which reduces hydrogen consumption in fuel cell power generation systems [26]. Additionally, a GM (1, N) power prediction strategy is proposed and compared with other strategies as a switch control strategy and logical threshold strategy and found that the power prediction strategy based on GM (1, N) is superior to the other two strategies in terms of output efficiency and fuel economy [27]. Thus, the grey system model has the advantages of providing an effective prediction with only a small amount of data and adapting to different operating conditions and environmental changes to be suitable for the study of power prediction.

Given the above, the real-time prediction method of fuel cell demand power is established based on the GM model in this study. With different sampling periods and prediction times, fuel cell power demand on commercial vehicles is estimated with three prediction strategies: single-step prediction, multi-step prediction and multiple prediction. Efficiency of these three strategies is compared under various working conditions, so as to achieve the high-precision prediction of power demand in dynamic changing environment.

2. Methodology

2.1. Fuel Cell Vehicle Power System Modelling

2.1.1. Fuel Cell Power System Topology

Fuel cell commercial vehicles were selected in this study with the lithium electronic battery as an auxiliary power source, forming an electric–electric hybrid power system with the fuel cell. This addition of power sources extends the working range of the power system and gives it more opportunities to work in the efficient area. As shown in Figure 1, the power of the vehicle is provided by the fuel cell and the power battery together, avoiding large fluctuations in the fuel cell load. The power battery can be used when additional power output is needed or when acceleration performance is temporarily improved. Additionally, the power battery can be used to recover the braking energy. By converting the braking energy into electric energy and storing it in the power battery, the energy utilization efficiency can be improved, and the life of the fuel cell can be extended. The main configuration information of the vehicle is shown in Table 1.

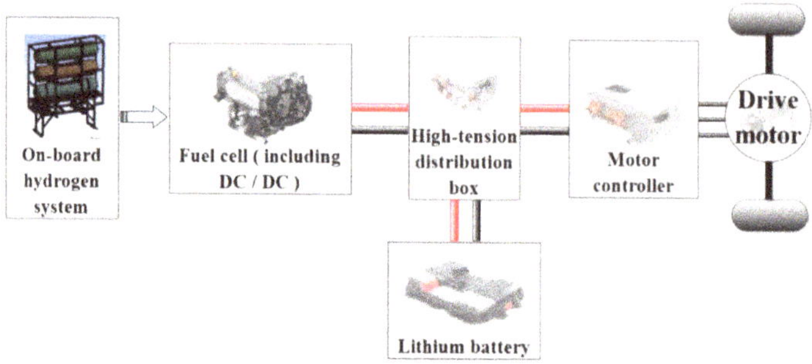

Figure 1. Structure of fuel cell and power battery.

Table 1. Power configuration of fuel cell vehicle.

Vehicle Type	49 Ton Fuel Cell Semi-Tractor
Rated power of drive motor (kW)	236
Peak power of drive motor (kW)	360
Fuel cell power rating (kW)	125
Power battery total energy (kWh)	141.312
Power battery nominal voltage (V)	614.4
Power battery nominal capacity (Ah)	230
Power battery pulse discharge ratio (C)	2
Power battery pulse discharge time (s)	30

2.1.2. Fuel Cell Variable Load Cycle

To optimize fuel cell and power battery collaboration, three key factors were considered as follows:

(i) Fuel cell durability: The durability of a fuel cell is affected by the frequency and magnitude of load changes. Frequent and large load changes will accelerate the accumulation of mechanical and chemical stresses inside the fuel cell, affecting its life. The high variable load speed of the fuel cell will cause the membrane electrode assembly (MEA) to undergo greater mechanical stress and a greater chemical degradation rate, especially when the load changes frequently and greatly. By setting the appropriate variable load period T_{load}, the frequency and amplitude of load change can be controlled. The variable load rate v_{load} is calculated as follows:

$$v_{load} = \frac{\Delta P}{T_{load}} \tag{1}$$

where ΔP is the variable load amplitude. Long T_{load} can reduce the rate of reducing each variable load, thereby reducing the impact on the fuel cell membrane electrode assembly and improving its durability.

(ii) Output characteristics of fuel cells: The output characteristics of fuel cells are soft and slow in response. In order to ensure that the fuel cell can supply the current steadily and avoid output instability or reduced efficiency due to frequent or drastic load changes, it is necessary to give the fuel cell sufficient time to adjust the output. The variable load cycle of the fuel cell should be long enough to ensure that its output can gradually adapt to new requirements when the load changes, avoiding the impact on system stability.

(iii) Pulse discharge time of the power battery: The power battery provides peak power when the car encounters high power demand (such as acceleration or climbing), and the pulse discharge time it can provide is usually 30–60 s. As an auxiliary energy source, the power battery can provide high currents in a short period of time, thereby reducing the instantaneous load demand of the fuel cell and extending the variable load cycle of the fuel cell to a certain extent. Therefore, in fuel cell commercial vehicles, during the peak power provided by the power cell, the fuel cell can gradually adjust the output to avoid drastic load changes affecting the durability of the fuel cell. For example, within 30 s of a power battery discharging, a fuel cell can gradually increase its output power without having to immediately reach maximum output in a short period of time.

Based on the above three considerations, the variable load period of the fuel cell can be set to 15 s on the fuel cell commercial vehicle. In this cycle, only the variable load rate of the fuel cell is greater than 10 kW/s, that is, only a lower variable load rate can meet the power needs of the vehicle. This cycle length is combined with the 30–60 s pulse discharge time of the power battery to ensure that the fuel cell can change load in time during each cycle and effectively match the pulse discharge time of the power battery. Specifically, the 15 s variable load cycle is moderate, neither too frequent to cause fuel cell life attenuation, nor too slow to affect the vehicle power performance, to ensure the stable output of the fuel cell at the same time, to avoid damage caused by large variable load.

2.1.3. Equivalent Circuit Model

In this study, an equivalent circuit model was constructed to simulate the internal polarization characteristics of the fuel cell by using circuit elements, including activation polarization, ohmic polarization, concentration polarization and a double-layer charge effect, to output the actual voltage and power of the fuel cell (shown in Figure 2).

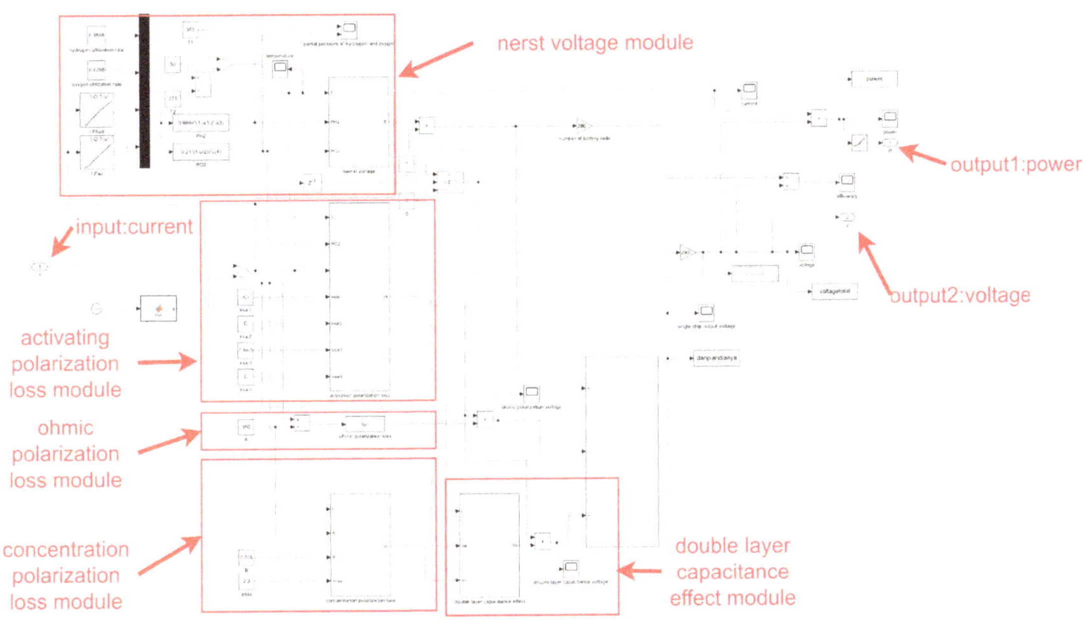

Figure 2. Fuel cell model structure.

The fuel cell polarization curve is a key index to evaluate the output performance of the fuel cell. In this study, the reliability and accuracy of the model were verified by comparing the polarization curves obtained under simulation and experimental conditions, in which the experimental conditions were run and tested on the fuel cell test bench.

In this study, the accuracy of the model was assessed by calculating the Mean Absolute Percentage Error (MAPE), a commonly used method to measure the accuracy of a predictive model, especially in the fields of statistics and machine learning. The advantage of this approach is that it generates a dimensionless metric that allows fair comparisons between the data of different sizes or units. In addition, the average relative error provides a visual percentage of error that helps us better understand how far the model is biased in its predictions.

2.2. Grey Prediction Modelling

The GM(1,1) model can convert non-stable data series into stable data series by the Accumulated Generating Operation, which can reveal the inherent regularity and development trend of data. The modelling steps were described in one of our previous papers [27]. GM(1,N) represents a first-order grey model with one dependent variable and $N-1$ independent variables. It is assumed that there are N sets of sequences composed of non-negative variables $x_i(i=1,2,\ldots N)$, and each set of sequences has m initial values:

$$x_i^{(0)} = \left\{x_i^{(0)}(1),\ \ x_i^{(0)}(2),\ \ \ldots\ \ x_i^{(0)}(m)\right\} \tag{2}$$

A new first-order cumulative sequence $x_i^{(1)}$ is calculated when each value of the initial sequence accumulates all the values as follows:

$$x_i^{(1)} = \left\{x_i^{(1)}(1),\ \ x_i^{(1)}(2),\ \ \ldots\ \ x_i^{(1)}(m)\right\} \tag{3}$$

where

$$x_i^{(1)}(t) = \sum_{j=1}^{t} x_i^{(0)}(j) \quad t = 1,2\ldots m \tag{4}$$

The grey differential equation of GM(1,N) is defined as follows:

$$\frac{dx_1^{(1)}(k)}{dt} + ax_1^{(1)}(k) = \sum_{i=2}^{n} b_i x_i^{(1)}(k) \tag{5}$$

where a is the system development parameter, and b_i is the driving parameter, representing the influence degree and polarity of the independent variable on the dependent variable. The dependent variable increases as the independent variable increases and vice versa. The grey differential equation can be defined as follows:

$$x_1^{(1)}(k) + az_1^{(1)}(k) = \sum_{i=2}^{n} b_i x_i^{(1)}(k) \tag{6}$$

where $z_1^{(1)}(k)$ is defined as

$$z_1^{(1)}(k) = \frac{x_1^{(1)}(k) + x_1^{(1)}(k-1)}{2} \tag{7}$$

The grey parameters of the grey model P_N represent the vector set of system development parameters and driving parameters, which can be calculated as follows:

$$P_N = (a, b_2, \ldots b_n)^T = B^{-1}Y_n \quad (m = N+1) \tag{8}$$

$$P_N = \left(B^T B\right)^{-1} B^T Y_n \quad (M > N+1) \tag{9}$$

$$P_N = B^T\left(B^T B\right)^{-1} Y_n \quad (M > N+1) \tag{10}$$

where

$$B = \begin{bmatrix} -z_1^{(1)}(2) & -x_2^{(1)}(2) & \cdots & -x_i^{(1)}(2) \\ -z_1^{(1)}(3) & -x_2^{(1)}(3) & \cdots & -x_i^{(1)}(3) \\ \vdots & \vdots & \vdots & \vdots \\ -z_1^{(1)}(m) & -x_2^{(1)}(m) & \cdots & -x_i^{(1)}(m) \end{bmatrix} \tag{11}$$

$$Y_n = \begin{bmatrix} x_1^{(0)}(2) \\ x_1^{(0)}(3) \\ \vdots \\ x_1^{(0)}(m) \end{bmatrix} \tag{12}$$

Then, the solution of Equation (5) can be obtained:

$$\hat{x}_1^{(1)}(k+1) = \left(x_1^{(0)}(1) - \sum_{i=2}^{n-1} \frac{b_i x_i^{(1)}(k+1)}{a} \right) e^{-at} + \sum_{i=2}^{n-1} \frac{b_i x_i^{(1)}(k+1)}{a} \tag{13}$$

Finally, the predicted value of the initial sequence can be obtained:

$$\hat{x}_1^{(0)}(k+1) = \hat{x}_1^{(1)}(k+1) - \hat{x}_1^{(1)}(k) \tag{14}$$

In this study, the GM (1,1) model is applied to predict the power demand at the future time (that is, after 15 s). At the same time, in order to explore the influence of different prediction steps and prediction times on the prediction accuracy, three predicting strategies are proposed, which are single-step prediction, multi-step prediction and multiple prediction.

2.2.1. Single-Step Grey Prediction

The process of the single-step grey prediction strategy is shown in Figure 3. The sample period of the power was set as 15 s. The whole power series obtained by sampling cover data of the current moment t_m and the three previous moments (t_{m-1}, t_{m-2}, t_{m-3}). Based on a prediction step size of 1, which is called single-step prediction, the prediction of the power demand at a t_{m+1} moment can be obtained.

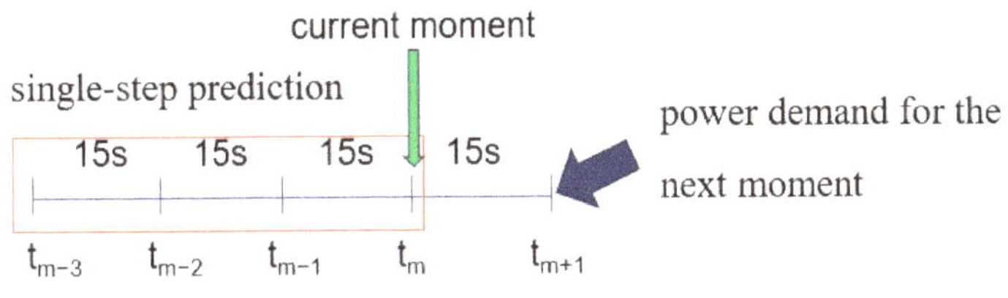

Figure 3. Process of single-step grey prediction.

2.2.2. Multi-Step Grey Prediction

According to the multi-step grey prediction strategy shown in Figure 4, compared with the single step prediction, the GM (1,1) model samples the power on a 5 s cycle, but the sequence of the input model also contains the power demand at four time points. Based on a prediction step size of 3 (i.e., multi-step prediction), the model predicts the power demand at time t_{m+3}.

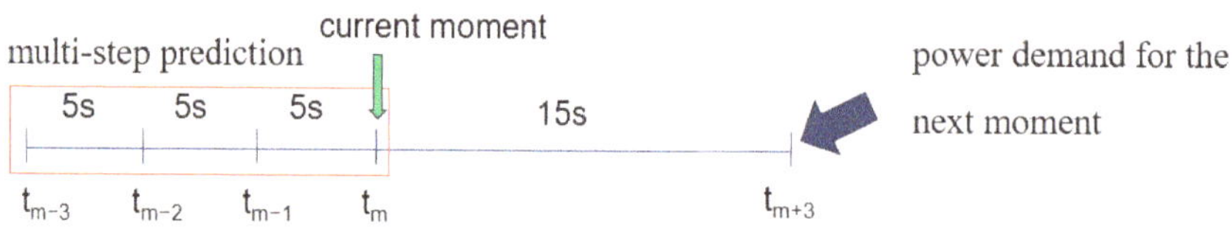

Figure 4. Process of multi-step grey prediction.

2.2.3. Multiple Grey Prediction

For the multiple prediction strategy shown in Figure 5, the sampling period is the same as that of multi-step prediction, which is set at 5 s. Compared with the first two predicting strategies, the remarkable feature of this strategy is that the input model sequence of this strategy covers the power demand data of 6 time points and expands the prediction number to 3 times; that is, the power demands of t_{m+1}, t_{m+2} and t_{m+3} are predicted. Specifically, the prediction of t_{m+1} is calculated based on a series of power demands at six time points from t_{m-5} to t_m. The power demand at 6 time points is selected for each step to ensure that the method is consistent with the above two strategies when predicting the power demand at t_{m+3} time and that there are definite data for 4 known moments.

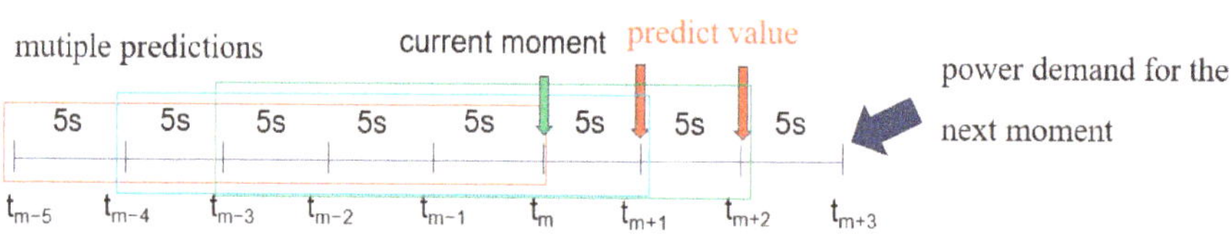

Figure 5. Process of multiple grey prediction.

3. Method Preprocessing

3.1. Fuel Cell Model Equivalent Circuit Model Evaluation

Figure 6 and Table 2 show the accuracy of the model evaluation in terms of fuel cell monolithic voltage and fuel cell stack power, respectively.

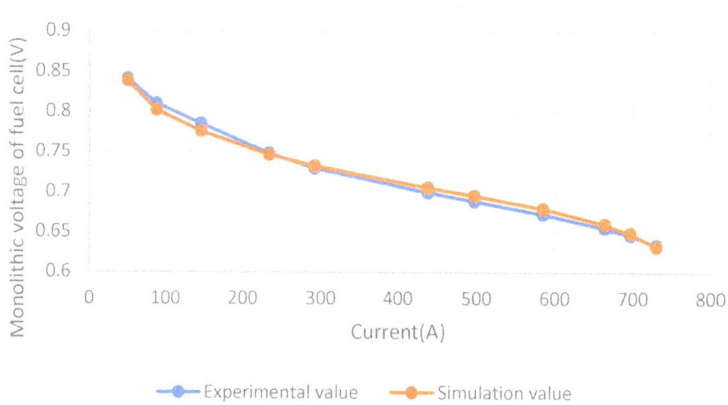

Figure 6. Comparison of fuel cell monolithic voltage.

Table 2. Comparison of fuel cell stack power at different currents.

Current (A)	Fuel Cell Stack Power (W)		Absolute Ratio Error (%)
	Experimental Value	Simulation Value	
50	13.9	13.8	0.72
88	23.5	23.3	0.85
146	37.8	37.4	1.06
234	57.7	57.6	0.17
292	70.2	70.1	0.14
438	101.1	101.9	0.79
497	112.8	113.9	0.98
585	129.7	131.1	1.08
664	143.4	144.6	0.84
697	148.5	149.2	0.47
730	152.7	152.3	0.26
Mean relative percentage error: 0.67%			

In the comparative analysis, the voltage and power performance of a single fuel cell under different current loads is investigated in detail. Figure 6 shows the downward trend of the monolithic voltage as the current increases over the range from 50 A to 730 A. The very low error range between the experimental and simulation data (the average error of 0.69%) indicates that the simulation model has a high accuracy in capturing the dynamic process of voltage changes with the current. Table 2 shows the stack output power performance of the fuel cell in the same current range. The simulation values showed a very low error (0.67%) in this section. In summary, these analysis results confirm the accuracy of the fuel cell model and can meet the needs of this study.

3.2. Model Input Condition

CHTC-HT is one of China's commercial vehicle-driving-condition standards, which plays an important role in vehicle performance evaluations for commercial vehicle manufacturing and supervision [28]. The working condition input process is shown in Figure 7. In this study, the vehicle dynamic model was used to obtain the vehicle power demand by setting the second-by-second speed data under the CHTC-HT working condition as input. Then, the vehicle power demand was set as the input variable of the fuel cell model, and the fuel cell power demand can be calculated as the output of the model.

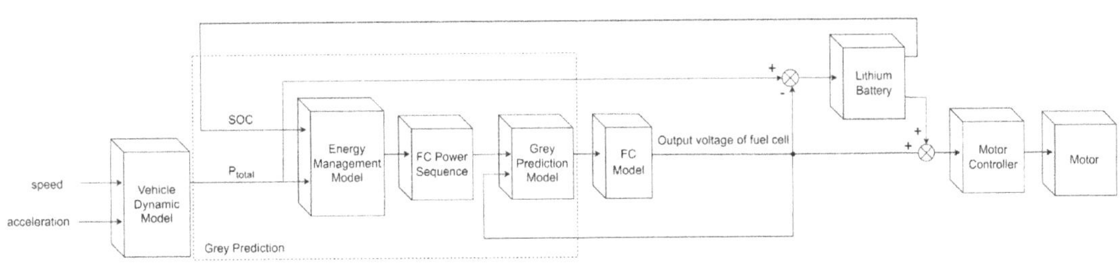

Figure 7. Process of model input condition.

Figure 8 shows the velocity and acceleration curves under the CHTC-HT condition. This working condition can be divided into three stages in chronological order. The first stage lasted 342 s as a low-speed operation. The second stage lasted 988 s as a medium-speed operation. The third stage lasted 470 s as a high-speed operation.

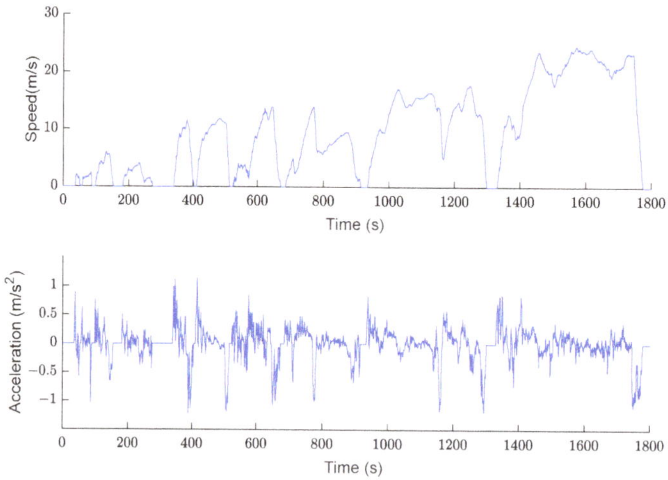

Figure 8. CHTC-HT working condition.

The obtained fuel–electric demand power curve is shown in Figure 9. It can be seen that, under the current working condition, the fuel–electric power demand has significant load changes or frequent load fluctuations. Therefore, in order to avoid the fuel cell performance attenuation caused by large load changes or frequent load changes, the power of the fuel cell would be estimated based on the grey GM(1,1) model.

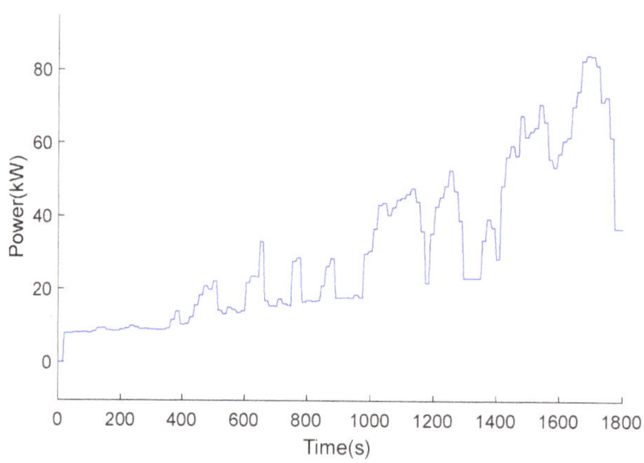

Figure 9. Fuel cell power demand under CHTC-HT condition.

4. Results

4.1. Under CHTC-HT Working Condition

Since the prediction of the GM(1,1) model requires a certain amount of historical data, in this study, the prediction started from the 500th second of the working condition. Then, the prediction of fuel cell power demand under the CHTC-HT working condition was carried out through the above three prediction strategies (i.e., single-step prediction, multi-step prediction and multiple prediction). Results of the simulation are shown in Figure 10.

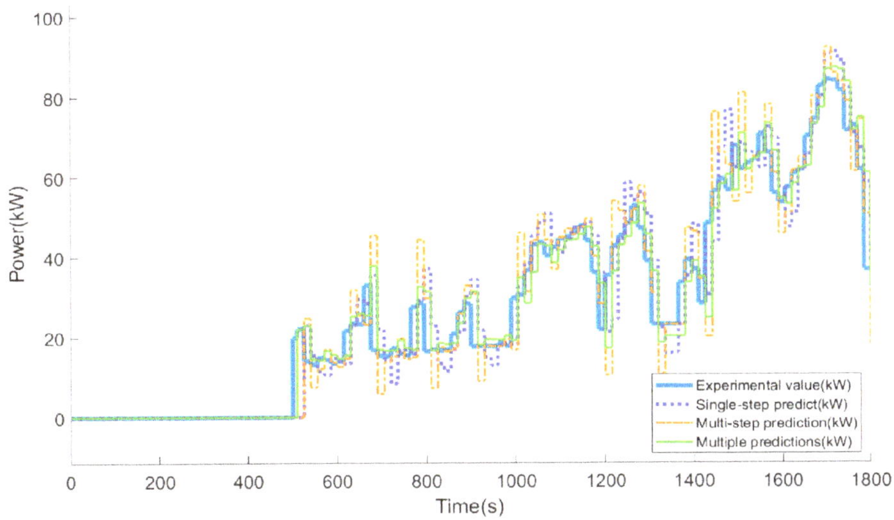

Figure 10. Prediction curves of fuel cell power demand with three strategies under CHTC-HT.

In this study, the average relative error is also used to evaluate the advantages and disadvantages of the predicting strategy. Table 3 listed the prediction errors of the three prediction strategies under the CHTC-HT condition. The errors of single-step prediction, multi-step prediction and multiple prediction are 22.957%, 23.366% and 16.944%, respectively. Under this condition, the prediction accuracy of the grey multiple prediction strategy is significantly higher than that of the other two prediction strategies.

Table 3. Average relative prediction errors of the three prediction strategies under CHTC-HT.

Grey Prediction Strategies	Average Relative Prediction Error (%)
Single-step prediction	22.957
Multi-step prediction	23.366
Multiple prediction	16.944

As can be seen from Figure 10 and Table 3, single-step prediction can track the experimental value change well at each step, but the performance in the area with large fluctuations may not be ideal, resulting in lag or an inaccurate prediction. When predicting multiple time steps, the accuracy may gradually decrease due to the accumulation of errors, especially in complex and frequently changing periods of time. For example, between 800 and 1400 s under the working condition, the power fluctuation is between 0 and 120 kW. Multiple prediction can smooth short-term fluctuations to a large extent and improve the stability of prediction results. For example, multiple predictions show better stability in regions with large power fluctuations, such as 600 s to 800 s and 1000 s to 1200 s.

4.2. Under Field-Testing Working Condition

In order to verify the applicability of the above three predicting strategies in real operating conditions, these three strategies were evaluated with the fuel cell operation data of fuel cell vehicles under actual driving conditions. The data used in this study were collected from the operation data of a 49-ton fuel cell heavy truck operating in Hebei

Province in China on March 3 in 2024 (shown in Figure 11), providing the power request sent by the vehicle to the fuel cell in real time.

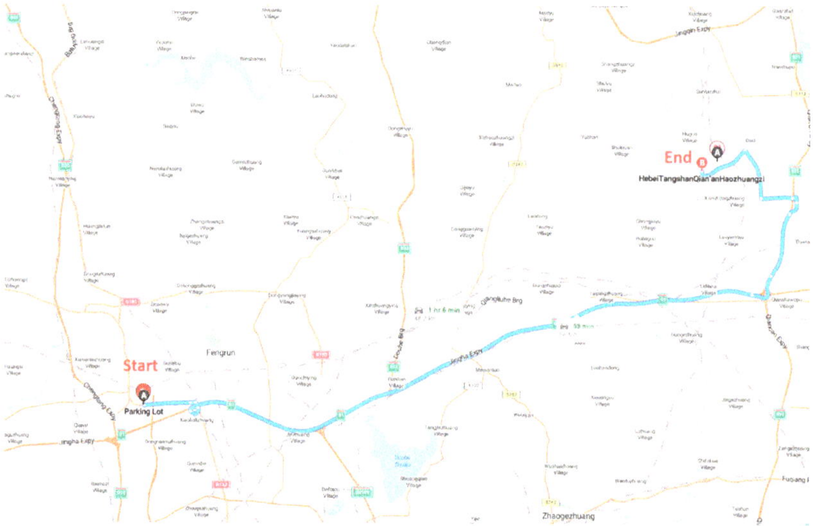

Figure 11. Truck route map in field testing.

The vehicle power system parameters are mainly used for logistic transportation. The performance and applicability of the proposed strategy under different environmental conditions can be evaluated by selecting the working conditions with different load variation degrees to verify the strategy. Thus, the stability and reliability of the strategy can be evaluated in practical application; further, the robustness can be verified by testing the performance of the strategy under different working conditions.

Two typical working conditions with different power request intensities covering a time span of 900 s in the running process of the vehicle are selected and marked as Working Condition A and Working Condition B, respectively. Under Working Condition A, the number of large load changes or frequent load changes is less, while the frequency of load changes is more frequent under Working Condition B. Figures 12 and 13 separately show the prediction curves of three strategies (single-step prediction, multi-step prediction, and multiple prediction) in field-testing Working Condition A and Working Condition B. And Tables 4 and 5 list the prediction errors of each strategy in detail.

Table 4. Average relative prediction errors of the three prediction strategies under Working Condition A in field testing.

Grey Prediction Strategies	Average Relative Prediction Error (%)
Single-step prediction	2.262
Multi-step prediction	2.502
Multiple prediction	2.169

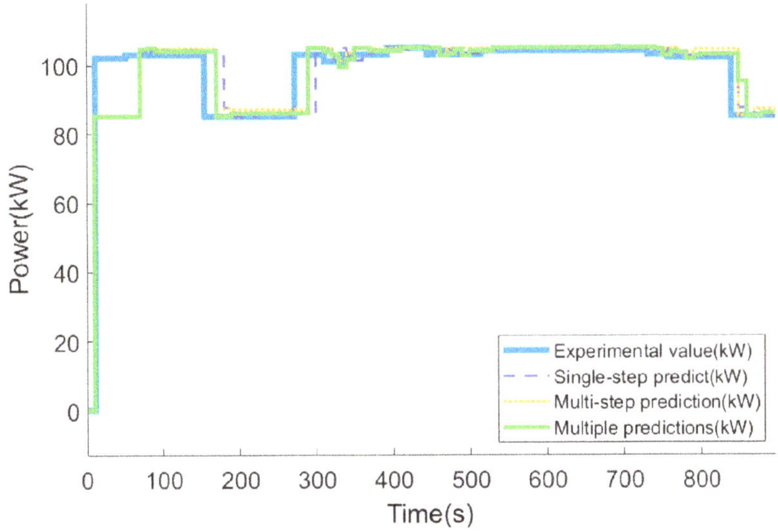

Figure 12. Prediction curves of fuel cell power demand with three strategies under Working Condition A in field testing.

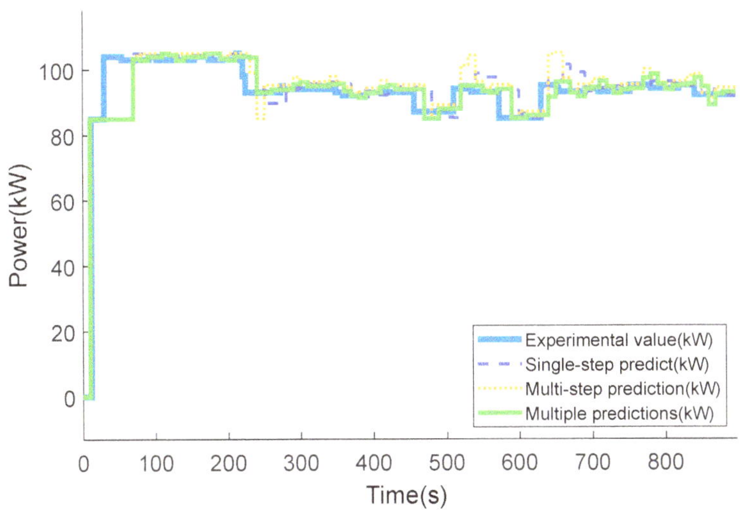

Figure 13. Prediction curves of fuel cell power demand with three strategies under Working Condition B in field testing.

Table 5. Average relative prediction errors of the three prediction strategies under Working Condition B in field testing.

Grey Prediction Strategies	Average Relative Prediction Error (%)
Single-step prediction	2.719
Multi-step prediction	3.151
Multiple prediction	1.930

The analysis results show that, both in Working Condition A with gentle fluctuation and Working Condition B with large fluctuation, the three predicting strategies all have large errors in the first 150 s of the working condition. Due to insufficient historical data at the beginning of the working condition, the three strategies all showed large errors of prediction. But after 150 s, the three strategies all showed better predicting performance.

In the prediction of Working Condition A, the average relative error of the multiple grey predicting strategy is 2.169%, which is the highest accuracy among the three strategies, followed by the single-step predicting strategy with 2.262% and the multi-step strategy with 2.502%. In the prediction of Working Condition B, the prediction error results of the three strategies showed similarly to Working Condition A. The prediction accuracy of the multiple grey predicting strategy was the highest with an average relative error of 1.930%, followed by the average relative error of the single-step strategy with an average relative error of 2.719% and the prediction effect of the multi-step strategy with an average relative error of 3.151%.

Combined with two conditions in field testing above, it can be concluded that, both under the working conditions with less load variation (Working Condition A) and frequent load variation (Working Condition B), the multiple grey strategy performed best in the fuel cell power demand prediction.

5. Conclusions

In this study, a fuel cell model was firstly constructed for commercial fuel cell vehicles, with the lithium electronic battery as an auxiliary power source and forming an electric–electric hybrid power system with the fuel cell. Power of the vehicle is provided by the fuel cell and the power battery. Fuel cell durability, output characteristics of fuel cells and pulse discharge time of the power battery were considered to determine the variable load period of the fuel cell (i.e., 15 s in this study) to optimize fuel cell and power battery collaboration on the fuel cell commercial vehicles.

In order to obtain the power demand of fuel cell, the GM (1,1) model was used based on the grey system method, which is an effective predicting tool to deal with small samples and uncertain information and is suitable for fuel cell systems with incomplete data information. Three different grey predicting strategies were proposed and tested in this study: single-step prediction, multi-step prediction and multiple prediction. In the single-step grey prediction strategy, the whole power series was obtained by sampling cover data of the current moment and previous moments and then was used to predict the power demand based on a step size of one. In multi-step grey prediction, the GM (1,1) model samples the power on a 5 s cycle and predicts the power demand based on a prediction step size of three. In the multiple grey prediction, the input model sequence of this strategy covers the power demand data of six time points and expands the prediction number to three times with a step-moved prediction time window.

Each strategy was simulated under CHTC-HT conditions to evaluate its effectiveness. The simulation results show that the grey multiple prediction strategy has the highest prediction accuracy when dealing with continuously changing dynamic conditions, indicated by the lowest error value of 16.944% when compared to the single-step prediction of 31.14% and multi-step prediction of 35.36%. In order to further verify the adaptability of the grey multiple prediction strategy in field testing, two different working conditions of fuel cell vehicles with more and less variable load conditions were selected (Working Condition A and Working Condition B). Results show that the multiple strategy showed the best prediction performance under stable conditions with less variable load (with the lowest error value of 2.169% compared to the single-step prediction of 2.262% and multi-step prediction of 2.502%) and dynamic conditions with frequent load changes (with the lowest error value of 1.930% compared to the single-step prediction of 2.719% and multi-step prediction of 3.151%), which can adjust the predicting model according to the real-time data and respond to the changes in external conditions in time.

In summary, the grey prediction model is an effective predicting tool in the power management of a fuel cell system, which enhances the adaptability and durability of the system. The power demand and supply of the fuel cell can be effectively predicted and managed based on the proposed multiple grey prediction model considering a dynamic time window. The system performance can further be optimized to improve the fuel efficiency and reliability of the overall operation of the vehicle.

Author Contributions: Conceptualization, J.L.; Methodology, Y.W.; Software, Y.L.; Validation, Y.W. and Y.L.; Formal analysis, Y.L.; Investigation, Y.L. and H.Z.; Resources, J.L. and H.Z.; Writing—original draft, Y.W. and Y.L.; Writing—review & editing, Y.W.; Supervision, J.L. and H.Z.; Funding acquisition, Y.W. and J.L. All authors have read and agreed to the published version of the manuscript.

Funding: This work was funded by [National Natural Science Foundation of China] grant number [52402434], [Zhejiang Provincial Nature Science Foundation of China] grant number [LY22E050018], and [Zhejiang Department of Transportation Science and Technology Plan] grant number [2024017]. And The APC was funded by [52402434].

Institutional Review Board Statement: Not applicable.

Informed Consent Statement: Not applicable.

Data Availability Statement: The original contributions presented in this study are included in the article. Further inquiries can be directed to the corresponding author.

Conflicts of Interest: The authors declare no conflict of interest.

References

1. Wei, T.; Yu, P.; Xue, B.; Zhang, Y.; Li, W. Modeling, Simulation and Characteristic Research of Proton Exchange Membrane Fuel Cell. *Agric. Equip. Veh. Eng.* **2022**, *60*, 41–46.
2. Wang, T.; Li, Q.; Yang, H.; Yin, L.; Wang, X.; Qiu, Y.; Chen, W. Adaptive Current Distribution Method for Parallel-connected PEMFC Generation System Considering Performance Consistency. *Energy Convers. Manag.* **2019**, *196*, 866–877. [CrossRef]
3. Qasem, N.A.; Abdulrahman, G.A. A Recent Comprehensive Review of Fuel Cells: History, Types, and Applications. *Int. J. Energy Res.* **2024**, *2024*, 7271748. [CrossRef]
4. Zheng, C.; Oh, C.; Park, Y.; Cha, S. Fuel Economy Evaluation of Fuel Cell Hybrid Vehicles Based on Equivalent Fuel Consumption. *Int. J. Hydrogen Energy* **2012**, *37*, 1790–1796. [CrossRef]
5. Yang, L. Research on Energy Management Strategy of Fuel Cell Vehicle Considering Battery Life Degradation. Master's Thesis, Jilin University, Jilin, China, 2022.
6. Hu, Z. Durability Modeling and State Estimation for Vehicular Fuel Cell System. Ph.D. Thesis, Tsinghua University, Beijing, China, 2019.
7. Huang, R.; Peng, Y.; Yang, J.; Xu, X.; Deng, P. Correlation analysis and prediction of PEM fuel cell voltage during start-stop operation based on real-world driving data. *Energy* **2022**, *260*, 124930. [CrossRef]
8. Tao, H.; Hu, Z.; Wang, Z.; Li, J.; Xu, L.; Ouyang, M. Working Conditions Analysis and Performance Degradation Estimation of Fuel Cell Buses Based on Real Road Data. In Proceedings of the 2023 7th CAA International Conference on Vehicular Control and Intelligence (CVCI 2023), Changsha, China, 27–29 October 2023.
9. Bahrami, M.; Martin, J.; Maranzana, G.; Pierfederici, S.; Weber, M.; Didierjean, S. Fuel Cell Management System: An Approach to Increase Its Durability. *Appl. Energy* **2022**, *306*, 118070. [CrossRef]
10. Bressel, M.; Hilairet, M.; Hissel, D.; Ould Bouamama, B. Remaining Useful Life Prediction and Uncertainty Quantification of Proton Exchange Membrane Fuel Cell under Variable Load. *IEEE Trans. Ind. Electron.* **2016**, *63*, 2569–2577. [CrossRef]
11. Jouin, M.; Gouriveau, R.; Hissel, D.; Pera, M.; Zerhouni, N. Prognostics of PEM Fuel Cell in a Particle Filtering Framework. *Int. J. Hydrogen Energy* **2014**, *39*, 481–494. [CrossRef]
12. Zhu, L.; Chen, J. Prognostics of PEM Fuel Cells Based on Gaussian Process State Space Models. *Energy* **2018**, *149*, 63–73. [CrossRef]
13. Javed, K.; Gouriveau, R.; Zerhouni, N.; Hissel, D. Prognostics of Proton Exchange Membrane Fuel Cells Stack Using an Ensemble of Constraints Based Connectionist Networks. *J. Power Sources* **2016**, *324*, 745–757. [CrossRef]

14. Liu, H.; Chen, J.; Hou, M.; Shao, Z.; Su, H. Data-based Short-term Prognostics for Proton Exchange Membrane Fuel Cells. *Int. J. Hydrogen Energy* **2017**, *42*, 20791–20808. [CrossRef]
15. Zeng, T. Research on Approach and Application of Short-Term Power Demand Prediction in Fuel Cell Vehicles. Ph.D. Thesis, Chongqing University, Chongqing, China, 2021.
16. Liu, S.; Tao, L.; Xie, N.; Yang, Y. On the New Model System and Framework of Grey System Theory. *J. Grey Syst.* **2016**, *28*, 1–15.
17. Lin, S.; Wang, R.; Li, M.; Zhou, J.; Deng, Y.; Luo, T.; Zhang, J.; Destech Publicat, I. Power Investment Prediction Based on the Improved GM(1,1) Model. In Proceedings of the International Conference on Electrical, Control, Automation and Robotics (ECAR), Xiamen, China, 16–17 September 2018.
18. Li, Y.; Mei, G.; Gaol, J.; Jiang, C.; Yuan, J. Application of Improved Failure Mode Recognition Algorithms and GM(1,1) Model on Fatigue Lifetime Prediction of Truss Structure. *J. Grey Syst.* **2014**, *26*, 85–95.
19. Ai, H.; Shi, Y. Application of GM (1,1) Model in PM2.5 Content Prediction. In Proceedings of the 2016 3rd International Conference on Education, Management and Computing Technology (ICEMCT), Hangzhou, China, 9–10 April 2016.
20. Sun, C. Research on Prediction of Visitors Quantity Based on GM(1,1)Method. In Proceedings of the 3rd Annual 2017 International Conference on Management Science and Engineering (MSE), Guilin, China, 18–20 August 2017.
21. Zheng, X. Application of Grey System Theory to Regional Logistics Demand Prediction. In Proceedings of the World Automation Congress 2012 (WAC), Puerto Vallarta, Mexico, 24–28 June 2012; IEEE: New York, NY, USA.
22. Ye, J.; Dang, Y.; Wang, J.; Yang, Y. Grey Prediction Model of Interval Grey Numbers Based on a Novel Compound Function Transformation. *J. Grey Syst.* **2017**, *29*, 155–176.
23. Wang, X.; Qi, L.; Chen, C.; Tang, J.; Jiang, M. Grey System Theory Based Prediction for Topic Trend on Internet. *Eng. Appl. Artif. Intell.* **2014**, *29*, 191–200. [CrossRef]
24. Chen, S.; Wang, D.; Liang, J. Sound Quality Analysis and Predicton of Vehicle Interior Noise Based on Grey System Theory. *Fluct. Noise Lett.* **2012**, *11*, 1250016. [CrossRef]
25. Sie, W.; Lian, R.; Lin, B. Enhancing Grey Prediction Fuzzy Controller for Active Suspension Systems. *Veh. Syst. Dyn.* **2006**, *44*, 407–430. [CrossRef]
26. Fu, Z.; Chen, Q.; Zhang, L.; Fan, J.; Zhang, H.; Deng, Z. Research on Energy Management Strategy of Fuel Cell Power Generation System Based on Grey-Markov Chain Power Prediction. *Energy Rep.* **2021**, *7*, 319–325. [CrossRef]
27. Wang, Y.; Lu, J.; Zhu, X.; Ye, J.; Kong, Y.; Hao, W. A GM-Based Energy Management Strategy of Hybrid Power System for Hydrogen Fuel Cell Buses. *J. Adv. Transp.* **2023**, *2023*, 6656612. [CrossRef]
28. China Automotive Technology Research Center Co., Ltd. *China Automotive Test Cycle-Part 2: Heavy-Duty Commercial Vehicles [Z]*; The State Administration for Market Regulation of China: Beijing, China, 2019.

applied sciences

MDPI

Article

High-Impedance Fault Detection in DC Microgrid Lines Using Open-Set Recognition

Ivan Grcić * and Hrvoje Pandžić

Faculty of Electrical Engineering and Computing, University of Zagreb, 10000 Zagreb, Croatia; hrvoje.pandzic@fer.hr
* Correspondence: ivan.grcic@fer.hr

Abstract: Detection of high-impedance faults in direct current microgrid lines presents a challenge for most conventional protection schemes because the magnitude of the fault current is similar to other transients that occur during normal operation. However, the waveform of high-impedance faults differs from that of other transients as it is characterized by a repetitive and nonlinear pattern caused by current reignition. Various methods have been proposed to exploit fault response waveforms for detecting high-impedance faults, including those based on deep discriminative intelligent classification. Different from previous works that focus on closed-set classification, this study frames fault detection as an open-set recognition problem, employing a neural network as the classifier. The resulting approach enables the detection of high-impedance faults as outliers from the normal operating states of microgrid lines with passive constant impedance loads and requires only the Fourier transform of the current signal as input to the neural network. Remarkably, the proposed solution eliminates the need for hard-to-model high-impedance faults in the training dataset and hence is more generally applicable. The proposed method consistently outperforms commercially available high-impedance fault detection systems, achieving high accuracy in fault detection.

Keywords: fault detection; fault protection; Fourier transform; machine learning; microgrids; neural networks

1. Introduction

The increasing integration of distributed energy resources (DERs) in the utility grid offers a higher reliability of power supply and lower energy costs for consumers [1]. DERs have different power ratings and, under certain conditions, can supply local consumers. This has led to the formation of local entities, called microgrids, which can operate independently of the utility grid [2].

Most renewable energy sources (RESs), battery energy storage and various loads use or produce direct current (DC) power. Coupling DC devices under the same microgrid avoids the need for alternating current (AC) to DC transformation. DC microgrids offer higher efficiency and simpler control as compared to their AC counterparts [3,4]. However, the protection of DC systems is complex, as faults cause fast-rising, high-magnitude currents. Although conventional protection methods are commonly used, various new approaches have been proposed to improve or solve problems associated to the protection of DC systems, such as signal-processing-based methods [5,6] and intelligent classification-based methods [7,8]. Similarly, this study focuses on faults that are not easily detected by conventional protection methods, requiring advanced techniques for their detection.

High-impedance faults (HIFs) present such a problem, as they cannot be detected by conventional system protection methods. However, the importance of HIF detection in DC systems is continuously stressed, as these systems are now prevalent in electric vehicles, photovoltaic plants, data centers, high-voltage DC transmission, and DC microgrids [9]. The only conventional system protection able to detect HIFs is differential protection, which has already been used for fault detection in cables in DC microgrids [10]. The method proved to be very effective; however, the differential protection is too expensive to implement on each microgrid line. In addition, the communication link is critical, as its failure will cause the protection not to work. It is also worth noting that other commercially available solutions developed specifically for HIF detection exhibit a relatively low successful action rate, with only 61.67% of the faults detected [11].

Various approaches have been adopted to solve HIF detection in DC microgrids. The authors in [12] observed incremental resistance as seen from the converter to detect an HIF. The method is simple, but how it behaves with different fault resistances is not investigated. The same problem arises with the method proposed in [13], where the impedance is predicted using an autoregressive model. In [14], the ground current is measured and when it reaches a certain threshold the operator is warned or, in case of a severe fault, the circuit breaker is tripped. In [15], the fault detection procedure is started when the current derivative exceeds a predefined value. In the next step, the resistance at both ends of the line segment is estimated using the least squares method. The sign of the resistances is then used to detect a fault and determine whether it is an external fault. An interesting approach for detecting partial discharge based on methods such as compressive sensing and phase diagrams is also presented [16,17]. These papers focus on two recent advancements in partial discharge detection: compressive sensing and phase-diagram-based analysis. These techniques are compared using real signals from an experimental setup designed to simulate realistic conditions, including load signals, reflections and propagation effects. The performance is evaluated based on the probability of detection and the false alarm ratio, providing insights into the effectiveness of these novel methods; however, they have yet to be applied to DC systems. In [18], the HIF detection is performed by injecting high-frequency oscillation at fault inception. However, the latter methods do not take into account the arcing nature of the HIF. In [19], the authors modeled the arcing nature of HIFs and used power spectral characteristic along with the current sag to detect them. They argued that a precise HIF model is crucial for the development of fault detection methods. Additionally, they noted that modeling is difficult due to the dependence of fault responses on operating conditions. The same is confirmed in [20], where it is noted that HIFs typically occur when a line comes into contact with a high-impedance surface, resulting in low-current arcs. These arcs occur intermittently and at random intervals because the contact area is not constant, posing a significant challenge for HIF modeling.

Although an HIF model cannot be asserted owing to its random properties, certain insights can be derived from the differences in data recorded during these failures and data recorded during normal operation. This can be achieved by applying machine learning models to find patterns or clusters or classify the data [21]. Several works have used machine learning models to distinguish an HIF from normal operation. In [22], the authors implemented an accurate HIF model with arcing, created a dataset and trained a k-nearest neighbors (kNN) classifier to detect HIFs. The dataset contains the difference in coefficient energy between the normal and faulty states obtained by the discrete wavelet transform (DWT) of the current signal and non-faulty samples. However, kNN is sensitive to noisy data, as is the case with HIF, and when the dimensionality of the data increases, its performance decreases. In [23], DWT was also used to pre-process the current measurement, but the support vector machine (SVM) classifier is used to detect HIFs. However, the arcing phenomenon is not considered when modeling HIFs and SVM is prone to problems similar to those of kNN.

Neural networks (NNs) offer certain advantages over classical machine learning algorithms such as kNN and SVM in terms of susceptibility to noise and the ability to work with high-dimensional data [24]. This is particularly advantageous for high-impedance fault (HIF) recognition because NNs can effectively process and classify the varying waveforms and patterns in power signals that traditional algorithms may struggle with. Furthermore, NN models are more adaptable, making them capable of handling data with complex, nonlinear relationships, which is a common characteristic in real-world fault detection scenarios. In [25], an NN is used in the process of HIF detection to predict the behavior of the power signal, but it is not used directly for classification. While this approach can successfully predict trends in the data, it lacks the ability to make definitive classification decisions, thus limiting its practical application in real-time fault detection. The method's reliance on predicting signal behavior highlights the importance of accurately modeling fault dynamics for future improvements. On the other hand, in [26], a recurrent neural network (RNN) was used to detect HIFs. The method uses a current measurement to detect HIFs in a radial DC microgrid. The RNN is used because its ability to work with time series has several advantages, such as taking into account previously seen samples without the need for fixed time windows. This temporal context enables the model to capture long-term dependencies and variations in the data, which is especially useful for dynamic fault detection in real-time environments. Furthermore, the RNN's structure is well suited to model the evolving nature of fault signals over time. HIFs are modeled as a nonlinear resistances that resemble fault current reignition. This approach, which models HIFs as dynamic systems, allows for a more realistic representation of fault behavior, contributing to more accurate detection when compared to static models. The modeling technique helps in distinguishing HIFs from other types of faults by focusing on their unique electrical characteristics. In [27], a pre-trained convolutional neural network (CNN) was fine-tuned to detect various types of faults in photovoltaic (PV) arrays, including arcing faults. The HIF is modeled as a time-varying resistance and is included in the dataset along with the responses of other faults. This method leverages the CNNs' ability to automatically extract features from raw data, significantly improving detection accuracy compared to manually engineered features. Similarly, a CNN was used for HIF detection in [28]. However, a complex multistep training procedure involves training a generative adversarial network (GAN) before training the CNN. This procedure also requires a substantial dataset, which in this case is obtained from a PV emulator. While this method can enhance model robustness, the reliance on GANs and large datasets introduces additional complexity and may limit its applicability in resource-constrained environments. A similar approach is taken in [29], where the wavelet transform of current during different events is used as input to the RNN. The accuracy achieved is higher than that of the previous method. However, the dataset used for training is significantly larger, comprising almost one thousand HIF samples. This increase in dataset size contributes to improved performance but also presents challenges in terms of data acquisition and storage, which may not be feasible for all applications. In [30], the NN is used to detect arcing faults in PV systems. The accuracy of the method is significant; however, the dataset size is tens of thousands of samples, and obtaining such a dataset requires significant effort. This emphasizes the trade-off between model performance and the practical challenges associated with collecting large datasets, which is a common limitation in many machine learning applications. The need for a huge dataset size and HIF modeling also persists in [31], where the authors used different HIF models to create the dataset for training CNN-based fault detection. While this approach improves the model's ability to generalize, it underscores the importance of having a comprehensive and diverse dataset to achieve reliable results. This can be particularly challenging in fields like power grid monitoring, where fault events may be rare or difficult to simulate. The same principle is used in [32], where the authors employed different HIF models to create a dataset for fault detection in DC microgrids. The use of various fault models provides a more robust foundation for training the system, but it also increases the complexity of dataset generation and may introduce biases if not carefully controlled. Such approaches highlight the ongoing challenge of balancing model robustness with dataset representativeness in fault detection tasks. In [33], the authors went a step further by generating arc faults on a real DC system to create a dataset for their intelligent classification method. This real-world dataset helps bridge the gap between theoretical models and practical applications, ensuring that the system can handle real fault events effectively. However, the effort involved in generating real-world data is resource-intensive and may not be feasible in all cases, especially for large-scale systems.

It is evident that intelligent fault detection methods heavily rely on HIF models in their datasets, requiring significant effort from the authors to accurately model HIFs in both simulations and real-world environments [34,35]. Hence, it is desirable to reduce or even completely eliminate the classifier's dependence on HIF modeling. One way to achieve this is by training the classifier to learn patterns that occur during normal operating conditions. Then, any deviation from these patterns is marked as an anomaly, or in this case, a fault. This approach is especially convenient for parts of the microgrid where the patterns of normal operating states are not diverse, as microgrid can have different topologies and operate in various states. However, the conventional training procedure for NNs does not enable this. Fortunately, there are specific methods for training NNs that allow them to detect outliers or anomalies. The problem of detecting outliers using machine learning models is called the open-set recognition (OSR) problem, formalized in [36], where the authors discussed how far the sample should be from the decision boundary to be considered an outlier. The same principle applies when employing NNs for classification [37]. Accordingly, various approaches have been proposed to enhance the OSR capabilities of NNs. These include methods that introduce different algorithms, metrics, outlier rejection rules, and modifications to the training procedure [38]. This work focuses on the latter, as it provides a straightforward and explainable procedure, which is introduced in Section 3.

Lines connecting the passive loads to the microgrid buses are typically unaffected by the microgrid's operating states, as their current is determined by the load impedance. Additionally, the number of events is minimal, only consisting of stationary states and load change events, simplifying the task of capturing all normal states. Given that there are methods allowing NNs to detect outliers, this paper proposes using that capability to identify HIFs in microgrid lines with passive loads as outliers from normal operating conditions. In this approach, the dataset contains only normal operating states, such as a stationary states or load changes. Since HIFs can be detected as outliers, there is no need to include them in the dataset, and more significantly, to model them. The benefits of this approach are manifold: (i) it eliminates the need of HIF modeling; (ii) the fault detection method is not entirely reliant on the selected HIF model; and (iii) detection is agnostic of the operating conditions under which an HIF occurs.

The contribution of this paper is summarized as follows:

- Modeling high-impedance fault detection as an open-set recognition problem;
- Development of a novel high-impedance fault detection method in DC microgrids based on neural network with open-set recognition capabilities;
- Simulation and detailed analysis of high-impedance faults in DC microgrid lines connecting passive loads.

The remainder of this paper is organized as follows. Section 2 provides a description of common types of faults that occur in microgrids, specifically low-impedance and high-impedance faults. Section 3 discusses the application of neural networks for classification, along with the OSR problem and its impact on classification performance. Furthermore, the proposed fault detection method is presented, emphasizing the utilization of the OSR framework for neural networks in fault detection. In Section 4, a case study demonstrating the effectiveness of the proposed method is presented. The DC microgrid model and dataset used in the study are described, and the parameters of the neural networks employed for detection are provided together with the results. Additionally, the algorithm used for the fault detection is outlined. Finally, Section 5 concludes the paper and summarizes the key findings and contributions of this research.

2. Faults in DC Microgrids

Protection of converter-based DC microgrids is an evolving topic that has been relatively poorly researched. The variety of power sources, possible operating points, converter types and control modes result in challenging protection system set ups. This section presents the two most common types of faults that the protection system deals with—low-impedance faults (LIF) and high-impedance faults (HIF), each with a different response.

2.1. Low-Impedance Faults

Low-impedance faults or short-circuit faults occur when a pole comes in contact with another pole or ground trough with low resistance, resulting in a significant voltage drop and high-magnitude current. In the case of an LIF, the contribution of the individual current sources to the fault depends mainly on the converter topology. Since DC microgrids are connected to the utility grid through a voltage source converter (VSC) and the largest contribution to the fault comes from the grid, its fault response is briefly described.

The VSC fault response is divided into three stages: capacitor discharge, diode freewheeling and grid-side current feeding [39]. During the capacitor discharge stage, its voltage drops from the nominal value to zero. Since the cable and fault resistances are low, the capacitor current reaches magnitudes that are about a hundred times higher than the rated current. When the capacitor is discharged, the diodes begin to conduct, marking the beginning of the diode freewheeling stage. The current remains high and the diodes are in danger of destruction. In the last stage, the grid starts feeding the fault, as the VSC behaves like a full-bridge rectifier. It is desirable to detect this fault before the fault current reaches its peak, and this can be achieved as its magnitude strongly deviates from that in normal operation.

2.2. High-Impedance Faults

High-impedance faults are difficult to detect because of their low current, which is similar in magnitude to that of the normal operating current. An additional problem is the arcing phenomenon caused by random reignition of the fault current. The randomness of the response makes the modeling of HIFs difficult, which is a prerequisite for the development of detection methods, especially those based on signal processing and intelligent classification. However, the HIF pattern can be well approximated with a model that includes a fault buildup, a shoulder, and a nonlinearity stage [40]. Such a model was proposed in [41]:

$$i_{j+1} = i_j - \frac{R \cdot i_j + k/i_j^{1.2} + 35 - V_{DC}\sin(\omega t)}{R - 1.2k/i_j^{2.2}}, \tag{1}$$

$$2n\pi + \pi/3 < \omega t < 2n\pi + 2\pi/3, n \in \mathbb{N}_0, \tag{2}$$

where k is the arc constant, i the HIF fault current, V_{DC} the nominal voltage, and R the equivalent fault resistance. The resulting HIF current is shown in Figure 1.

This model was used in [42] to develop an HIF detection method for DC aircraft systems based on the second derivative of the voltage and a comb filter. It is also used for HIF detection in DC distribution networks, where a mathematical-morphology-based fault detection method is proposed [43].

To implement HIF in Matlab/Simulink, the electrical model from [44] is used. The model consists of impedance that is used to simulate the buildup stage and diodes with sources connected in series to create repetitive nonlinear behaviour.

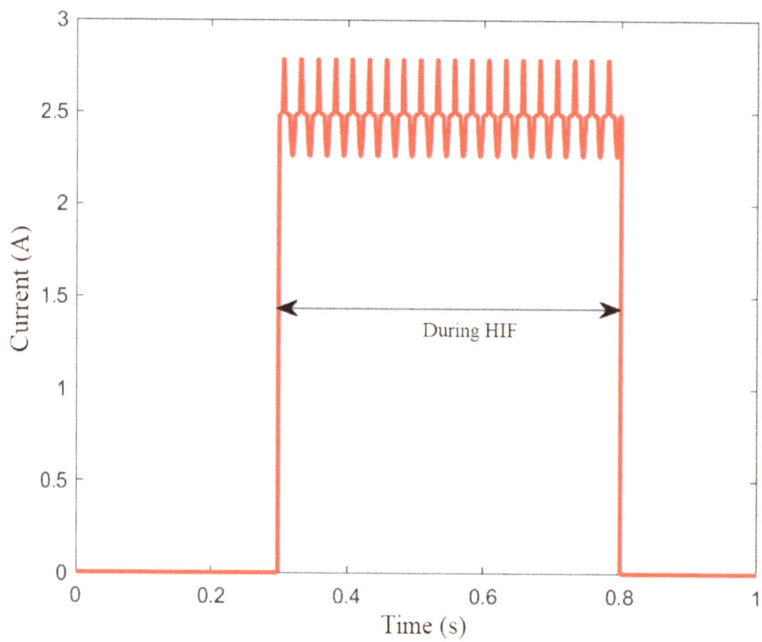

Figure 1. Fault current of the HIF model. The fault buildup starts at t = 0.3 s and lasts until the current reaches its maximum magnitude and is followed by the shoulder. The nonlinear stage lasts until the fault is cleared at t = 0.8 s.

3. Fault Detection

Protection system is essential for the safe and effective operation of electrical systems. Its task is to eliminate faults in the shortest possible time to avoid danger to people and damage to equipment. At the core of any protection system is a fault detection system that determines whether there is a fault in the system. However, as described in Section 2, not all fault types are easy to detect, nor do they cause the same damage. LIFs, which are characterized by high currents and significant voltage drops, are easily detected by the protection system. Conventional protection schemes such as overcurrent are effective, but new methods such as those presented in [45] are often proposed to further increase the level of safety.

As far as HIFs are concerned, there is no fully effective system of fault detection other than differential protection. However, the cost of differential protection is high and depends heavily on the communication link, so it is not an option for microgrid protection. However, various HIF detection methods have been proposed, many of which are based on intelligent classification, since fault detection is essentially a classification task. As mentioned in the introduction, this work focuses on HIF detection with NNs trained in a way that considers an open-set recognition problem. Therefore, this section describes classification with NNs, the OSR problem and how the NNs are adapted to solve this problem.

3.1. Classification with Neural Networks

The aim of the classification is to assign an input sample to one of the predefined classes based on its features. The importance of the features is determined during the training phase of the classifier, where the most salient features will have a greater impact on the classification decision. In supervised learning, the dataset contains samples with corresponding class labels. The training procedure minimizes the loss between the output of the classifier and label by shifting the parameters of the classifier. In the case of neural networks, these are the weights and biases of individual layers. The basic feed-forward neural network (FFNN) defines a mapping $\mathbf{z} = g(\mathbf{x};\boldsymbol{\theta})$, where $\mathbf{x}$ is an input, $\mathbf{z}_k, k \in K$ is a predicted score (logit) for each of K classes, and $\boldsymbol{\theta}$ are parameters learned during the training. The logits vector $\mathbf{z}$ is normalized by the softmax function to obtain the probabilities for each class:

$$\hat{\mathbf{y}}_k = \frac{\exp(\mathbf{z}_k)}{\sum_j^K \exp(\mathbf{z}_j)}. \quad (3)$$

Each element $\hat{\mathbf{y}}_k$ is greater than 0, and the sum of elements of vector $\hat{\mathbf{y}}$ is equal to 1.

The training procedure minimizes the loss function $\mathcal{L}(\hat{\mathbf{y}}, \mathbf{y}; \boldsymbol{\theta})$, which measures the loss between the output $\hat{\mathbf{y}}$ and the ground truth labels $\mathbf{y}$. To find the optimal parameters, the following optimization problem is solved:

$$\underset{\boldsymbol{\theta}}{\operatorname{argmin}}\ \mathcal{L}(\hat{\mathbf{y}}, \mathbf{y}; \boldsymbol{\theta}) = \underset{\boldsymbol{\theta}}{\operatorname{argmin}} \sum_i^n \ell(\hat{\mathbf{y}}, \mathbf{y}_i; \boldsymbol{\theta}), \quad (4)$$

where n is the number of samples. This optimization problem is nonlinear, so its solution requires the use of a gradient descent optimizer. The optimizer searches for the minimum of the loss function and the performance of the classifier is tested on the test dataset. If it appears that the classifier is starting to overfit, the process is stopped to preserve the generalization ability [46]. A standard choice for classification tasks is the cross-entropy loss:

$$\ell(\hat{\mathbf{y}}, \mathbf{y}) = -\sum_{i=1}^{K} \hat{\mathbf{y}}_i \log \mathbf{y}_i \quad (5)$$

The function is constructed to significantly increase the loss value when the predicted value is far from the ground truth.

The NN's performance depends on its architecture, as the capability of a model increases with the number of layers and neurons per layer. However, there are NN architectures that are adapted to the data type.

Convolutional NNs are commonly used for image classification and signal processing. The convolutional layers of a CNN act as filters to extract features from the input, with each layer consisting of multiple kernels representing different feature maps. The second type of layer in CNNs is the pooling layer, which is usually placed between adjacent convolutional layers to reduce their output size. Finally, the fully connected layer produces the output [47].

On the other hand, RNNs are used for sequential data because they have an additional recurrent layer that has a memory effect. That is, the hidden state $\mathbf{h}$ of the previous step(s) of the RNN is taken into account when generating the output. Formally, the output $\hat{\mathbf{y}} = g(\mathbf{x}, \mathbf{h}^{t-1}; \boldsymbol{\theta})$ depends not only on the current input but also on the previously seen inputs. Therefore, RNN models are best suited for working with time series, either for prediction or classification tasks [48].

3.2. Open-Set Recognition

3.2.1. Open-Set Recognition Problem

The open-set recognition problem might be difficult to comprehend when described directly for NNs, so a simple example showing binary classification in Figure 2 is provided. The decision boundary separating the positive ("+") and negative ("o") samples lays in the margin plane A, which is maximized during the training of the classifier. Although this is an efficient approach to distinguish between these two classes, it leads to an overgeneralization for the OSR problem. In this setting, the unknowns ("?") and samples from unknown classes ("△") would be labeled as positive samples. The reason for this is that these samples are not considered when creating the boundary—it is only determined by the positive and negative samples, so there is nothing to limit the propagation of the positive label [36]. The suggested approach is to introduce the plane Ω that bounds the space of positive samples. Now, the samples that are far away in the positive half-space can be rejected by the classifier.

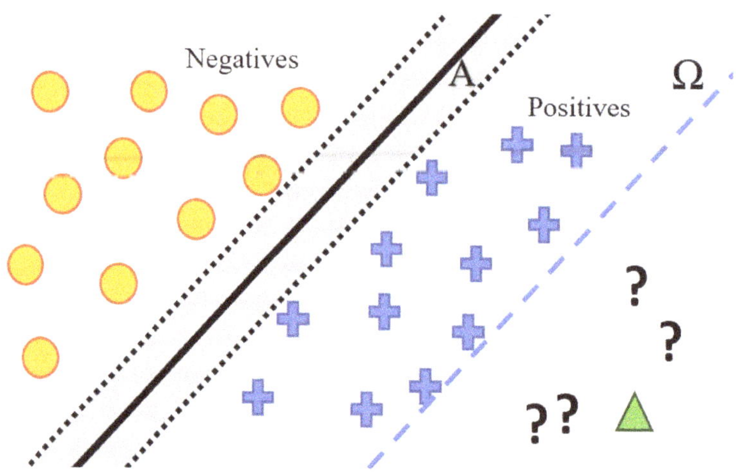

Figure 2. Illustration of the open-set recognition problem. Plane A separates the sample space into two half-spaces, so that unknowns ("?") and samples from unknown classes ("△") are part of the positive sample space. The second plane Ω is added to limit the open space risk by restricting the positive sample space. In case of fault detection, negative samples are steady states and positive samples are load changes. The unknowns and unknown classes are HIFs and other disturbances that occur in the microgrid.

The formal definition of the OSR problem is given in [36] and presented here in a short form. First, the measurable recognition function $f : \mathbb{R}^d \mapsto \mathbb{N}_0$ is defined, which maps the d-dimensional input x to the label y. Next, S_o is defined as a ball containing the positively labeled open space O and all positive training samples. The open space risk $R_O(f)$ is now defined as:

$$R_O(f) = \frac{\int_O f(x)\,dx}{\int_{S_0} f(x)\,dx}, \quad (6)$$

measuring the ratio of positively labeled open space and the overall measure of positively labeled space. Before defining the OSR problem, some instances have to be introduced. Positive training data are defined as $\hat{V} = \{v_1, \ldots, v_m\}$ from $\mathcal{P}$, and samples from other known classes $\mathcal{K}$ negative training data $\hat{K} = \{k_1, \ldots, k_n\}$. $\mathcal{U}$ is the set of all other unknown classes seen only in testing, hence the test set is defined as $\mathcal{T} = \{t_1, \ldots, t_z\}$, where $t_i \in \mathcal{P} \cup \mathcal{K} \cup \mathcal{U}$. The goal is to find function $f \in \mathcal{H}$ ($\mathcal{H}$ is a class of functions that separate the hyperplanes), trained on training data $\hat{V} \cup \hat{K}$ with given open space risk function $R_O(f)$ and empirical risk function R_ε, such that it minimizes the open set risk:

$$\underset{f \in \mathcal{H}}{\operatorname{argmin}}\{R_O(f) + \lambda_r R_\varepsilon(f(\hat{V} \cup \hat{K}))\}, \tag{7}$$

where λ_r is the regularization constant. Now, not only is the empirical risk of the training dataset reduced, but also the open space risk associated with unknown classes $\mathcal{U}$.

3.2.2. Open-Set Recognition Using NNs

The same problem persists in classification with neural networks, as samples from unknown classes are often classified with high confidence [37]. The reason for this behavior is due to the closed-set nature of NNs, i.e., the input sample is always assigned to one of the known classes. Since the final layer of the NN is softmax, the probability of the input belonging to each class is determined and the one with the highest probability is selected. However, it has been shown that there are "fooling" inputs that can lead to a high score in the classification even though they are significantly different from the rest of the samples belonging to that class [49]. Much worse, adversarial samples can also be created. These samples visually resemble those in the training dataset but result in high confidence output that is incorrect.

To address this issue, several approaches have been proposed [37]. However, one of the recent approaches that became very popular shows that if a stronger classifier is used, its open-set performance is on par with other OSR approaches. The authors of [50] concluded that a classifier's open-set performance correlates positively with its closed-set performance across different datasets, model architectures, and objectives. However, there are certain additions to the classifier's training procedure that allow for such performance improvement. The first is to augment the dataset with more data, as this increases the accuracy. The same effect is achieved by warming up the learning rate and smoothing the labels. The former helps with numerical instabilities at the beginning of the training, since the parameters (weights and biases) are random values that can be far from their optimal values. Hence, the learning process is started with a learning rate lower than the initial one (η) and gradually increased until it equals the initial rate [51]. In this work, a linear increase is used.

Since the one-hot encoded representation of the label vector $\mathbf{y}$ is equal to 1 at the location $i = y$ and 0 otherwise, label smoothing affects the labels by decreasing the value of the true label for a small constant ϵ and increasing the other element values to $\epsilon/(K-1)$, where K is the number of classes:

$$\hat{\mathbf{y}}_i = \begin{cases} 1-\epsilon, & \text{if } i = y, \\ \epsilon/(K-1), & \text{otherwise.} \end{cases} \tag{8}$$

This leads to a change in the optimal solution of the loss function:

$$\mathbf{z}_i^* = \begin{cases} \log((K-1)(1-\epsilon)/\epsilon) + \alpha, & \text{if } i = y, \\ \alpha, & \text{otherwise,} \end{cases} \tag{9}$$

where α is any real number. Constant ϵ affects the solution by making it finite, which is shown to increase the generalization capability [52].

It is also beneficial to train the model longer as compared to the normal setting by using the cosine learning rate decay. This decay reduces the learning rate from the initial value to zero according to the cosine law, which benefits the training progress [52]. The expression for the decay is as follows:

$$\eta_t = \frac{1}{2}\left(1 + \cos\left(\frac{t\pi}{T}\right)\right)\eta, \tag{10}$$

where η_t is the learning rate at epoch t, T is the total number of epochs, and η is the initial learning rate.

Finally, the maximum logit score (MLS) is used as an open-set indicator instead of the normalized softmax score (note that MLS is used as an indicator, but softmax is used during the training and testing phase of the classifier). Logits are outputs of the penultimate layer of the NN that are given as inputs to the softmax layer. When they are normalized, the information about their magnitude, which proves to be a good indicator, is lost. Therefore, open-set detection is performed directly on logits, and in this case specifically on the logit with the highest value. Formally, MLS is defined as:

$$\text{MLS}(\mathbf{z}) = \max_{1 \le i \le K} \mathbf{z}_i, \tag{11}$$

where $\mathbf{z}$ is a vector with prediction scores (logits) and K the number of classes.

3.3. Proposed Fault Detection Method

Considering NN classification and the problem of open-set recognition and its impact on neural networks, the proposed method uses the OSR paradigm to detect a worst-case disturbance in microgrids that cannot be detected by most conventional protection systems—a high-impedance fault. In this work, the focus is on lines connecting constant current passive loads, as this load type represents the majority in practical scenarios. The idea behind the method is to identify normal operating states of a microgrid line that connects passive load: its steady states and transients, which denote the change between the steady states, i.e., load changes. Other disturbances, in this case destructive HIFs, which should be removed from the system in the shortest time possible, are detected using an NN. However, the OSR paradigm allows these to be detected as outliers without having to model them and include them in the training dataset. Since HIFs behave differently depending on the different conditions (surface material and humidity, voltage levels, etc.), this method helps significantly in their detection [53].

The proposed HIF detection method is examined for the simplest form of NNs, the FFNN. The classifier is first trained using a standard hyperparameter tuning procedure, where the architecture of the FFNN is varied until satisfactory results are achieved and the optimizer parameters are adjusted for an efficient learning procedure. Then, the classifier is trained using the same architecture but with additional steps, as proposed in [50]. Finally, the OSR performance of the method is investigated.

The current is used as the input as transients can be detected from its waveform. It should be noted that the predominant component of DC current is the 0 Hz component, which is not useful indicator since it changes in case of both load change and HIF events. However, the difference becomes apparent when the signal is decomposed in the frequency domain using the fast Fourier transform (FFT) algorithm. The FFT algorithm is used as it is a fast solution for obtaining harmonic decomposition, especially when computation time is a crucial aspect [54]. The HIF event contains many more high-frequency components than the load change event. Hence, frequencies higher than 0 Hz are used as indicators to distinguish between the events [5]. The method should ideally distinguish between a static operating point (no transients) and load changes, flagging HIFs as outliers. The diagram of the method is provided in Figure 3.

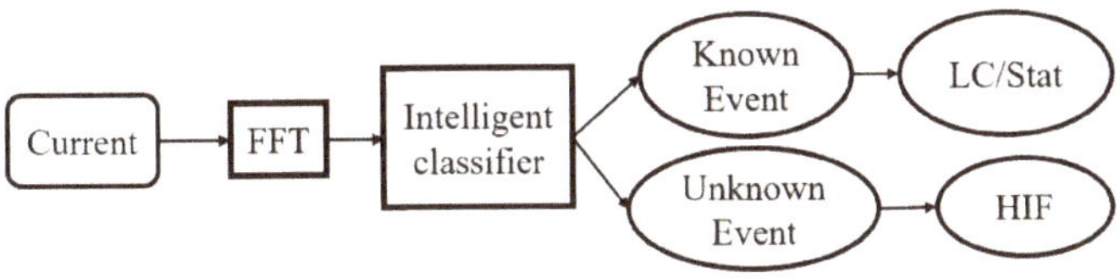

Figure 3. Overview of the proposed fault detection method, where, based on the FFT of the current, an intelligent classifier decides whether the event is known (and to which class it belongs) or not.

4. Case Study

4.1. System Description

The DC microgrid system considered in this study is a renewable energy-based microgrid that includes inverter-interfaced power sources and energy storage, inspired by the DC microgrid testbed developed at the City College of New York [55]. The microgrid considered has a radial topology (a common bus) to which a photovoltaic system, battery energy storage system (BESS) and a constant current passive load are connected, as shown in Figure 4.

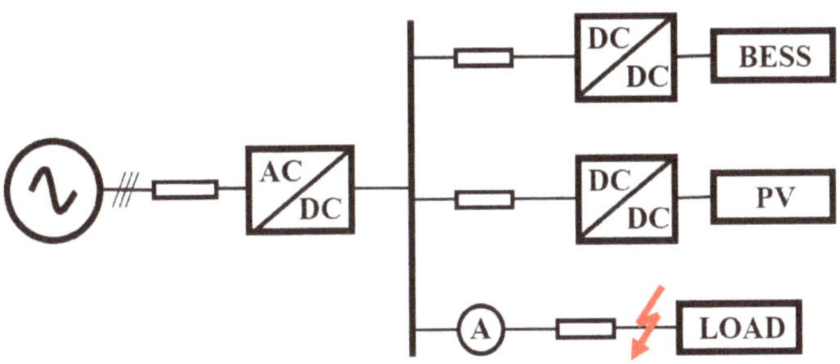

Figure 4. Simulation model of a DC microgrid. Model includes battery energy storage system, PV module, passive load, and utility grid connection. The fault location is on the line that connects load to the bus.

The PV system generates a maximum power of 6 kWp at a temperature of 25 °C. It is connected to the bus via a DC-DC boost converter with maximum power point tracking (MPPT) control, ensuring the maximum electricity output. The 12 kWh capacity BESS ensures that the excess power is stored and used when needed. The bidirectional DC-DC converter is in current control mode, enabling an effective bidirectional operation. The passive load is rated at 5 kW. The VSC connects the DC microgrid to the AC utility grid and regulates the voltage at the bus. All parameters of the microgrid are summarized in Table 1.

The current measurement for fault detection is taken at the line's inception, denoted by the ammeter symbol, with the fault location represented by the red arrow sign. The fault location is varied along the line, ranging from 10% to 90% of the line length.

Table 1. Parameters and rated values of bus, battery energy storage system, voltage-source converter, PV system, load and microgrid lines.

Parameter	Rated Value
Bus voltage	500 V
VSC rated power	15 kW
BESS capacity	12 kWh
PV peak power	6 kW @ 25 °C
Load	5 kW
Line parameters	20 mΩ, 62 μH

4.2. Dataset

The dataset includes various types of events, each represented by time series of current measurements, with a length of 1000 samples. These measurements are taken at the beginning of the line connecting passive load to the bus. Since the sampling frequency used is 10 kHz, the time duration of each time series is 100 ms, which is sufficient to capture the transients of interest. The elementary state of the microgrid is a steady state in which the current is flat with ripples originating from converter operations, and this state is included in the dataset. Load changes, such as the increase/decrease in generation or load, are also included. Finally, the responses of the microgrid to HIFs with different parameters are collected for testing of the method. The first two classes are used for training the classifier. The unknown class, in this case the HIF, is also included to test the OSR performance of the classifiers, but they are not seen by the classifier during the training procedure. Gaussian noise with zero mean and standard deviation of 0.01, is added to better approximate the real-world response. The entire dataset, without noise, can be found in the online repository [56]. The dataset contains 200 waveforms of normal operation and 21 HIF responses. A representative of each event is depicted in Figure 5. If the load is in its stationary state, the current remains flat throughout the entire window. Non-stationary events occur at $t = 0$. After the load change, the current reaches a new steady state. However, during HIF, even though a new steady state is achieved, the arcing phenomenon is clearly visible in time and frequency domains (Figure 6). The amplitudes that HIF achieves at high frequencies surpass those of stationary state and load change, thereby revealing a distinct difference.

It is worth noting that standards for arc fault protection in PV systems exist, each defining the arc type, conductor size, fault location, and other requirements [57]. Even though subject of this research is a line that connects a passive load, the authors created dataset according to the IEC 63027 standard while also incorporating additional fault locations and current levels, as suggested in [58]. However, this aspect will be emphasized in future work, where the method will undergo experimental testing.

During data preprocessing, each event occurrence is shifted in time to better represent the different times the event could appear and to augment the training data. Subsequently, each sequence is transformed from the time domain to the frequency domain using the fast Fourier transform (FFT), and the absolute value is taken. To simplify, the values at negative frequencies are added to their positive counterparts, and the dominant zero-frequency component is removed by zero multiplication. This reduces the input dimensionality to 500. Finally, the dataset is standardized. The two classes representing the stationary state and load change are encoded with numbers 0 and 1, respectively. The dataset is designed to serve as an example of the OSR problem but can also be used to test any HIF detection method.

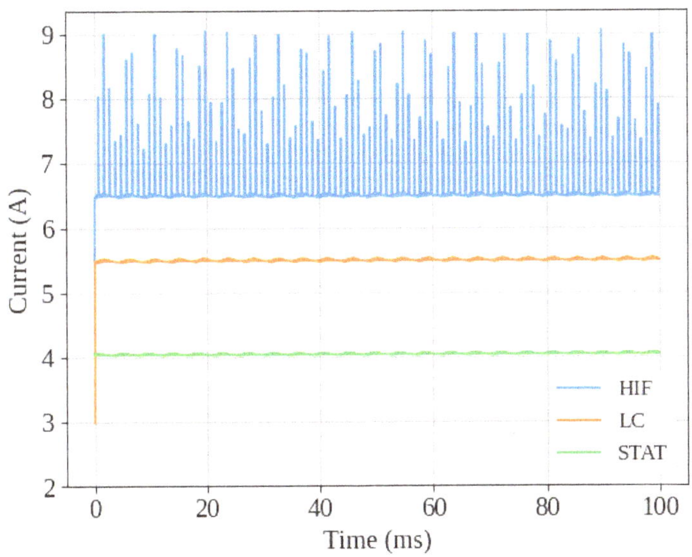

Figure 5. Representatives of dataset classes (steady state, load change) and HIF in the time domain. The stationary current is flat with ripples, while the load change current transitions to another steady state after a transient period. Notably, the arcing phenomenon is clearly visible during the HIF.

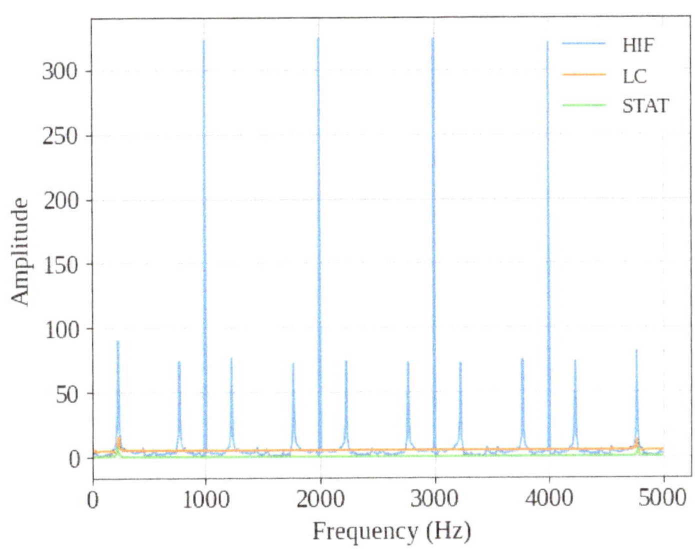

Figure 6. Representatives of dataset classes (steady state, load change) and HIF in the frequency domain. Amplitudes of high-frequency components of the current signal are relatively low during the stationary state and load changes. However, during the HIF event, these amplitudes significantly increase, indicating a distinct pattern.

4.3. Results

The FFNN is used because its architecture is simple and the training procedure is relatively straightforward. The FFNN is first trained to achieve the best possible accuracy (defined as ratio of correctly classified samples and total number of samples) on the test dataset. The second FFNN, with the same architecture, is trained but with improvements that aim to increase its open-set performance. The FFNN is set to take 500 points as inputs, as this is the size of the transform of the time sequence that describes an event, as explained before. The input layer is followed by three linear layers with sizes 2000, 2000, and 400, each of which has a Rectified Linear Unit (ReLU) nonlinearity at its output. The number of nodes in the first hidden layer is selected to increase the dimensionality of the input. The second hidden layer retains the same number of nodes, as iterative parameter selection indicated that further increases in dimensionality did not enhance the neural network's performance. The third hidden layer reduces dimensionality before the final reduction in the output layer. The number of nodes in this layer is reduced by a factor of five compared to the previous hidden layer, as the iterative parameter selection process demonstrated that this achieves an optimal balance between the neural network's complexity and accuracy. The final linear layer reduces the size from 400 to a vector of size 2, with each element representing a class. The ReLU nonlinear function is applied at the output of each hidden layer, as it is computationally efficient and avoids the vanishing gradient problem encountered with other nonlinear functions, such as the sigmoid function. A batch size of 32 samples is used, as it provides a good balance between convergence speed and the neural network's generalization ability. This same rationale guided the selection of the learning rate and number of epochs. Additionally, RMSprop is employed as the optimizer due to its ability to ensure stable convergence, even in cases with sparse gradients. Both the learning rate warmup and label smoothing were determined iteratively in alignment with the OSR framework. The architectures for both classifiers, as well as other parameters and the optimizer used, can be found in Table 2.

Table 2. FFNN parameters. The neural networks used share the same architecture; however, one is trained using the standard procedure, while the other follows the OSR framework (incorporating cosine learning rate, learning rate warmup and label smoothing).

	FFNN	FFNN + OSR
No. hidden layers	3	3
Nodes per layer	[500, 2000, 2000, 400, 2]	[500, 2000, 2000, 400, 2]
Nonlinearity	ReLU	ReLU
Batch size	32	32
Optimizer	RMSprop	RMSprop
Learning rate	1×10^{-4}	Cosine (1×10^{-3} to 1×10^{-4})
Learning rate warmup	-	10% of epochs (init. 1×10^{-4})
Epochs	50	50
Label smoothing	-	0.1
Accuracy	99.99%	99.99%

As can be seen from the table, the only difference between the classifiers is in the improvements related to the OSR performance. Instead of a fixed value, the cosine learning rate with warmup was used, which means that the learning rate is increased from 1×10^{-4} to 1×10^{-3} during the first 10% of the total number of epochs. After it reaches its maximum, it is reduced for the remaining number of epochs according to the cosine law. The accuracy of the classifier is not reduced after applying all OSR rules, and the learning process converges as shown in Figure 7.

The reason for using MLS becomes clearer when dealing with unknown samples that are not part of the dataset, in this case HIFs. The standard approach is that for each input sample, the classifier produces a vector of size equal to the number of classes. The vector elements are normalized using the softmax function to obtain probabilities of the input belonging to each class; i.e., each vector element is in the range [0, 1] and their sum is equal to 1. However, this approach does not take into account the vector amplitude information, which is extremely useful when working with unknowns. Therefore, the MLS approach is used to detect HIFs. The output of the classifier is not softmaxed, but the maximum element of this vector is used to represent the class.

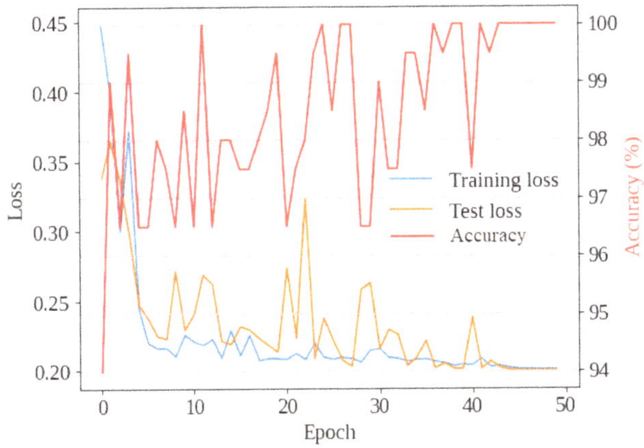

Figure 7. Loss and accuracy plots for FFNN with OSR improvements. Training and test loss converge, indicating a stable and successful training procedure. Accuracy rises rapidly and approaches near-perfect score.

The FFNN and the FFNN with OSR enhancements are fed the same test and HIF samples. Figures 8 and 9 show the maximum logits of the test samples and the HIFs. (blue dots and orange triangles, respectively).

The MLSs of the steady-state class (0) for FFNN without improvements are closely spaced, but for the load change class (1) they are widely scattered (Figure 8). The latter discrepancy arises from differences in the directions in which load changes occur, as well as variations in the magnitudes of load changes. The features characterizing steady states are similar, with the only difference being the 0 Hz component (representing different loading levels), which is removed during preprocessing. As expected, HIFs are misclassified because the NN always assigns a class to the inputs. Furthermore, the NN assigns HIFs to both classes, even though the HIF response is more similar to that of load change. Moreover, the MLSs of known classes and the HIF overlap, indicating that the classifier struggles to handle unknowns effectively.

On the other hand, the results for the FFNN trained according to the OSR suggestions are shown in Figure 9. The scale of the logits shrank on the y-axis, but now the samples of both classes are grouped. The NN decided correctly that these samples are closer to the load change event than to the steady state, assigning all the HIFs to the class 1. More importantly, the MLSs of the HIFs are far from the rest of the MLSs, indicating that the input is from the unknown class. Thus, the HIF can be identified as an outlier and rejected by the classifier in a trivial manner.

The OSR approach proves valuable in detecting outliers. The neural network, with these additions to the training procedure, demnostrates an enhanced ability to detect outliers in a straightforward manner.

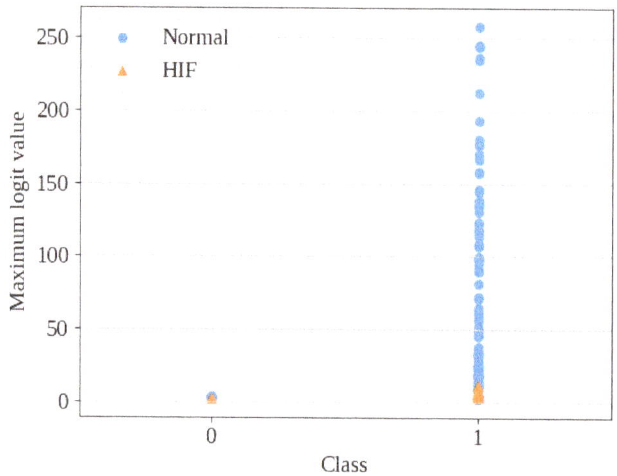

Figure 8. Maximum logit score values of FFNN fed with normal test samples (steady state and load change) and HIF samples indicate that the FFNN assigns HIF samples to both classes, making them indistinguishable from other normal samples.

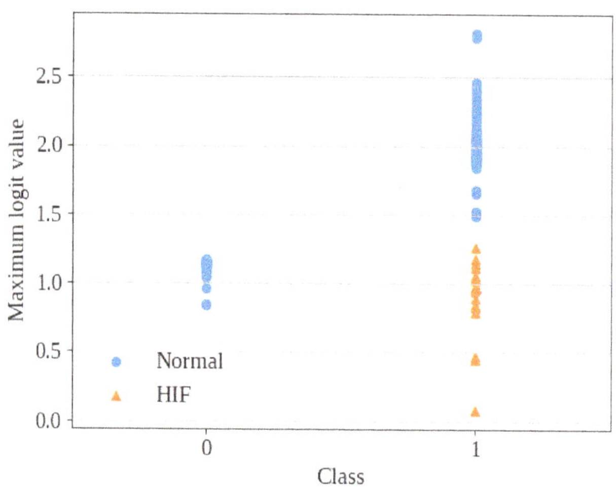

Figure 9. Maximum logit score values of FFNN with OSR improvements, when fed with normal test samples and HIF samples, reveal that the FFNN assigns HIF samples exclusively to class 1 (the class with transients). More importantly, it clusters HIFs separately from other normal samples.

4.4. Comparison with Existing NN-Based Methods

This section compares the proposed method with other neural network-based approaches, evaluating their performance in terms of complexity, computational efficiency, and detection accuracy. The comparison is limited to NN-based methods, as they closely align with the framework of the proposed approach. This focused evaluation enables a meaningful assessment of the relative strengths and limitations. Furthermore, it highlights the advancements and contributions of the proposed method within this specific context.

Table 3 presents several approaches based on different types of neural networks, all tested on the same dataset used to evaluate the proposed method. The proposed method achieves the highest detection accuracy (99.99) among all methods, demonstrating its performance while maintaining low complexity. Both the proposed method and the approach in [32] exhibit high computational efficiency; however, the proposed method significantly outperforms [32] in terms of the detection accuracy (99.99% vs. 70.09%). In terms of complexity, while the CNN-based method in [31] achieves high detection accuracy (96.79%), it suffers from high complexity, making it less suitable for resource-constrained or real-time applications. The back-propagation neural network (BPNN)-based method in [30] offers a balance between medium complexity and high efficiency but falls short in accuracy (95.31%) compared to the proposed method. Finally, unlike all other methods, the proposed method supports OSR, making it robust and adaptable to scenarios involving unknown or unexpected faults. This capability provides a significant advantage over other approaches.

Table 3. Comparison of the proposed method with existing methods.

Method	Type of NN	Complexity	Detection Accuracy	OSR Capability
[30]	BPNN	Medium	95.31%	No
[31]	CNN	High	96.79%	No
[32]	FFNN	Low	70.09%	No
Proposed	FFNN	Low	**99.99%**	Yes

4.5. Fault Detection Algorithm

The OSR improved performance of the NN, allowing it to be used not only as a classifier but also as an outlier detector (HIFs in this case). Algorithm 1 summarizes the fault detection method. The offline part of the algorithm consists of the dataset creation, the classifier training, and determining the bounds of the samples per class. The bounds can be chosen in any manner that separates outliers. In this work, a fixed bound for both classes is proposed, but other methods can be used instead. The goal of the bounds is to capture the range in which the normal samples for that class occur. In the case shown in Figure 9, the bound for both classes could be any number below the lowest scoring sample of each class. Anything under this limit is considered an outlier.

Algorithm 1 High-impedance fault detection

OFFLINE
1. Obtain dataset with normal operating conditions (steady state, load changes)
2. Transform data from time to frequency domain using FFT algorithm
3. Train NN-based classifier C with OSR suggestions
4. Obtain responses to HIFs
5. Find MLS for normal samples (test set) and HIFs
6. Determine boundaries that enclose normal samples of each class

ONLINE
1. Take N consecutive samples of current measurement and create vector $\mathbf{X}$
2. Find **MLS** and assigned class k from $C(\mathbf{X})$
3. **if MLS** outside boundaries for class k:
   HIF detected.
   else return to step 1.

The online part describes how fault detection works. The method requires passing a vector of length N, in this case 500, to the pre-trained classifier. The output of the classifier contains logits from which the maximum (MLS) is determined, along with the class that the classifier assigned to the input. The assigned class specifies the bounds against which the MLS is to be compared. If the MLS is outside the bounds, it is declared an outlier, in this case an HIF. If it is not, $\mathbf{X}$ is updated.

4.6. Discussion

The goal of this study is to develop an HIF detection method that is based on an NN trained according to the OSR suggestions. These suggestions allow unknowns (inputs that are not seen by the classifier during its training—in this case HIFs) to be treated as outliers. The first advantage of this approach is that the HIF responses do not need to be included in the dataset, which relieves the user of the need to model HIFs and reduces complexity of the classifier (NN). This is achieved because both the FFNN and the CNN classifiers have simple architectures but can still properly handle unknowns, making it possible to set bounds around the correctly classified samples. Consequently, the detection of outliers (HIFs) is simple—it is enough to check whether the MLS of an input is outside the bounds of the class assigned by the classifier. It should be noted that it cannot be known a priori which class an input will be assigned to, so the MLS is compared to the bounds of the class into which the classifier has placed it. This is the general case, but it may happen that the same bounds can be applied to both classes, so it is not necessary to know which class the classifier has assigned to the input. The bounds do not have to be fixed, as any strategy can be used to detect outliers. In this work, the fixed bound is used for simplicity.

Speaking of disturbances other than HIFs, some characteristics could be included in the dataset, but it is difficult to know and model all the disturbances that occur in a microgrid. It is assumed here that the HIF response is specific and different from that of other disturbances. Moreover, 100 ms is long enough that temporary disturbances do not trigger the trip signal, while HIFs, on the other hand, persist. At the time of writing, the authors are not aware that a dataset for HIF detection exists. Since the dataset is publicly available, it can be used as a basis for further research. Any improvements to the dataset and method are encouraged. Future work can be based on any of the above: enhancing the dataset by including more normal operating points or adding real-world HIF responses, improving the classifiers or the training procedure, or treating outliers differently. Testing the method on real-world data would increase pressure on classifier's discriminative capabilities, potentially raising the need for input filtering, using different signal decomposition methods (e.g., wavelet transform), or incorporation of additional input features. One of the research directions could cover adaptation of the proposed method for lines that connect power sources, energy storage systems or active loads, as it is currently limited to lines connecting passive loads. Another research direction could investigate applying the method to long DC lines, such as those used in HVDC transmission systems.

It is also worth noting that hardware could be a limiting factor, as the method requires implementation of the neural network on the processor. This involves handling fast matrix multiplication and implementation of nonlinear functions, which must be carefully managed to minimize computational overhead and ensure fast execution. Both factors impact the neural network's classification performance, which, in turn, directly affects the fault detection performance [59,60].

5. Conclusions

Detecting high-impedance faults is a challenging task for conventional protection systems due to their low current magnitude. However, advanced intelligent classifiers are capable of detecting repetitive patterns that characterize such faults. In this work, we present an HIF detection method based on intelligent classification, using a feed-forward neural network as the classifier. The FFNN is trained to distinguish normal operating points (knowns) from unknowns, which in this case are identified as HIFs and treated as outliers. The key conclusions drawn from this study are as follows: (i) the number of classes in the dataset can be reduced by focusing on normal operating states; (ii) HIFs can be detected as outliers using a neural network trained according to open-set recognition principles; (iii) the architecture of the neural network does not need modification, only the training procedure; (iv) FFNNs are highly effective for HIF detection, with an accuracy of 99.99%. Additionally, the dataset used in this study is made publicly available, as no similar datasets exist in the literature. This method offers a promising approach for improving fault detection in power grids, especially in microgrids and other low-current scenarios.

Author Contributions: Conceptualization, I.G. and H.P.; methodology, I.G. and H.P.; validation, I.G.; formal analysis, I.G.; resources, H.P.; data curation, I.G.; writing—original draft preparation, I.G.; writing—review and editing, H.P.; visualization, I.G.; supervision, H.P. All authors have read and agreed to the published version of the manuscript.

Funding: This research received no external funding.

Institutional Review Board Statement: Not applicable.

Informed Consent Statement: Not applicable.

Data Availability Statement: The data used to support the findings of this study are available from the corresponding author upon request.

Conflicts of Interest: The authors declare no conflicts of interest.

Abbreviations

The following abbreviations are used in this manuscript:

AC	Alternating current
BESS	Battery energy storage system
BPNN	Back-propagation neural network
CNN	Convolutional neural network
DC	Direct current
DER	Distributed energy sources
DWT	Discrete wavelet transform
FFNN	Feed-forward neural network
FFT	Fast Fourier transform
GAN	Generative adversarial network
HIF	High-impedance fault
kNN	k-Nearest neighbors
LIF	Low-impedance fault
MLS	Maximum logit score
MPPT	Maximum power point tracking
NN	Neural network
OSR	Open-set recognition
PV	Photo-voltaic
ReLU	Rectified linear unit
RNN	Recurrent neural network
RES	Renewable energy sources
SVM	Support vector machine
VSC	Voltage source converter

References

1. Muhtadi, A.; Pandit, D.; Nguyen, N.; Mitra, J. Distributed Energy Resources Based Microgrid: Review of Architecture, Control, and Reliability. *IEEE Trans. Ind. Appl.* **2021**, *57*, 2223–2235. [CrossRef]
2. Hatziargyriou, N.; Asano, H.; Iravani, R.; Marnay, C. Microgrids. *IEEE Power Energy Mag.* **2007**, *5*, 78–94. [CrossRef]
3. Papadimitriou, C.N.; Zountouridou, E.I.; Hatziargyriou, N.D. Review of hierarchical control in DC microgrids. *Electr. Power Syst. Res.* **2015**, *122*, 159–167. [CrossRef]
4. Ansari, S.; Chandel, A.; Tariq, M. A Comprehensive Review on Power Converters Control and Control Strategies of AC/DC Microgrid. *IEEE Access* **2021**, *9*, 17998–18015. [CrossRef]
5. Grcić, I.; Pandžić, H.; Novosel, D. Fault Detection in DC Microgrids Using Short-Time Fourier Transform. *Energies* **2021**, *14*, 277. [CrossRef]
6. Čuljak, M.; Pandžić, H.; Havelka, J. Mathematical Morphology-Based Fault Detection in Radial DC Microgrids Considering Fault Current from VSC. *IEEE Trans. Smart Grid* **2023**, *14*, 2545–2557. [CrossRef]
7. Grcić, I.; Pandžixcx, H. Fault Detection in DC Microgrids using Recurrent Neural Networks. In Proceedings of the 2021 International Conference on Smart Energy Systems and Technologies (SEST), Vaasa, Finland, 6–8 September 2021; pp. 1–6.
8. Ali, Z.; Terriche, Y.; Abbas, S.Z.; Hassan, M.A.; Sadiq, M.; Su, C.L.; Guerrero, J.M. Fault Management in DC Microgrids: A Review of Challenges, Countermeasures, and Future Research Trends. *IEEE Access* **2021**, *9*, 128032–128054. [CrossRef]
9. Yao, X.; Wang, J.; Schweickart, D.L. Review and recent developments in DC arc fault detection. In Proceedings of the 2016 IEEE International Power Modulator and High Voltage Conference (IPMHVC), San Francisco, CA, USA, 6–9 July 2016; pp. 467–472.
10. Dhar, S.; Patnaik, R.K.; Dash, P.K. Fault Detection and Location of Photovoltaic Based DC Microgrid Using Differential Protection Strategy. *IEEE Trans. Smart Grid* **2018**, *9*, 4303–4312. [CrossRef]
11. Wang, Y.; Hou, L.; Paul, K.C.; Ban, Y.; Chen, C.; Zhao, T. ArcNet: Series AC Arc Fault Detection Based on Raw Current and Convolutional Neural Network. *IEEE Trans. Ind. Inform.* **2022**, *18*, 77–86. [CrossRef]
12. Paz, F.; Ordonez, M. High-Impedance Fault Detection Method for DC Microgrids. In Proceedings of the 2019 IEEE 10th International Symposium on Power Electronics for Distributed Generation Systems (PEDG), Xi'an, China, 3–6 June 2019; pp. 787–792.
13. Taheri, B.; Hosseini, S.A. Detection of High Impedance Fault in DC Microgrid Using Impedance Prediction Technique. In Proceedings of the 2020 15th International Conference on Protection and Automation of Power Systems (IPAPS), Shiraz, Iran, 30–31 December 2020; pp. 68–73.
14. Sheikh, A.A.; Wakode, S.A.; Deshmukh, R.R.; Ballal, M.S. Detection of High Impedance Fault in DC Microgrid. In Proceedings of the 2020 IEEE International Conference on Power Electronics, Drives and Energy Systems (PEDES), Jaipur, India, 16–19 December 2020; pp. 1–4.
15. Rao, G.K.; Jena, P. Fault Detection in DC Microgrid Based on the Resistance Estimation. *IEEE Syst. J.* **2022**, *16*, 1009–1020. [CrossRef]
16. Stanescu, D.; Digulescu, A.; Ioana, C.; Serbanescu, A.; Zhang, Y.D.; Senjyu, T.; So-In, C.; Joshi, A. On the Existing and New Potential Methods for Partial Discharge Source Monitoring in Electrical Power Grids. In *Smart Trends in Computing and Communications*; Lecture Notes in Networks and Systems; Springer: Singapore, 2022; Volume 286.
17. Stanescu, D.; Digulescu, A.; Ioana, C.; Serbanescu, A. Entropy-Based Characterization of the Transient Phenomena—Systemic Approach. *Mathematics* **2021**, *9*, 648. [CrossRef]
18. Srivastava, C.; Tripathy, M. Novel Grounding and Protection Strategy for DC Microgrid Restraining Fault Current. *IEEE Trans. Power Deliv.* **2024**, *39*, 2182–2193. [CrossRef]
19. Miao, W.; Wang, Z.; Wang, F.; Lam, K.H.; Pong, P.W.T. Multicharacteristics Arc Model and Autocorrelation-Algorithm Based Arc Fault Detector for DC Microgrid. *IEEE Trans. Ind. Electron.* **2023**, *70*, 4875–4886. [CrossRef]
20. Valero, A. High Impedance Fault Detection Method in Multi-Grounded Distribution Networks. Ph.D. Dissertation, École polytechnique de Bruxelles Université Libre de Bruxelles, Brussels, Belgium, 2012.
21. Masa, A.V.; Werben, S.; Maun, J.C. Incorporation of data-mining in protection technology for high impedance fault detection. In Proceedings of the 2012 IEEE Power and Energy Society General Meeting, San Diego, CA, USA, 22–26 July 2012; pp. 1–8.
22. Subramaniam, K.; Illindala, M.S. High Impedance Fault Detection and Isolation in DC Microgrids. In Proceedings of the 2019 IEEE/IAS 55th Industrial and Commercial Power Systems Technical Conference (I&CPS), Calgary, AB, Canada, 5–8 May 2019; pp. 1–8.
23. Wakode, S.A.; Kumar, S.; Ballal, M.S. A Flexible Protection Approach for DC Microgrid Using Wavelet based Machine Learning Technique. In Proceedings of the 2021 IEEE 2nd International Conference on Smart Technologies for Power, Energy and Control (STPEC), Bilaspur, Chhattisgarh, India, 19–22 December 2021; pp. 1–5.
24. Bramareswara Rao, S.N.V.; Kumar, Y.V.P.; Amir, M.; Muyeen, S.M. Fault detection and classification in hybrid energy-based multi-area grid-connected microgrid clusters using discrete wavelet transform with deep neural networks. *Electr. Eng.* **2024**. [CrossRef]

25. Taheri, B.; Hosseini, S.A.; Salehimehr, S.; Razavi, F. A Novel Approach for Detection High Impedance Fault in DC Microgrid. In Proceedings of the 2019 International Power System Conference (PSC), Tehran, Iran, 9–11 December 2019; pp. 287–292.
26. Grcić, I.; Pandžić, H. Artificial Neural Network for High-Impedance-Fault Detection in DC Microgrids. In Proceedings of the 2023 IEEE PES Conference on Innovative Smart Grid Technologies—Middle East (ISGT Middle East), Abu Dhabi, United Arab Emirates, 12–15 March 2023; pp. 1–5.
27. Aziz, F.; Haq, A.U.; Ahmad, S.; Mahmoud, Y.; Jalal, M.; Ali, U. A Novel Convolutional Neural Network-Based Approach for Fault Classification in Photovoltaic Arrays. *IEEE Access* **2020**, *8*, 41889–41904. [CrossRef]
28. Lu, S.; Sirojan, T.; Phung, B.T.; Zhang, D.; Ambikairajah, E. DA-DCGAN: An Effective Methodology for DC Series Arc Fault Diagnosis in Photovoltaic Systems. *IEEE Access* **2019**, *7*, 45831–45840. [CrossRef]
29. Veerasamy, V.; Wahab, N.I.A.; Othman, M.L.; Padmanaban, S.; Sekar, K.; Ramachandran, R.; Hizam, H.; Vinayagam, A.; Islam, M.Z. LSTM Recurrent Neural Network Classifier for High Impedance Fault Detection in Solar PV Integrated Power System. *IEEE Access* **2021**, *9*, 32672–32687. [CrossRef]
30. Yang, J.; Wang, Y. Identification and Detection of DC Arc Fault in Photovoltaic Power Generation System. In Proceedings of the 2020 International Conference on Intelligent Transportation, Big Data & Smart City (ICITBS), Vientiane, Laos, 11–12 January 2020; pp. 440–444.
31. Omran, A.H.; Said, D.M.; Hussin, S.M.; Mirsaeidi, S.; Abid, Y.M. An Intelligent Classification Method of Series Arc Fault Models Using Deep Learning Algorithm. In Proceedings of the 2020 IEEE International Conference on Power and Energy (PECon), Penang, Malaysia, 21–24 July 2020; pp. 44–48.
32. Patil, D.D.; Bindu, S.; SushilThale. Arc Fault Detection in DC Microgrid Using Deep Neural Network. In Proceedings of the 2021 4th Biennial International Conference on Nascent Technologies in Engineering (ICNTE), NaviMumbai, India, 15–16 January 2021; pp. 1–6.
33. Cai, X.; Wai, R.-J. Intelligent DC Arc-Fault Detection of Solar PV Power Generation System via Optimized VMD-Based Signal Processing and PSO–SVM Classifier. *IEEE J. Photovolt.* **2022**, *12*, 1058–1077. [CrossRef]
34. Lu, S.; Sahoo, A.; Ma, R.; Phung, B.T. DC Series Arc Fault Detection Using Machine Learning in Photovoltaic Systems: Recent Developments and Challenges. In Proceedings of the 2020 8th International Conference on Condition Monitoring and Diagnosis (CMD), Phuket, Thailand, 25–28 October 2020; pp. 416–421.
35. Mansouri, M.; Trabelsi, M.; Nounou, H.; Nounou, M. Deep Learning-Based Fault Diagnosis of Photovoltaic Systems: A Comprehensive Review and Enhancement Prospects. *IEEE Access* **2021**, *9*, 126286–126306. [CrossRef]
36. Scheirer, W.J.; de Rezende Rocha, A.; Sapkota, A.; Boult, T.E. Toward Open Set Recognition. *IEEE Trans. Pattern Anal. Mach. Intell.* **2013**, *35*, 1757–1772. [CrossRef]
37. Bendale, A.; Boult, T.E. Towards Open Set Deep Networks. In Proceedings of the 2016 IEEE Conference on Computer Vision and Pattern Recognition (CVPR), Las Vegas, NV, USA, 27–30 June 2016; pp. 1563–1572.
38. Geng, C.; Huang, S.-J.; Chen, S. Recent Advances in Open Set Recognition: A Survey. *IEEE Trans. Pattern Anal. Mach. Intell.* **2021**, *43*, 3614–3631. [CrossRef] [PubMed]
39. Beheshtaein, S.; Cuzner, R.M.; Forouzesh, M.; Savaghebi, M.; Guerrero, J.M. DC Microgrid Protection: A Comprehensive Review. *IEEE J. Emerg. Sel. Top. Power Electron.* 2019, *early access*.
40. Bayati, N. Fault Detection and Location of DC Microgrids. Ph.D. Dissertation, Faculty of Engineering and Science, Aalborg University, Aalborg, Denmark, 2020.
41. Li, J.; Kohler, J.L. New insight into the detection of high-impedance arcing faults on DC trolley systems. *IEEE Trans. Ind. Appl.* **1999**, *35*, 1169–1173.
42. Bayati, N.; Savaghebi, M. A high-impedance fault detection scheme for DC aircrafts based on comb filter and second derivative of voltage. *Green Energy Intell. Transp.* **2023**, *2*, 100073. [CrossRef]
43. Oh, Y.; Kim, C.; Gwon, G.; Noh, C.; Bukhari, S.; Haider, R.; Gush, T. Fault detection scheme based on mathematical morphology in last mile radial low voltage DC distribution networks. *Int. J. Electr. Power Energy Syst.* **2019**, *106*, 520–527. [CrossRef]
44. Andrea, J.; Schweitzer, P.; Tisserand, E. A New DC and AC Arc Fault Electrical Model. In Proceedings of the 2010 56th IEEE Holm Conference on Electrical Contacts, Charleston, SC, USA, 4–7 October 2010; pp. 1–6.
45. Yadav, N.; Tummuru, N.R. Short-Circuit Fault Detection and Isolation Using Filter Capacitor Current Signature in Low-Voltage DC Microgrid Applications. *IEEE Trans. Ind. Electron.* **2022**, *69*, 8491–8500. [CrossRef]
46. Goodfellow, I.; Bengio, Y.; Courville, A. *Deep Learning*; MIT Press: Cambridge, MA, USA, 2016.
47. Gu, J.; Wang, Z.; Kuen, J.; Ma, L.; Shahroudy, A.; Shuai, B.; Liu, T.; Wang, X.; Wang, G.; Cai, J.; et al. Recent advances in convolutional neural networks. *Pattern Recognition* **2018**, *77*, 354–377. [CrossRef]
48. Hüsken, M.; Stagge, P. Recurrent neural networks for time series classification. *Neurocomputing* **2003**, *50*, 223–235. [CrossRef]
49. Nguyen, A.; Yosinski, J.; Clune, J. Deep neural networks are easily fooled: High confidence predictions for unrecognizable images. In Proceedings of the 2015 IEEE Conference on Computer Vision and Pattern Recognition (CVPR), Boston, MA, USA, 7–12 June 2015; pp. 427–436.

50. Vaze, S.; Han, K.; Vedaldi, A.; Zisserman, A. Open-Set Recognition: A Good Closed-Set Classifier is All You Need. In Proceedings of the Tenth International Conference on Learning Representations (ICLR), Virtual Event, 25–29 April 2022; pp. 1–26.
51. He, K.; Zhang, X.; Ren, S.; Sun, J. Deep Residual Learning for Image Recognition. In Proceedings of the 2016 IEEE Conference on Computer Vision and Pattern Recognition (CVPR), Las Vegas, NV, USA, 27–30 June 2016; pp. 770–778.
52. He, T.; Zhang, Z.; Zhang, H.; Zhang, Z.; Xie, J.; Li, M. Bag of Tricks for Image Classification with Convolutional Neural Networks. In Proceedings of the 2019 IEEE/CVF Conference on Computer Vision and Pattern Recognition (CVPR), Long Beach, CA, USA, 15–20 June 2019; pp. 558–567.
53. Ghaderi, A.; Ginn, H.L.; Mohammadpour, H.A. High impedance fault detection: A review. *Electr. Power Syst. Res.* **2017**, *143*, 376–388. [CrossRef]
54. Saribulut, L.; Teke, A.; Tumay, M. Fundamentals and literature review of Fourier transform in power quality issues. *J. Electr. Electron. Eng. Res.* **2013**, *5*, 9–22.
55. Saleh, M.; Esa, Y.; Mhandi, Y.; Brandauer, W.; Mohamed, A. Design and implementation of CCNY DC microgrid testbed. In Proceedings of the 2016 IEEE Industry Applications Society Annual Meeting, Portland, OR, USA, 2–6 October 2016; pp. 1–7.
56. Grcić, I. Dataset HIF Detection [Data Set]. Available online: https://github.com/ivangrcic/dataset-hif-detection (accessed on 15 November 2024.).
57. Putzke, J.L.; Michels, L.; Bellinaso, L.V. Electric Arcs in Photovoltaic Systems: A Comparative Analysis of IEC 63027, UL 1699B, and GB-t 39750 Standards. In Proceedings of the 2023 IEEE 8th Southern Power Electronics Conference and 17th Brazilian Power Electronics Conference (SPEC/COBEP), Florianopolis, Brazil, 26–29 November 2023; pp. 1–5.
58. Wang, Y.; Zhou, J.; Paul, K.C.; Zhao, T.; Sheng, D. Explainability Approach-Based Series Arc Fault Detection Method for Photovoltaic Systems. *IEEE Access* **2024**, *12*, 45530–45542. [CrossRef]
59. Zhang, Z.; Li, J. A Review of Artificial Intelligence in Embedded Systems. *Micromachines* **2023**, *14*, 897. [CrossRef] [PubMed]
60. Capra, M.; Bussolino, B.; Marchisio, A.; Shafique, M.; Masera, G.; Martina, M. An Updated Survey of Efficient Hardware Architectures for Accelerating Deep Convolutional Neural Networks. *Future Internet* **2020**, *12*, 113. [CrossRef]

applied sciences

MDPI

Article

A Bi-Objective Optimization Strategy of a Distribution Network Including a Distributed Energy System Using Stepper Search

Suliang Ma [1,*], Zeqing Meng [1], Yilin Cui [2] and Guanglin Sha [3]

1 School of Electrical and Control Engineering, North China University of Technology, Shijingshan District, Beijing 100144, China; 15721627501@163.com
2 Shandong Electric Power Company Haiyang Power Supply Company, Haiyang 265100, China; cyl_yilin_03@163.com
3 Distribution Technology Center, China Electric Power Research Institute, Haidian District, Beijing 100192, China; guanglinsha@163.com
* Correspondence: msl13811581885@ncut.edu.cn

Abstract: The optimal scheduling of DES is to solve a multi-objective optimization problem (MOP) with complex constraints. However, the potential contradiction between multiple optimization objectives leads to the diversity of feasible solutions, which has a serious impact on the selection of optimal scheduling strategies. Therefore, a stepper search optimization (SSO) method has been proposed for a bi-objective optimization problem (BiOP). Firstly, a constrained single-objective optimization problem (CSiOP) has been established to transform a BiOP and describe an accurate pareto front curve. Then, based on the characteristics of pareto front, the rate of the pareto front is analyzed by the SSO, and the best recommended solution of the BiOP is obtained. Finally, in the IEEE 33 with a DES simulation, by comparing other methods, the SSO method can better control the bi-objective optimization results to be 1–2.5 times as much as the optimal result under each single optimization objective and avoid the imbalance between the two optimization objectives. Additionally, the optimization speed of the SSO method is more than 10 times faster than that of the non-dominated sorting genetic algorithm (NSGA). Further, the SSO method will provide a novel idea for solving MOP.

Keywords: multi-energy system; bi-objective optimization problem (BiOP); stepper search optimization (SSO); pareto front; tolerance control

1. Introduction

In modern power systems, power generation systems and power consumption units with different characteristics have entered the power grid on a large scale, which puts forward higher requirements for the power balance and stable operation of the power grid [1]. In addition, under the premise of ensuring the stable operation of power supply, transmission and consumption, the optimal scheduling of a distributed energy system (DES) needs to take into account multiple objectives such as its economy and application performance [2,3], but there are often contradictions between these objectives, which results in a complex and difficult optimization process of scheduling strategies [4]. Therefore, how to balance the conflicts between multiple optimization objectives and form a reasonable optimal scheduling strategy is a key issue that restricts the further application of the distributed energy system.

1.1. Literature Review

The optimal scheduling problem of a DES is a typical multi-objective optimization problem (MOP) with constraints, characterized by the optimal control of various operating indexes under the constraints of the actual operating boundary of the system. Regarding the solution of the MOP, some experts from various countries have carried out a large number of studies [5–7]. In [8], a non-dominated sorting genetic algorithm (NSGA) is proposed to achieve the optimization of operating costs, labor efficiency and the penalty cost of pollutant gas emission in an integrated park scenario, obtaining an effective non-dominated solution set. Reference [9], after applying an NSGA to obtain the set of non-dominated solutions, proposes to utilize an approximate ideal solution sorting method to obtain scheduling instructions including wind farm (WF), photovoltaic station (PV), battery energy storage system (BESS) and load. Similarly, reference [10] evaluates the optimal capacity allocation scheme of an energy storage system (ESS) from two aspects of balancing the reliability and economy of the power grid by using the method of approximate ideal solution sorting. Reference [11] combines the improved artificial bee colony (ABC) algorithm with the entropy weight method to recommend an optimal allocation scheme for the energy storage system. Reference [12] uses the genetic algorithm (GA) for a multi-objective optimization of the three parameters in combined cycle power plants to determine and recommend optimal values. The characteristic of these methods is to leverage the global optimization ability of swarm intelligence optimization algorithms and, after obtaining numerous non-dominated solutions, use evaluation methods to recommend the final result. The disadvantage of this approach is that non-dominated solutions tend to be limited and few in number, while recommended solutions are directly affected by different evaluation methods.

Another way to solve an MOP is to combine multiple optimization objectives into a joint optimization objective, transforming it into a single-objective optimization problem (SiOP), of which one of the most common ways is to combine multiple optimization objective functions using summation [13,14]. For example, in the economic optimization of electric power, reference [15] equates the power loss to an operating cost and then sums it with the economy of other operating units in the system to form an SiOP. Reference [16] transforms the aging damage of the charging and discharging power process of the BESS into the operating cost of the BESS, accomplishing the union of multiple optimization objectives. References [17,18] equate the amount of discarded new energy as an economic loss for electricity applications, transforming the problem of new energy consumption into an optimization problem for economic objectives. Similarly, in references [19,20], an economic benefit of the reserve power from the power generation unit has been established to decide the operation status of the power grid and achieve the optimization of scheduling strategies for multiple types of energy. Other than converting various optimization objectives into the economic indicators mentioned above, some studies have achieved the reasonable addition of multiple optimization objectives by eliminating the physical meaning of each optimization objective function or allocating weights [21,22]. For example, references [23,24] combine multiple optimization objective functions through a weight allocation method, but the subjective weight values defined by users are easy to cause a lack of objectivity. By analyzing the indicators from the application scenarios, references [25,26] have assigned the objective function weight and completed the MOP transformation based on the analytic hierarchy process and fuzzy theory, respectively. Although the above method can simply and effectively transform the MOP into a constrained single-objective optimization problem (CSiOP), there is a certain deviation in the economic equivalent accuracy of the optimization objective. Additionally, the way of the weight allocation has a great influence on the optimization results, and the current objective weight allocation method is not perfect enough.

A summary of the above studies is listed in Table 1. Several comments are concerned. (1) In a MOP, a solution set for multi-objective optimization problems can be obtained by non-dominated swarm intelligence optimization, but selecting a recommended solution is difficult. There are few studies exploring the pareto frontier morphology. (2) Few studies implement or explore the pareto frontier morphology. (3) The method of transforming a MOP into a CSiOP has been extensively studied, but there is often a lack of discussion on the effectiveness of the transformation approach.

Table 1. Relevant research review.

Reference	System Components					Grid	Number of Objective Functions	Methodology
	DG	WF	PV	BESS	Other			
[5]	✓			✓		✓	2	PSO-TLBO
[6]	✓	✓		✓		✓	2	ECA + TOPSIS
[8]			✓		✓	✓	3	MO-NSGA-II
[10]	✓			✓		✓	3	MC + TOPSIS
[11]		✓	✓	✓	✓	✓	3	EW + ABC
[13]				✓	✓	✓	11	Weighted summation
[14]				✓	✓		3	MPC + adaptive weight
[16]		✓	✓	✓	✓	✓	2	Weighted summation
[17]		✓		✓	✓	✓	2	B-L optimization model
[19]		✓	✓	✓	✓	✓	3	TSDRO
[21]	✓			✓		✓	3	Multi-OPF model
[26]			✓	✓		✓	3	MOMUS

1.2. Contributions and Paper Organization

Therefore, in the MOP exhibited by the optimal scheduling of a distributed energy system, there are two key issues that need to be solved urgently. One is how to effectively obtain a pareto front or a non-dominated solution set for multiple optimization objectives, and the other is how to obtain a reasonable and unique recommended solution in the non-dominated solution set. To solve the above problems, a bi-objective optimization method is proposed based on stepper search optimization (SSO) from the perspective of pareto front form and the changing rate. This has provided an effective and recommended solution based on the transformation and the search of the bi-objective optimization problem (BiOP). The main contributions of this paper are summarized as follows.

(1) The SSO method has transformed a BiOP into a CSiOP by changing the boundary conditions. It can effectively describe the pareto front of the BiOP, which helps to solve the problem of how to choose the recommended solution on the pareto front.

(2) For the first time, the SSO method adopts stepper search on the pareto front to form the best-recommended solution of a BiOP based on the change rate of each non-dominated solution, with a novel approach being provided for selecting the recommended solution of a BiOP.

The organization of the manuscript is as follows. In the first section, the mathematical model of a DES is established, including the constraints of the operation process and two optimization objective functions. Then, the problems in the process of bi-objective optimization are analyzed, and the analysis process and implementation flow of the SSO method are shown. In the third section, based on the simulation example of IEEE 33 defined by the Institute of Electrical and Electronics Engineers (IEEE), the SSO method is compared and discussed with other methods. Finally, in the fourth section, the relevant conclusions and future research directions of this paper are given.

2. Mathematical Model of Distribution Network

2.1. The Operation Constraints of Distribution Network

Figure 1 shows a DES under an IEEE 33 distribution network including two photovoltaic power stations, two wind power generation stations, distributed generation (DG) and a BESS. It can be seen that there are multiple energy units scattered throughout the IEEE 33 distribution network. The distributed photovoltaic and wind-power generation stations provide green energy for the DES, while DG and ESS are used to ensure the energy balance and economic of the DES. In the application of the DES, certain operation constraints should be guaranteed, including the interactions of power, current and voltage among power nodes in the system, which conform to some basic physical laws such as energy conservation and node voltage equations [27]. These constraints for operation can be roughly divided into two categories, equality constraints and inequality constraints.

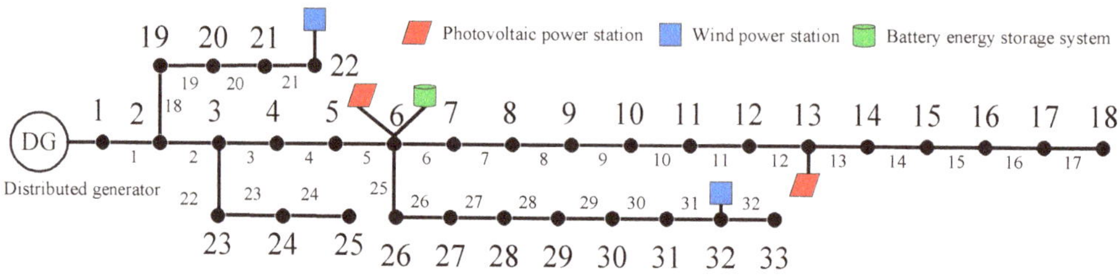

Figure 1. The DES under the IEEE 33 distribution network.

Firstly, in the operation process of the DES mentioned above, the equation constraints satisfied are the power-balance equation as shown in Equation (1). At time t, $P_{\mathrm{DG},i}(t)$, $P_{\mathrm{ESS},i}(t)$ and $P_{\mathrm{New},i}(t)$ respectively represent the active power of the DG, ESS and new energy station on the i-th power node. Additionally, $Q_{\mathrm{DG},i}(t)$ and $Q_{\mathrm{ESS},i}(t)$ are the reactive power of the DG and ESS. Similarly, the branch current and active and reactive power have been described as $I_{i,j}(t)$, $P_{i,j}(t)$ and $Q_{i,j}(t)$, respectively.

$$\begin{cases} P_{\mathrm{DG},i}(t) + P_{\mathrm{ESS},i}(t) + P_{\mathrm{New},i}(t) - P_{\mathrm{LOAD},i}(t) = \sum\limits_{j=1}^{N_b} P_{i,j}(t) - \sum\limits_{j=1}^{N_b} I_{i,j}(t) R_{i,j} \\ Q_{\mathrm{DG},i}(t) + Q_{\mathrm{ESS},i}(t) - Q_{\mathrm{LOAD},i}(t) = \sum\limits_{j=1}^{N_b} Q_{i,j}(t) - \sum\limits_{j=1}^{N_b} I_{i,j}(t) X_{i,j} \end{cases} \quad (1)$$

where $P_{\mathrm{LOAD},i}(t)$ and $Q_{\mathrm{LOAD},i}(t)$ are the active and reactive power of the load. $R_{i,j}$ and $X_{i,j}$ are the resistance and reactance between the j-th and the i-th power node. The number of power nodes is N_b.

Then, Ohm's law equation for each node in the power grid is considered as shown in Equation (2). $U_i(t)$ is the voltage of the i-th power node.

$$U_i(t) = U_j(t) + I_{i,j}(t) \times \left(R_{i,j}^2 + X_{i,j}^2\right) - 2 \times \left(P_{i,j}(t) \times R_{i,j} - Q_{i,j}(t) \times X_{i,j}\right) \quad (2)$$

Meanwhile, to balance the energy of the ESS during each typical day, Equation (3) represents the relationship between the energy state of the ESS and the charging and discharging power. At time t, the state of energy (SOE), the charge and discharge active power of the ESS on the i-th power node are $C_{\mathrm{SOE},i}(t)$, $P_{\mathrm{ESS\_c},i}(t)$ and $P_{\mathrm{ESS\_d},i}(t)$. In a typical day, the starting and ending SOEs of the ESS are $C_{\mathrm{SOE},i}(0)$ and $C_{\mathrm{SOE},i}(T)$.

$$C_{\mathrm{SOE},i}(t+\Delta t) = C_{\mathrm{SOE},i}(t) - \left(P_{\mathrm{ESS\_d},i}(t)/\eta_{\mathrm{d},i} + P_{\mathrm{ESS\_c},i}(t) \times \eta_{\mathrm{c},i}\right) \times \Delta t \times 100 / S_{\mathrm{ESS},i}^{\mathrm{N}} \quad (3)$$

where, on the i-th power node, the initial SOE, rated power, rated capacity and charging and discharging efficiency of the ESS have been described as $C_{\mathrm{SOE},i}^{\mathrm{init}}$, $P_{\mathrm{ESS},i}^{\mathrm{N}}$, $S_{\mathrm{ESS},i}^{\mathrm{N}}$, $\eta_{\mathrm{c},i}$ and $\eta_{\mathrm{d},i}$.

Equation (4) shows the energy balance relationship and initial state of charge in each typical day. The starting and ending SOEs of the ESS are $C_{\mathrm{SOE},i}(0)$ and $C_{\mathrm{SOE},i}(T)$.

$$C_{\mathrm{SOE},i}(0) = C_{\mathrm{SOE},i}(T) = C_{\mathrm{SOE},i}^{\mathrm{init}} \quad (4)$$

The energy storage output power has been defined by Expression (5).

$$P_{\mathrm{ESS},i}(t) = P_{\mathrm{ESS\_d},i}(t) + P_{\mathrm{ESS\_c},i}(t) \quad (5)$$

In addition to meeting the above equation constraints, the actual operation of the distributed energy system needs to meet some inequality constraints [28,29]. Firstly, the actual power must be less than or equal to its rated power for DG as shown in Inequality (6). The maximum and minimum active power of the i-th DG are $P_{\mathrm{DG},i}^{\min}$ and $P_{\mathrm{DG},i}^{\max}$, respectively. Then, $Q_{\mathrm{DG},i}^{\min}$ and $Q_{\mathrm{DG},i}^{\max}$ are the maximum and minimum active power.

$$\begin{cases} P_{\mathrm{DG},i}^{\min} \leq P_{\mathrm{DG},i}(t) \leq P_{\mathrm{DG},i}^{\max} \\ Q_{\mathrm{DG},i}^{\min} \leq Q_{\mathrm{DG},i}(t) \leq Q_{\mathrm{DG},i}^{\max} \end{cases} \tag{6}$$

Similarly, the Inequality (7) represents that the actual power is less than or equal to its rated power for ESS, and the SOE range of the i-th ESS is $[C_{\mathrm{SOE},i}^{\mathrm{L}}, C_{\mathrm{SOE},i}^{\mathrm{D}}]$. At time t, the binary variables indicating the charging and discharging statuses of the i-th ESS have been defined by $u_i^{\mathrm{c}}(t)$ and $u_i^{\mathrm{d}}(t)$. They have limited the ability of the ESS to charge and discharge energy at the same time.

$$\begin{cases} 0 \leq P_{\mathrm{ESS\_d},i}(t) \leq u_i^{\mathrm{d}}(t) \times P_{\mathrm{ESS},i}^{\mathrm{N}} \\ -u_i^{\mathrm{c}}(t) \times P_{\mathrm{ESS},i}^{\mathrm{N}}(t) \leq P_{\mathrm{ESS\_c},i}(t) \leq 0 \\ u_i^{\mathrm{c}}(t) + u_i^{\mathrm{d}}(t) \leq 1 \quad u_i^{\mathrm{c}}(t) = 0 or 1 \quad u_i^{\mathrm{d}}(t) = 0 or 1 \\ C_{\mathrm{SOE},i}^{\mathrm{L}} \leq C_{\mathrm{SOE},i}(t) \leq C_{\mathrm{SOE},i}^{\mathrm{U}} \end{cases} \tag{7}$$

In terms of new energy generation station, the output power must be less than or equal to the maximum power that can be output from the new energy field station at the current moment as shown in Inequality (8). The maximum active power of the i-th new energy generation station has been described as $P_{\mathrm{New},i}^{\mathrm{ref}}(t)$.

$$0 \leq P_{\mathrm{New},i}(t) \leq P_{\mathrm{New},i}^{\mathrm{ref}}(t) \tag{8}$$

Meanwhile, considering the requirements of power grid operation, it is necessary to limit the voltage of power grid nodes and branch currents, as shown in Inequality (9) where the maximum value of the node voltage and the branch current have been defined as $U_{\max}$ and $I_{\max}$, respectively. The minimum value of the node voltage is $U_{\min}$.

$$\begin{cases} U_{\min} \leq U_i(t) \leq U_{\max} \\ 0 \leq I_{i,j}(t) \leq I_{\max} \end{cases} \tag{9}$$

In addition to the above common inequality constraints, the distributed power supply and energy storage system can output both active and reactive power, and the actual output of active and reactive power shows a quadratic inequality relationship with the rated power of DG and ESS, as shown in Inequality (10) where the rated power of the DG and ESS on the i-th power node are $S_{\mathrm{DG},i}^{\mathrm{N}}$ and $P_{\mathrm{ESS},i}^{\mathrm{N}}$, respectively. $||\cdot||_2$ represents the 2-norm of the vector.

$$\begin{cases} \left\| P_{\mathrm{DG},i}(t) \quad Q_{\mathrm{DG},i}(t) \right\|_2 \leq S_{\mathrm{DG},i}^{\mathrm{N}} \\ \left\| P_{\mathrm{ESS},i}(t) \quad Q_{\mathrm{ESS},i}(t) \right\|_2 \leq P_{\mathrm{ESS},i}^{\mathrm{N}} \end{cases} \tag{10}$$

The last class of inequality constraints derives from a relaxation process for nonlinear equality constraints, such as the i-th node power balance equation $P^2_{i,j}(t) + Q^2_{i,j}(t) = I_{i,j}(t) \times U_i(t)$. This type of constraint exhibits the nonlinearity of the multiplication between the variables $I_{i,j}(t)$ and $U_i(t)$, and this equation will be relaxed as a quadratic inequality form shown in Inequality (11).

$$\left\| \begin{matrix} 2 \times P_{i,j}(t) \\ 2 \times Q_{i,j}(t) \\ I_{i,j}(t) - U_i(t) \end{matrix} \right\|_2 \leq \left(I_{i,j}(t) + U_i(t) \right) \tag{11}$$

The following can be seen from the above constraints: (1) The coupling relationship between the components of a DES is significant, and the output of each subsystem will have a direct impact on the operating range of other subsystems. (2) The actual operation of a DES contains a large number of equations, inequalities and quadratic constraints, and the actual operating boundaries of the system are relatively complex. (3) There are many discrete variables of type 0–1, such as u_i^c and u_i^d, and continuous variables in the constraints forming a mixed variable optimization problem. The continuous variables have included $P_{DG,i}$, $P_{ESS,i}$, $P_{New,i}$, $Q_{DG,i}$, $Q_{ESS,i}$, $I_{i,j}$, $P_{i,j}$, $Q_{i,j}$, U_i, $C_{SOE,i}$, $P_{ESS\_c,i}$ and $P_{ESS\_d,i}$.

2.2. The Objective Function of Distribution Network

Section 2.1 shows the operational constraints of the DES, and this section will establish the optimization objective function of the DES. There are many elements in a DES, and a variety of different optimization objectives are established from the perspectives of grid performance and economy, such as grid loss, node voltage deviation, new energy utilization rate, system operating cost and so on. However, the SSO method in this paper is more friendly for the BiOP. Therefore, this paper takes the two optimization objectives of grid loss and system operating cost as examples to carry out the subsequent analysis. It should be noted that these two objective functions are only listed for the purpose of illustrating the methodology of this paper, which does not mean that the SSO method is limited to optimizing the following functions.

(1) To reduce the power line loss of the DES, the first optimization objective function J_1 is established, as shown in Expression (12), and is a key index that effects the transmission efficiency of the power grid.

$$\min \quad J_1 = \frac{1}{2}\sum_{t=1}^{T}\sum_{i=1}^{N_b}\sum_{j=1}^{N_b}\left(i_{i,j}^2(t) \times \sqrt{R_{i,j}^2 + X_{i,j}^2}\right) \tag{12}$$

(2) To reduce the operating costs of the DES, the second optimization objective function J_2 is established including the cost of DG, ESS and the new energy generation station, as shown in Expression (13), where $S_{DG,i}$, $S_{ESS,i}$ and $S_{New,i}$ respectively represent the electricity consumption of the i-th DG, ESS and new energy during time period Δt. $S_{DG,i} = P_{DG,I} \times \Delta t$, $S_{ESS,i} = P_{ESS,I} \times \Delta t$ and $S_{New,i} = P_{New,I} \times \Delta t$. In this paper, $\Delta t = 1$ h. Then, where $k_{1,I}$, $k_{2,I}$ and $k_{3,i}$ respectively represent the cost coefficient of DG [30,31], ¥/kWh^2, ¥/kWh, ¥. Significantly, $k_{1,i}$ represents the cost coefficient of the square of the active power of DG. Therefore, the unit of coefficient $k_{1,i}$ is ¥/kWh^2. The levelized cost coefficient of the ESS and the new energy generation station have been defined as $k_{4,i}$ and $k_{5,i}$, ¥/kWh.

$$\min \quad J_2 = \sum_{t=1}^{T}\sum_{i=1}^{N_b}\left(k_{1,i}S_{DG,i}^2(t) + k_{2,i}S_{DG,i}(t) + k_{3,i} + k_{4,i}S_{ESS,i}(t) + k_{5,i}S_{New,i}(t)\right) \tag{13}$$

The following can be seen from the above optimization objectives: (1) The physical dimensions of the two optimization objectives are different, so it is difficult to directly compare and analyze the importance between them, and it is not convenient to transform them into a single optimization objective by adding weights. (2) The two optimization objectives are used to evaluate the two aspects of the operation of the DES and do not have dominance, which means that the two optimization objective functions are equivalent in the distributed energy system. (3) Both objective functions exhibit quadratic form, which implies that the optimization and control of the DES is a multi-objective quadratic planning problem.

3. The Problem and Solution of the BiOP

3.1. The Problem of the BiOP

The first section shows that the optimal control problem of the DES can be described as a bi-objective quadratic programming problem with a large number of complex constraints. This mathematical optimization problem can be described as Expressions (14) and (15), where $X = [x_1, x_2, \ldots, x_m]$ is a vector composed of optimization variables x_i, $i = 1, 2, \ldots, m$, while $f_1(X)$ and $f_2(X)$ respectively represent two optimization objective functions, $g_k(X)$ represents the k-th inequality constraint, $h_l(X)$ represents the l-th equality constraint and x_i^{lb} and x_i^{ub} respectively represent the lower and upper bounds of the optimization variable x_i.

$$\underset{X}{\text{minimize}} \quad F(X) = [f_1(X), f_2(X)] \tag{14}$$

$$\text{Subject to} \quad \begin{cases} g_k(X) \leq 0, k = 1, 2, \ldots, K \\ h_l(X) \leq 0, l = 1, 2, \ldots, L \\ x_i^{\mathrm{lb}} \leq x_i \leq x_i^{\mathrm{ub}}, i = 1, 2, \ldots, m \end{cases} \tag{15}$$

In a MOP, there is generally no solution to ensure that both optimization objectives are the most global optimal. The way to solve such optimization problems is to use the pareto optimization method to obtain a series of non-dominated solutions, and there is no definite dominance relationship between different optimization solutions rather than a global optimal solution. Figure 2 shows the results of analyzing pareto optimality using two optimization objectives as examples.

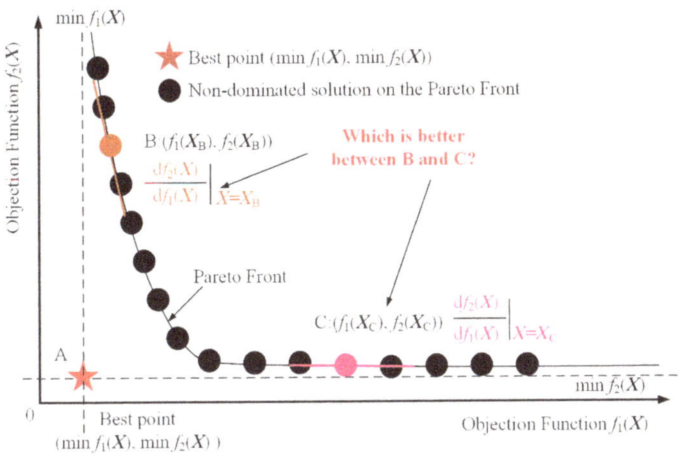

Figure 2. Block diagram for describing a bi-objective optimization problem.

From Figure 2, it can be observed that if each objective is independently optimized, a global optimum can be obtained. The points marked by these global optimum values are defined as the optimal points in the multi-objective function space, such as point A in the figure. However, this optimal point is not achievable in multi-objective optimization problems. Pareto optimization is used to obtain a series of non-dominated solutions that form a convex surface in the space of multi-objective functions, which is the pareto front. Obviously, it is difficult to choose the most reasonable solution among many non-dominated solutions on the pareto front. However, it is worth thinking about in this paper that non-dominated solutions on the pareto front, such as points B and C, show great differences in their slopes $\mathrm{d}f_1(X)/\mathrm{d}f_1(X)\,|_{X\,=\,X_B}$ or X_C. In the neighborhood of point B, when $f_1(X)$ changes slightly, $f_2(X)$ will change dramatically. In the neighborhood of point C, on the contrary, when $f_1(X)$ changes greatly, $f_2(X)$ hardly changes. It is from this phenomenon that this paper carries out a study from the perspective of the degree of the change of the two optimization objective functions and proposes a bi-objective optimization method based on stepper search to complete the problem of selecting the non-dominated solutions on the pareto front.

Based on the above analysis, it can be seen that the method proposed in this paper needs to address the following two issues. (1) How to define the change in the slope of the pareto front. (2) How to find the point with the largest slope change. It is very unfortunate that pareto front parsing expressions cannot be built in the currently available methods, and the above problems can only be solved by means of numerical analysis. Therefore, in the SSO method, the most crucial part is to transform the description form of the BiOP, and the transformed optimization problem is described as shown in Expression (16).

$$\begin{array}{ll}\text{minimize} & J(\mathbf{X}) = r_2|r_1 = c \\ \text{Subject to} & \left\{\begin{array}{l} g_k(\mathbf{X}) \leq 0, k = 1,2,\ldots,K \\ h_l(\mathbf{X}) \leq 0, l = 1,2,\ldots,L \\ x_i^{\text{lb}} \leq x_i \leq x_i^{\text{ub}}, i = 1,2,\ldots,m \\ f_1(\mathbf{X}) \leq |\min(f_1(\mathbf{X}))|r_1 + \min(f_1(\mathbf{X})) \\ f_2(\mathbf{X}) \leq |\min(f_2(\mathbf{X}))|r_2 + \min(f_2(\mathbf{X})) \end{array}\right.\end{array} \quad (16)$$

Expression (16) has introduced two optimization variables r_1 and r_2, which respectively represent the deviation margins between the two optimization objectives and their respective optimal values, and transforms the optimization problem described in Expression (14) into a constraint inequality. The optimization goal $J(\mathbf{X}) = r_2 \mid r_1 = c$ means that the problem of minimum optimization deviation margin r_2 is solved under the condition of $r_1 = c$, where c represents a constant value that can be changed. Obviously, the optimization result of Expression (16) will converge to $\min(J_1(\mathbf{X}))$ when the parameter $c \rightarrow 0$, while the optimization result of Expression (16) will converge to $\min(J_2(\mathbf{X}))$ when the parameter $c \rightarrow \infty$. According to the optimization model described in Expression (16), when the parameter c changes from 0 to ∞, the pareto front of the dual optimization objectives described in Expressions (14) and (15) can be formed.

3.2. A Bi-Objective Optimization Method Based on SSO

According to the mathematical model described in Expression (14), this paper has proposed a SSO method for solving the BiOP. Figure 3 shows the stepper search process of the SSO method. Suppose the current search point is a, whose coordinates are obtained by using the single-objective optimization model described in the interior-point method optimization Expression (14) under the condition $r_1 = c$, that is, point $a = \min(r_2 \mid r_1 = c)$.

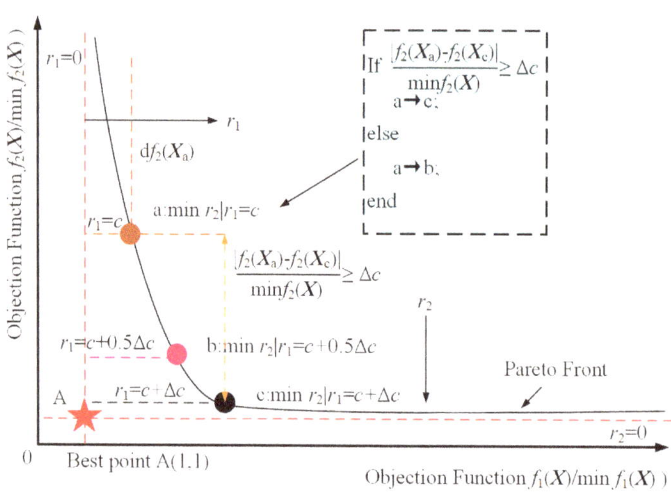

Figure 3. Schematic diagram of the SSO method.

Modifying the value of r_1 in the direction of the objective function $f_1(\mathbf{X})$ with a step size Δc, $r_1 = r_1 + \Delta c$, it can be seen that the point a will move to the point c, $\min(r_2 \mid r_1 = c + \Delta c)$. At this point, if the absolute value of the change in the direction of the objective function $f_2(\mathbf{X})$ is greater than or equal to $\min(f_2(\mathbf{X})) \times \Delta c$, which means that the change in the optimization objective function $f_2(\mathbf{X})$ is greater under the change in the objective function $f_1(\mathbf{X})$, then the point c will be the starting point of the next search. If the absolute value of the change in the direction of the objective function $f_2(\mathbf{X})$ is less than $\min(f_2(\mathbf{X})) \times \Delta c$, then this means that the change in the optimization objective function $f_2(\mathbf{X})$ is less than the change in the objective function $f_1(\mathbf{X})$, and the search step should be narrowed down (the dichotomy can be utilized to change Δc), that is, $r_1 = r_1 + 0.5 \times \Delta c$. Additionally, the SiOP described in Expression (14) is optimized once again by using the interior point method, and point a will move to $\min(r_2 \mid r_1 = c + 0.5 \times \Delta c)$ at point b. Again, compare the amount of change in the optimization objective functions $f_1(\mathbf{X})$ and $f_2(\mathbf{X})$. By continuously completing the above process, when the deviation value of the variation of the optimization objective functions $f_1(\mathbf{X})$ and $f_2(\mathbf{X})$ is small or reaches the set maximum number of iterative searches, the search process will be stopped, and the result of the last search will be the final optimization result. Based on the stepper search approach depicted in Figure 3, the pseudo-code of the SSO algorithm has been listed in Algorithm 1.

Algorithm 1 Pseudo code of the SSO method.

Input: In the DES described in this article, the BiOP has been shown in mathematical relation (1)~(13), and the mathematical model for transforming the BiOP of the DES is shown in Expression (16). The initial step parameter Δc and the deadline of stepper search are defined. The deadline of the stepper search is the maximum number of iterations $N_{\max}$ and stop threshold λ. Assuming that the iteration N is 0,

Optimize two single optimization objective functions independently. The minimum value of each optimization objective is $\min(J_1(\mathbf{X}))$ and $\min(J_2(\mathbf{X}))$, respectively;
$r_1^N \leftarrow \Delta c$ optimize Expression (16), and $\min(r_2^N \mid r_1^N)$ is obtained.
Calculate $J_1^N(\mathbf{X})$ and $J_2^N(\mathbf{X})$ under the above r_1^N and r_2^N condition,
While $N < N_{\max}$
$r_1^{N+1} \leftarrow r_1^N + \Delta c$ optimize Expression (16), and $\min(r_2^{N+1} \mid r_1^{N+1})$ is obtained.
Calculate $J_1^{N+1}(\mathbf{X})$ and $J_2^{N+1}(\mathbf{X})$ under the above r_1^{N+1} and r_2^{N+1} condition;
if $\mathrm{abs}(\mid J_1^{N+1}(\mathbf{X}) - J_1^N(\mathbf{X}) \mid / \mid \min(J_1(\mathbf{X})) \mid - \mid J_2^{N+1}(\mathbf{X}) - J_2^N(\mathbf{X}) \mid / \mid \min(J_2(\mathbf{X})) \mid) \leq \lambda$ or $N > N_{\max}$
break;
else
if $\mid J_1^{N+1}(\mathbf{X}) - J_1^N(\mathbf{X}) \mid / \mid \min(J_1(\mathbf{X})) \mid \leq \mid J_2^{N+1}(\mathbf{X}) - J_2^N(\mathbf{X}) \mid / \mid \min(J_2(\mathbf{X})) \mid$
$\Delta c \leftarrow \Delta c$;
$N \leftarrow N + 1$;
else
$N \leftarrow N$;
$\Delta c \leftarrow \Delta c/2$;
end
end
end
Output: $(J_1^N(\mathbf{X}), J_2^N(\mathbf{X}))$ and $\mathbf{X}$ under the r_1^N and r_2^N condition.

4. Case and Discussion

4.1. Case Condition

To analyze and validate the SSO methodology, this paper has carried out the analysis based on the DES in the IEEE 33 distribution network as shown in Figure 1. And the lithium-ion battery energy storage system has been adopted as the BESS in this paper. Table 2 has represented the case parameters and condition of the power grid, DG, photovoltaic power station, wind power station and BESS. The computational environment of this example is Matlab 2020a under Win10 system, the computer memory is 32 G, the CPU is Intel Core i7, the main frequency is 2.3 GHz and the number of cores is 16. Figure 4 shows the power curves of the power load and photovoltaic and wind power stations in a typical day system. The time interval Δt = 1 h, and the typical day T = 24 h.

Table 2. Parameters and conditions.

Case Parameter	Numerical Value	Case Parameter	Numerical Value
Capacity of power grid	10 MVA	Rated power of DG	10 MVA
Nominal voltage	12.66 kV	Cost coefficient k_1 of DG	40,000 ¥/kWh^2
The voltage range	±10%	Cost coefficient k_2 of DG	650 ¥/kWh
The maximum branch current	456 A	Cost coefficient k_3 of DG	8 ¥
Charge/discharge efficiency of BESS	95%, 92%	Rated power of photovoltaic at #6 node	4 MW
SOE range of BESS	90%, 10%	Rated power of photovoltaic at #13 node	2 MW
Initial SOE of BESS	50%	Levelized cost of photovoltaic	0.293 ¥/kWh
Rated power of BESS	3.3 MW	Rated power of wind farm at #22 node	3 MW
Rated capacity of BESS	13.2 MWh	Rated power of wind farm at #22 node	2 MW
Levelized cost of BESS	0.574 ¥/kWh	Levelized cost of wind farm	0.3 ¥/kWh
Initial stepper parameter Δc	0.1	The stop threshold λ	1×10^{-4}

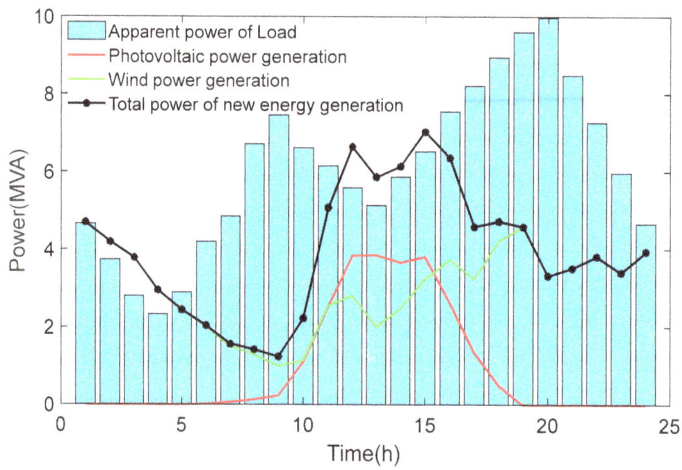

Figure 4. Electricity generation and consumption curve from new energy stations and load under typical day.

In Table 2 and Figure 4, it can be seen that the apparent power of the load is less than 10 MVA, and the total rated power of new energy stations is greater than 10 MVA. Additionally, the apparent power of the load is much greater than the actual maximum output power of the new energy stations at 5–11 h and 16–24 h, respectively. It means that there is an imbalance between power supply and consumption in the distribution network system and a reasonable control of BESS and DG is necessary.

Meanwhile, this paper compares the SSO method with other common multi-objective optimization methods. Method 1 is an equal-weight summation method, method 2 is a Euclidean distance minimization method, and the optimization objective functions are shown in Expressions (17) and (18), respectively. In addition, the constraint conditions for method 1 are shown in Expression (15), and the constraint conditions for method 2 are shown in Expression (16). Further, as the method 3, the traditional NSGA method combined with TOPSIS [9] has been used for a comparison of the proposed methods. The maximum iterative number is 30, and the number of individuals is 20 in the population.

$$\text{minimize} \quad J(\mathbf{X}) = \frac{J_1(\mathbf{X})}{|\min(J_1(\mathbf{X}))|} + \frac{J_2(\mathbf{X})}{|\min(J_2(\mathbf{X}))|} \tag{17}$$

$$\text{minimize} \quad J(\mathbf{X}) = r_1^2 + r_2^2 \tag{18}$$

4.2. Result Analysis

With the above simulation parameters, a single objective optimization is performed based on the two objective functions of the DES mathematical model described in Section 1 to obtain the minimum values $\min(J_1) = 0.0294$ and $\min(J_2) = 11{,}584.32$. Table 3 demonstrates the comparison of the two objective function values under optimization with a single objective, method 1, method 2, and the SSO method. Figure 5 shows the optimization results under three different methods and the stepper search process of the SSO method. Figure 6 shows the differences of the optimization scheduling results between three different methods from four perspectives: distributed power, new energy generation power stations, energy storage system and grid node voltage.

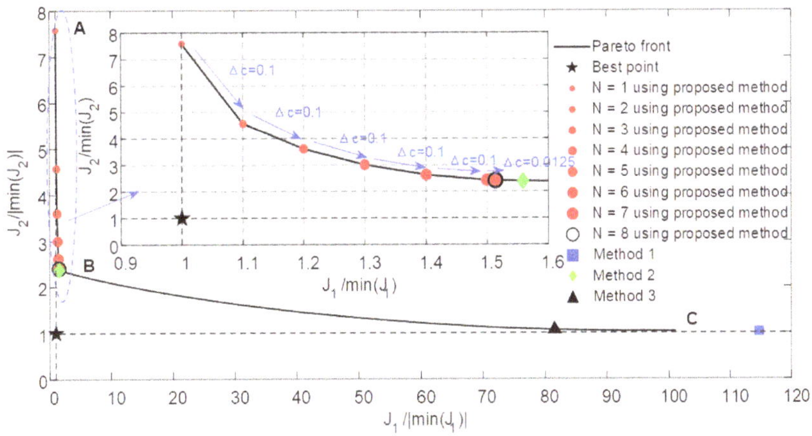

Figure 5. The distribution of optimization results ($J_1/|\min(J_1)|$, $J_2/|\min(J_2)|$) from the different methods at the pareto front.

Table 3. J_1 and J_2 values under different methods.

	min J_1	min J_2	Method 1	Method 2	Method 3	SSO Method
$J_1/|\min(J_1)|$	1	140.7114	114.7497	1.5586	81.4999	1.5141
$J_2/|\min(J_2)|$	7.7836	1	1.0248	2.3756	1.0637	2.4047
Computing time	18.44 s	18.67 s	21.94 s	26.22 s	3907.16 s	280.44 s

Table 3 presents the following: (1) Optimizing with a single objective guarantees that the results obtained will be optimal for that metric, but the other metric will be so far from the optimal position that it will not be possible to balance the two optimization objectives, and the results obtained will be very poor. (2) Compared to single-objective optimization, the optimization result obtained by using method 1 and 3 increases only by 0.0248 and 0.0637 in the $J_2/\min(J_2)$ direction but decreases by 25.9617 and 59.2115 in the $J_1/\min(J_1)$ direction, respectively. It means that the optimization result is better than the single-objective optimization approach in terms of the whole system but still exhibits poorer performance in the $J_1/\min(J_1)$ direction. (3) Compared to method 1, although method 2 and the SSO method decrease by 1.3508 and 1.3799, respectively, in the $J_2/\min(J_2)$

direction, they increase by 113.1911 and 113.2356, respectively, in the $J_1/\min(J_1)$ direction, improving approximately 80 times. It means that the scheduling strategies based on these two methods will be more conducive to the overall performance of the system. (4) Compared to method 2, the SSO method decreases by 0.0291 in the $J_2/\min(J_2)$ direction and increases by 0.0445 in the $J_1/\min(J_1)$ direction. Although this numerical change is small, it achieves approximately 1.5 times the benefit, which means that the SSO method is slightly better than method 2. (5) In terms of solving time, the proposed method is superior to method 3 and reduces the solution time by more than 10 times, but it is inferior to other methods. The reason for the above results is that the proposed method and method 3 have iterative processes, and there are more iterations required in method 3.

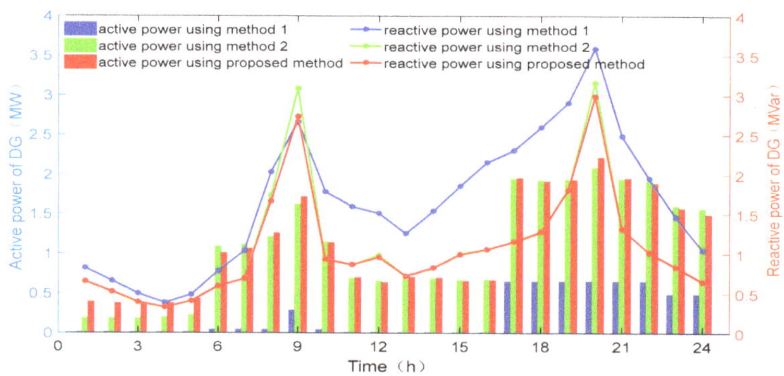

(**a**) The operating curve of the DG.

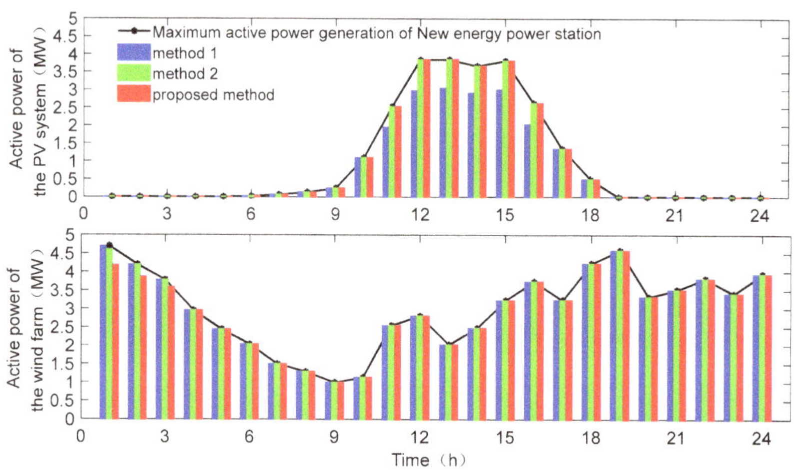

(**b**) The operating curve of the new energy power stations.

Figure 6. *Cont.*

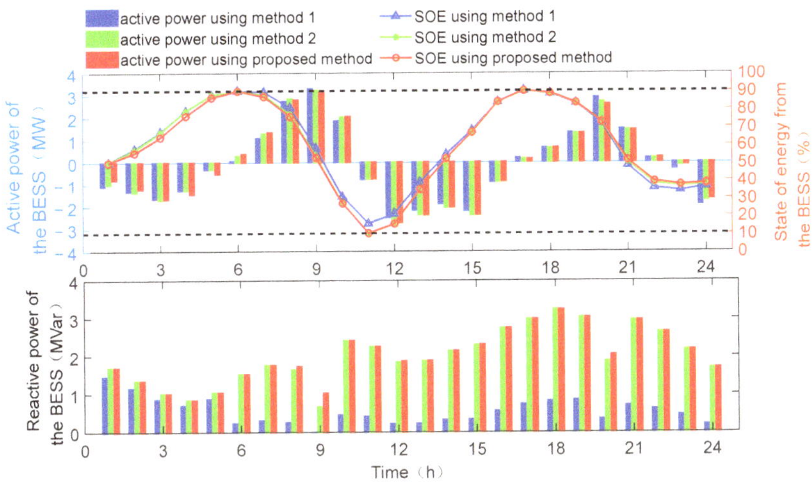

(**c**) The operating curve of the BESS.

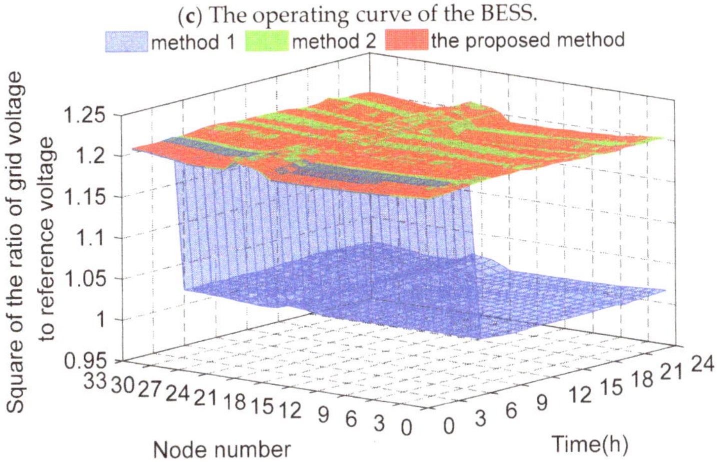

(**d**) The operating curve of the grid voltage.

Figure 6. Comparison results of operational performance of different distribution network methods.

Figure 5 shows the distribution characteristics of different methods on the pareto front and the stepper search process of the SSO method in the coordinate system composed of $(J_1/|\min(J_1)|, J_2/|\min(J_2)|)$. From the figure, it can be observed that (1) the two optimization objectives show significant contradiction, that is, as the value of one optimization objective decreases, the value of the other optimization objective increases. (2) In the coordinate system composed of $(J_1/|\min(J_1)|, J_2/|\min(J_2)|)$, the pareto front shows two parts with significant slope differences, namely, the AB section and BC section. In the AB section, as J_1 slowly increases, J_2 rapidly decreases, which means that by alleviating the requirements for target J_1, greater benefits can be obtained on target J_2. On the other hand, in the BC section, as J_1 increases significantly, J_2 slowly decreases, which means that the substantial reduction of the requirements for the target J_1 has limited improvement in the direction of the target J_2. Therefore, the optimization results of method 1 and 3 are located in the BC section and are far from point B, which is bad for the whole system. Meanwhile, the search results of both method 2 and the SSO method are close to point B, the point of the slope mutation of the pareto front, which means that the results of these two methods are better for the overall system. (3) The search process of the SSO method takes $\Delta c = 0.1$

as the step, and, after updating five times, it gradually decreases for three times, resulting in $\Delta c = 0.0125$, and then $\text{abs}(J_1/|\min(J_1)| - J_2/|\min(J_2)|) < \lambda$ is searched. This implies that the stepper search method proposed in this paper is effective and can search for the recommended solution quickly.

From the operating curves of the DG presented in Figure 6a, the curves of method 2 and the SSO method are basically the same, and the trend of the output power during a typical day is basically consistent with the load power demand, whereas in method 1, in order to significantly reduce power loss (optimization objective J_1), the distributed power supply continuously outputs a large amount of reactive power and reduces the active power output. From the comparison of the operating curves from PV and wind power generation system shown in Figure 6b, the actual power of wind power generation has been well consumed based on the three scheduling strategies, but there is a phenomenon of photovoltaic power abandonment in the application of method 1, which is concentrated in the peak of PV power generation and the low valley of load power consumption between 12~15 h. Additionally, it can be seen that method 2 and the SSO method are better than method 1. Figure 6c shows the output active power curve and energy state of the BESS. Between 0:00 and 6:00, the BESS takes the initiative to charge and dissipate the wind power generation that is larger than the load demand. Between 6:00 and 11:00, the load active power is greater than the maximum output power of the new energy, and the BESS discharges to support the load power demand. Between 11:00 and 16:00, due to more photovoltaic power generation, the new energy generation is greater than the load demand, and the BESS is reasonably charged again, ready to cope with the next decline in photovoltaic power generation caused by the difference between the load power and the new energy power supply. The BESS has solved the power balance problem and ensured the economic operation of the DES. Figure 6d shows the voltage curves of each node from the power grid at different times under different optimization methods. It can be seen that compared with method 1, the voltage of each node from the power grid is close to 1.21 (the square of 110% of the reference voltage) in the whole time period under method 2 and the SSO method, thus reducing the line current and power loss.

4.3. Discussion

Firstly, this section discusses the influence of the step parameter Δc of the SSO method on the search stability and speed. Table 4 shows the calculation time and optimization target value of the SSO method under the step parameter Δc, and Figure 7 shows the change process of the corresponding curve.

Table 4. Optimization results under different parameters Δc using the SSO method.

Δc	0.02	0.03	0.05	0.08	0.1	0.15	0.18	0.2	0.25
Search time	688.67	442.67	358.85	325.88	280.44	309.58	316.85	330.09	263.99
$J_1/|\min(J_1)|$	1.52	1.51	1.51	1.52	1.51	1.53	1.54	1.60	1.52
$J_2/|\min(J_2)|$	2.40	2.40	2.40	2.40	2.40	2.40	2.38	2.37	2.40

Table 4 and Figure 7 show that (1) with the decrease in step parameter Δc, the solution time of the SSO method increases; especially when $\Delta c \leq 0.03$, the solution time almost doubles. However, when the step parameter Δc is large, $\Delta c \in [0.08, 2.5]$, the solution time of the SSO method fluctuates in a small range between 250~350 s. (2) With the decrease in step parameter Δc, the fluctuation ranges of $J_1/\min(J_1)$ and $J_2/\min(J_2)$ are 1.51~1.60 and 2.37~2.40, respectively. This means that the accuracy of the SSO method is insensitive to the step parameter Δc. Therefore, in the SSO method, a larger step size can be selected for calculation by selecting the step parameter Δc.

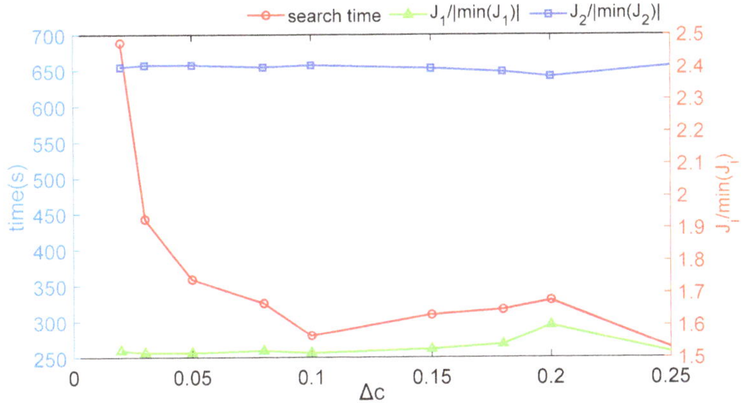

Figure 7. Search time and optimization results under different parameter Δc using the SSO method.

Then, to verify the applicability of the SSO method, this section establishes two optimization objectives J_3 and J_4 from the aspects of node voltage and new energy station in the DES, which are defined as shown in Expressions (19) and (20). At time t, the i-th node voltage and the power have been defined as $V_i(t)$ and $P_{\mathrm{New},i}(t)$, respectively. The duration of the typical day and the number of grid nodes have been set to T and N_{b}. The voltage reference value is V_{base}.

$$\min \quad J_3 = \max_{t=1}^{T}\left(\max_{i=1}^{N_{\mathrm{b}}}\left(\mathrm{abs}\left(V_i^2(t)/V_{\mathrm{base}}^2 - 1\right)\right)\right) \tag{19}$$

$$\min \quad J_4 = -\sum_{t=1}^{T}\sum_{i=1}^{N_{\mathrm{b}}}\left(P_{\mathrm{New},i}(t)\right) \tag{20}$$

Table 5 represents the comparison results under the dual objectives of (J_1, J_3) with single-objective optimization, methods 1, 2 and 3 and the SSO method, respectively. Figure 8 shows the optimization results under four different methods and the stepper search process of the SSO method.

Table 5. J_1 and J_3 values under different methods.

	min J_1	**min J_3**	**Method 1**	**Method 2**	**Method 3**	**SSO Method**
$J_1/\mid(\min(J_1)\mid$	1	65.9354	1.4879	1.5170	1.0161	1.5031
$J_3/\mid\min(J_3)\mid$	78.7690	1	1.5731	1.5448	71.4381	1.5581
Computing time	18.44 s	18.61 s	17.34 s	21.42 s	7715.89 s	263.52 s

In Table 5, comparing the results of single-objective optimization, methods 1 and 2 and the SSO method can effectively balance the contradiction between the two objectives, but method 3 is not. Method 3 focuses more on the objective function J_1, resulting in an excessive loss of J_3. Obviously, it is not good. Compared to the other methods, using the best results of each objective function as a baseline, methods 1 and 2 decrease by 0.4879 and 0.5170 in the $J_1/\mid\min(J_1)\mid$ direction and improve by 0.5731 and 0.5448 in the $J_3/\mid\min(J_3)\mid$ direction. Then, the SSO method reduces 0.5031 in the $J_1/\min(J_1)$ direction and enhances 0.5581 in the $J_3/\mid\min(J_3)\mid$ direction. Apparently, the SSO method is more balanced. In addition, the solving time of the proposed method is superior to method 3 and reduces the solution time by about 30 times, but it is inferior to the other methods. This phenomenon and its causes are consistent with Table 3.

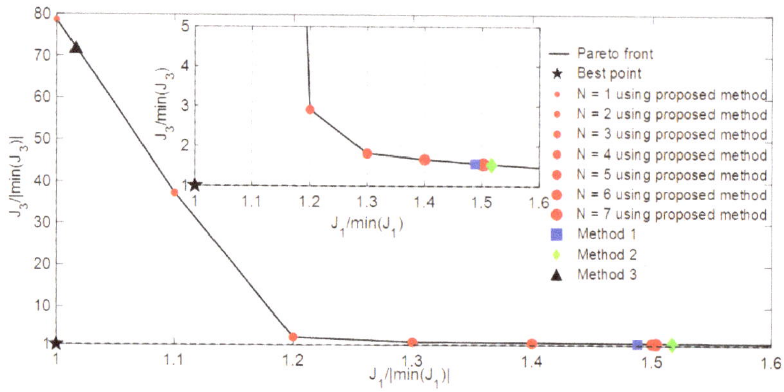

Figure 8. The distribution of optimization results $(J_1/|\min(J_1)|, J_3/|\min(J_3)|)$ at the pareto front.

Figure 8 has represented the distribution characteristics of the different methods on the pareto front and the stepper search process of the SSO method under the coordinate system consisting of $(J_1/|\min(J_1)|, J_3/|\min(J_3)|)$ and $(J_1/|\min(J_1)|, J_4/|\min(J_4)|)$, respectively. With the first optimization objective value decreasing, the second optimization objective value rises. Further, the pareto front exhibits two components with significant differences in the slope. However, the slope changes exhibited by the pareto front are different, which means that if the above multiple objectives are comprehensively considered to form a multi-objective optimization problem, the slope changes in different objective function directions will be different.

Table 6 and Figure 9 exhibit the comparison results under the dual objectives of (J_1, J_4) with single-objective optimization, methods 1, 2 and 3 and the SSO method, respectively.

Table 6. J_1 and J_4 values under different methods.

	min J_1	**min J_4**	**Method 1**	**Method 2**	**Method 3**	**SSO Method**
$J_1/|(\min(J_1)|$	1	49.5009	1.0906	1.1398	1.5947	1.1016
$J_4/|\min(J_4))|$	−0.5872	−1	−0.7795	−0.8228	−1.0000	−0.7900
Computing time	18.44 s	14.03 s	11.96 s	13.15 s	2251.62 s	109.45 s

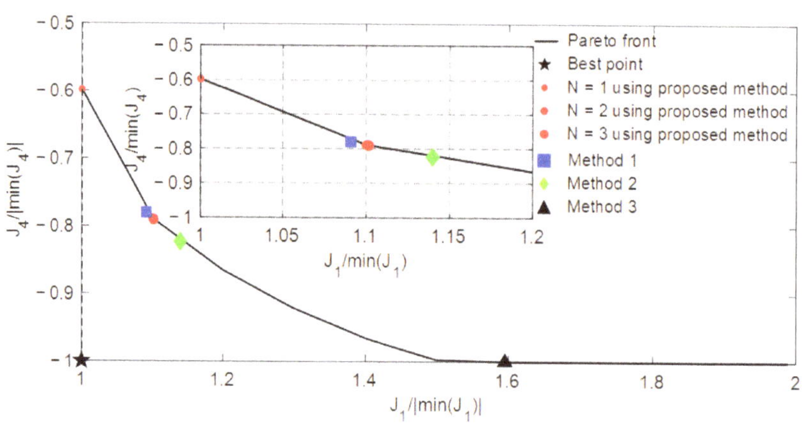

Figure 9. The distribution of optimization results $(J_1/|\min(J_1)|, J_4/|\min(J_4)|)$ at the pareto front.

In Table 6, the optimization results from methods 1, 2 and 3 and the SSO method have achieved effective balance in two objective function directions. It is different from

Table 6. However, the results of the solving time are the same. Methods 1, 2 and 3 decrease by 0.0906, 0.13982 and 0.5947 in the $J_1/|\min(J_1)|$ direction and improve by 0.2205, 0.1772 and 0 in the $J_4/\min(J_4)$ direction. Then, the SSO method reduces 0.1016 in the $J_1/|\min(J_1)|$ direction and enhances 0.21 in the $J_4/|\min(J_4)|$ direction. The above results show that the SSO method can find equitable recommended solutions more consistently than other methods.

Compared to Figures 5 and 8, the pareto front curve in Figure 9 has more stages of slope variation. The optimization results of method 3 differ significantly from other methods and is almost equivalent to the results of the single optimization objective J_4. It has caused an imbalance between the two optimization objectives' functions. Then, the result of method 1 is at the intersection of two slopes. It means that the changes in the two optimization objective functions will vary. Combining the values of Table 6, it can be seen that the optimization effect of the SSO method is better. Additionally, based on the results in Figures 5, 8 and 9, it can be seen that the SSO method can be applied to different forms of pareto front format. The adaptability of the method is good.

5. Conclusions

In the optimization scheduling of the DES, the SSO method has been proposed. By transforming a BiOP into a CSiOP, the pareto front of the BiOP is accurately depicted, and, according to the changing characteristics of the pareto front, the stepper search method is introduced to solve the problem of the difficulty of choosing among many feasible solutions in the MOP, obtaining a more reasonable and unique recommended solution and effectively balancing the contradiction between two optimization objectives. Furthermore, the SSO method has fewer parameters, and the parameters have less impact on the accuracy of the optimization results. The method is also highly stable. Additionally, when facing different bi-objective optimization problems, the SSO method can search for excellent recommended solutions and show good adaptability to bi-objective optimization problems.

At present, the limitation of this study is that the proposed method can only be applied to bi-objective optimization problems. Therefore, expanding the application of the SSO method to more optimization objective problems is the focus of future research. In addition, the diversity of pareto frontier morphology changes will be the core link of subsequent research.

Author Contributions: Conceptualization, S.M.; Methodology, S.M.; Validation, Z.M. and G.S.; Formal analysis, Z.M. and Y.C.; Writing—original draft, Z.M.; Writing—review & editing, S.M., Y.C. and G.S.; Visualization, Y.C.; Funding acquisition, G.S. All authors have read and agreed to the published version of the manuscript.

Funding: This work was supported in part by the Beijing Nova Program (Z211100002121081).

Institutional Review Board Statement: Not applicable.

Informed Consent Statement: Not applicable.

Data Availability Statement: The data used to support the findings of this study are available from the corresponding author upon request.

Conflicts of Interest: Author Yilin Cui was employed by the company Shandong Electric Power Company Haiyang Power Supply Company. The remaining authors declare that the research was conducted in the absence of any commercial or financial relationships that could be construed as a potential conflict of interest.

References

1. Li, Q.; Wang, J.; Chen, J.; Ding, T.; Gu, C. A hierarchical multi-area capacity planning model considering configuration ratios of renewable energy and energy storage systems with multi-area coordination. *IET Gener. Transm. Distrib.* **2023**, *17*, 3658–3677. [CrossRef]
2. Zeng, L.; Xu, J.; Wang, Y.; Liu, Y.; Tang, J.; Wen, M.; Chen, Z. Day-ahead interval scheduling strategy of power systems based on improved adaptive diffusion kernel density estimation. *Electr. Power Energy Syst.* **2023**, *147*, 108850. [CrossRef]

3. Zuo, H.; Teng, Y.; Cheng, S.; Sun, P.; Chen, Z. Distributed multi-energy storage cooperative optimization control method for power grid voltage stability enhancement. *Electr. Power Syst. Res.* **2023**, *216*, 109012. [CrossRef]
4. Ahmadi, B.; Ceylan, O.; Ozdemir, A.; Fotuhi-Firuzabad, M. A multi-objective framework for distributed energy resources planning and storage management. *Appl. Energy* **2022**, *314*, 118887. [CrossRef]
5. Cheng, T.; Chen, M.; Wang, Y.; Li, B.; Hassan, M.A.; Chen, T.; Xu, R. Adaptive Robust Method for Dynamic Economic Emission Dispatch Incorporating Renewable Energy and Energy Storage. *Complexity* **2018**, *2018*, 2517987. [CrossRef]
6. Di, H.; Xin, Y.; Lin, G.; Zhang, C.; Chen, H. Interval economic dispatch of multi-objective cooperative game considering wind power and energy storage connected to power grid. *Control Theory Appl.* **2021**, *38*, 1061–1070.
7. Sandgani, M.R.; Sirouspour, S. Coordinated Optimal Dispatch of Energy Storage in a Network of Grid-connected Microgrids. *IEEE Trans. Sustain. Energy* **2017**, *8*, 1166–1176. [CrossRef]
8. Wu, X.; Liao, B.; Su, Y.; Li, S. Multi-objective and multi-algorithm operation optimization of integrated energy system considering ground source energy and solar energy. *Int. J. Electr. Power Energy Syst.* **2023**, *144*, 108529. [CrossRef]
9. Liu, Z.; Xiao, Z.; Wu, Y.; Hou, H.; Xu, T.; Zhang, Q.; Xie, C. Integrated optimal dispatching strategy considering power generation and consumption interaction. *IEEE Access* **2021**, *9*, 1338–1349. [CrossRef]
10. Ma, S.; Wu, Y.; Li, J.; Hou, X.; Li, D. Optimal economic planning method of energy storage system to improve grid reliability. *Acta Energiae Solaris Sin.* **2024**, *45*, 251–262.
11. Jiang, W.; Chen, Z. Optimization of microgrid energy storage system capacity based on improved artifical bee colony algorithm. *J. Shanghai Univ. Electr. Power* **2021**, *37*, 415–421+427.
12. Mishra, A.; Arora, B.B.; Arora, A. Multi-Objective Optimization of an Inlet Air-Cooled Combined Cycle Power Plant. *J. Therm. Sci. Eng. Appl. Trans. ASME* **2023**, *15*, 071005. [CrossRef]
13. Liu, Y.; Liu, Q.; Guan, H.; Li, X.; Bi, D.; Guo, Y.; Sun, H. Optimization Strategy of Configuration and Scheduling for User-Side Energy Storage. *Electronics* **2021**, *11*, 120. [CrossRef]
14. Sun, J.; Li, Z.; Li, J.; Wu, G.; Xia, Y. Hybrid power system with adaptive adjustment of weight coefficients multi-objective model predictive control. *Int. J. Electr. Power Energy Syst.* **2023**, *153*, 109296. [CrossRef]
15. Li, L.; Liu, P.; Li, Z.; Wang, X. A multi-objective optimization approach for selection of energy storage systems. *Comput. Chem. Eng.* **2018**, *115*, 213–225. [CrossRef]
16. Terlouw, T.; Gabrielli, P.; AlSkaif, T.; Bauer, C.; McKenna, R.; Mazzotti, M. Optimal economic and environmental design of multi-energy systems. *Appl. Energy* **2023**, *347*, 121374. [CrossRef]
17. Fan, W.; Tan, Q.; Zhang, A.; Ju, L.; Wang, Y.; Yin, Z.; Li, X. A Bi-level optimization model of integrated energy system considering wind power uncertainty. *Renew. Energy* **2023**, *202*, 973–991. [CrossRef]
18. Pourghaderi, N.; Fotuhi-Firuzabad, M.; Moeini-Aghtaie, M.; Kabirifar, M.; Lehtonen, M. Optimal energy and flexibility self-scheduling of a technical virtual power plant under uncertainty: A two-stage adaptive robust approach. *IET Gener. Transm. Distrib.* **2023**, *17*, 3828–3847. [CrossRef]
19. Qiu, Y.; Li, Q.; Ai, Y.; Chen, W.; Benbouzid, M.; Liu, S.; Gao, F. Two-stage distributionally robust optimization-based coordinated scheduling of integrated energy system with electricity-hydrogen hybrid energy storage. *Prot. Control Mod. Power Syst.* **2023**, *8*, 33. [CrossRef]
20. Luo, Y.; Gao, Y.; Fan, D. Real-time demand response strategy base on price and incentive considering multi-energy in smart grid: A bi-level optimization method. *Int. J. Electr. Power Energy Syst.* **2023**, *153*, 109354. [CrossRef]
21. Zhang, Y.; Zhong, K.; Deng, W.; Cheng, R.; Chen, M.; An, Y.; Tang, Y. An optimal scheduling of renewable energy in flexible interconnected distribution networks considering extreme scenarios. *IET Renew. Power Gener.* **2023**, *17*, 2531–2541. [CrossRef]
22. Yang, H.; Wu, J.; Du, X. Thermo-economic analysis and multi-objective optimization of solar aided pumped thermal electricity storage system. *J. Energy Storage* **2023**, *70*, 107994. [CrossRef]
23. Zhao, J.; Ju, L.; Luo, W.; Zhao, J. Reactive voltage control model and method considering partitioned dynamic reactive power reserve. *Electr. Power Autom. Equip.* **2015**, *35*, 100–105.
24. Sagawa, D.; Tanaka, K. Machine Learning-Based Estimation of COP and Multi-Objective Optimization of Operation Strategy for Heat Source Considering Electricity Cost and On-Site Consumption of Renewable Energy. *Energies* **2023**, *16*, 4893. [CrossRef]
25. Zhong, Y.; Huang, M.; Ye, C. Multi-objective optimization of microgrid operation based on dynamic dispatch of battery energy storage system. *Electr. Power Autom. Equip.* **2014**, *34*, 114–121.
26. Liao, X.; Qian, B.; Jiang, Z.; Fu, B.; He, H. Integrated Energy Station Optimal Dispatching Using a Novel Many-Objective Optimization Algorithm Based on Multiple Update Strategies. *Energies* **2023**, *16*, 5216. [CrossRef]
27. Ma, S.; Wu, Y.; Li, J.; Xiong, J.; Zeng, W. Research on optimal configuration of centralized battery energy storage for multiple service objectives. *High Volt. Appar.* **2023**, *59*, 75–86.
28. Ma, G.; Li, J.; Zhang, X.P. Energy Storage Capacity Optimization for Improving the Autonomy of Grid-Connected Microgrid. *IEEE Trans. Smart Grid* **2023**, *14*, 2921–2933. [CrossRef]
29. Chen, Y.; Feng, L.; Li, X.; Zoghi, M.; Javaherdeh, K. Exergy-economic analysis and multi-objective optimization of a multi-generation system based on efficient waste heat recovery of combined wind turbine and compressed CO_2 energy storage system. *Sustain. Cities Soc.* **2023**, *96*, 104714. [CrossRef]

30. Ma, S.; Wu, Y.; Jiang, Y.; Li, Y.; Sha, G. Research on two-stage optimization control method for energy storage systems based on multi service attribute utility evaluation. *J. Energy Storage* **2024**, *46*, 3041–3060. [CrossRef]
31. Lu, M.; Guan, J.; Wu, H.; Chen, H.; Gu, W.; Wu, Y.; Ling, C.; Zhang, L. Day-ahead optimal dispatching of multi-source power system. *Renew. Energy* **2022**, *183*, 435–446. [CrossRef]

applied sciences

MDPI

Article

Zero Carbon Emissions Due to Ultra-High RES Penetration in Interconnected Island

Emmanuel Karapidakis [1], Ioannis Mozakis [1], Marios Nikologiannis [2,*] and Antonios Tsikalakis [1]

[1] Electrical and Computer Engineering Department, Hellenic Mediterranean University, 71410 Heraklion, Greece; karapidakis@hmu.gr (E.K.); ioannismozakis@gmail.com (I.M.); tsikalant@hmu.gr (A.T.)

[2] Institute of Energy, Environment and Climate Change, Hellenic Mediterranean University, 71410 Heraklion, Greece

* Correspondence: marios_nikolog@protonmail.com

Featured Application: The described methodology can be utilized to simulate and optimize microgrids with similar or desired designs, with the objective of assessing the logistical and environmental performance indicators of their operations.

Abstract: European islands have been leading the charge in renewable energy innovation. Yet, the intermittent nature of sources like solar and wind poses challenges such as grid saturation and frequency variations. Limited interconnection with mainland grids exacerbates these issues, necessitating backup from conventional power sources during low-production periods. Until 2021, Crete operated independently, but new infrastructure now integrates it with the Greek mainland grid, facilitating swift energy transfers. This integration enables surplus power from Crete's solar and wind systems to be transmitted to the mainland and vice versa. However, reliance on remote power production exposes the island to market fluctuations and distant disruptions, impacting electricity production. Storage technologies offer a solution, enhancing renewable energy penetration while reducing carbon emissions. Green hydrogen, a rising storage method, shows promise in offsetting carbon emissions. Its clean-burning nature minimizes environmental impact and reduces reliance on costly and harmful conventional sources. This study aims to evaluate the feasibility of achieving carbon-neutral electricity production in Crete, Greece, using hydrogen storage to offset annual carbon emissions in a financially viable and sustainable manner. Hydrogen's clean-burning properties reduce environmental impact and lessen dependence on expensive and environmentally harmful conventional sources. The methodology prioritizes the independence of the Cretan electricity system, utilizing electrolysis to produce green hydrogen and proton-exchange membrane (PEM) fuel cells for energy generation. It investigates the optimal expansion of renewable energy systems, including photovoltaic (PV) and wind turbine (WT) parks, alongside the installation of hydrogen storage, under specific assumptions. This proposed installation aims to achieve both island independence and profitability, requiring an additional expansion in PV capacity of 2.13 GW, WT capacity of 3.55 GW, and a hydrogen system with electrolyzer and fuel cell capacities totaling 278.83 MW each, along with a hydrogen tank capacity of 69.20 MWh. The investment entails a capital expenditure (CAPEX) of 6,516,043,003.31 EUR for a nearly zero net present value (NPV) over 20 years. However, carbon neutrality cannot be attained through this optimal solution alone, as relying solely on carbon sequestration from olive groves, the primary crop cultivated on the island, is insufficient as a carbon sink method. The annual net carbon emissions from electricity production, island transport, residential heating, and carbon sequestration are estimated at 94,772.22 tCO_2.

Keywords: renewable energy sources; zero carbon emissions; electricity storage; Cretan energy system; green hydrogen; interconnected islands

1. Introduction

The increasingly evident climatic shifts present a formidable challenge for the global scientific community. With every facet of existence affected, from weather patterns to ecosystem dynamics, thorough study and analysis are imperative for a transition toward sustainable and environmentally secure living conditions. Scientists worldwide are diligently investigating these changes, striving to comprehend their complexities and devise innovative solutions. Through collaborative efforts and interdisciplinary approaches, they aim to pave the way for a future where humanity coexists harmoniously with the planet, fostering habitats that prioritize ecological integrity and resilience.

At present, considerable attention is being directed towards exploring the vast potential of renewable energy sources (RES) across the globe. Various alternative methods of energy acquisition are being meticulously evaluated, with a focus on deploying cutting-edge technologies to maximize efficiency while minimizing environmental impact. From solar and wind to hydroelectric and geothermal power, the spectrum of renewable energy options offers a promising avenue for reducing reliance on fossil fuels and mitigating the adverse effects of climate change. Research and development efforts are increasingly geared towards refining existing technologies and pioneering new approaches to harnessing renewable energy on a large scale. By embracing these advancements and fostering a global culture of innovation, societies can aspire towards a future powered by sustainable, clean energy solutions that benefit both people and the planet.

The traditional electrical grid is experiencing substantial augmentation to ensure system dependability and enhance both energy efficacy and ecological sustainability. A conventional grid is transforming into a smart grid (SG) that includes a diversified, heterogeneous mix of operational measures such as smart appliances, meters, and renewable energy resources. The SG is a new design for the electric power grid since it produces more efficient outcomes and is more reliable. The SG is one of the most promising innovations as a solution to the energy dilemma. A major advancement involves the developments in power electronics technologies that have contributed to the development of power conversion systems used in RES and other energy applications. The benefits include an increase in reliability, resilience of power sources, and electromagnetic compatibility between different power sources applied to microgrids [1]. Nonetheless, it is intricate, and decentralizing it might be of immense advantage [2].

From the beginning of the Industrial Revolution, the amount of carbon dioxide (CO_2) equivalent in the atmosphere has approximately doubled, according to various reports [3]. As a result, this has led to a rise in world average temperature, which is having an adverse effect on the climate on a global scale [4]. Recent advancements in a variety of renewable energy technologies have improved their reliability while simultaneously minimizing their cost, making them suitable for usage in a wider variety of scenarios. Significant improvements have been achieved in both renewable energy systems and hydrogen technologies, resulting in increased efficiency and productivity [5,6]. The Mediterranean area is distinguished by its abundance of indigenous solar and wind energy resources, which have the potential to be used in the creation of power, heat, and cooling [7].

1.1. Islands Pioneers in RES Integration

In the context of European legislation and the shift to green energy technologies [8], the extensive literature has been established regarding the application of ecologically friendly technologies to island power systems. Juswanto and Ali [9] conducted a thorough analysis of recent advancements in the field of renewable energy in the Pacific islands. According to the authors, the present contribution of renewable energies such as solar and wind power to their overall energy balance is between 20 and 40 percent.

Meschede et al. [10] composed an overview of the 100% usage of renewable energy on islands. A total of 97 papers were analyzed on islands, with a focus on various regions like the Atlantic Ocean, Mediterranean islands, North and Baltic Sea, Caribbean Sea, Pacific Ocean, and Indian Ocean. The majority of papers in their study aimed to assess renewable energy systems, with a distinction between studies focusing on achieving 100% renewable energy supply and those analyzing different renewable energy systems on islands. According to the authors, Solar PV and wind technologies are the primary technologies for achieving 100% renewable energy systems on islands. On the other hand, there is a need to seriously consider area limitations on islands for all renewable energy sources based on full energy system studies and area demand. The International Renewable Energy Agency (IRENA) [11] has reported on international uses of RES in island settings. The paper included a number of case studies dealing with the use of renewable energy sources for the production of electricity on a variety of islands located all over the globe. There have been reports of the use of RES on European islands [12]. The possibility of islands fulfilling all their energy needs with renewable energy sources was discussed in the study, with specific reference to the island of Samso. It focuses on organizational and technological challenges as well as on local considerations affecting the availability and usage of RES. It offers a variety of case studies about the exploitation of RES in EU islands.

Marczinkowski et al. [13] conducted research on the topic of converting an island's energy systems to be 100% reliant on RES by the year 2030. The authors investigated how three different islands in Europe—Samso, Orkney, and Madeira—made the transition to the total use of renewable energy sources in order to achieve energy independence. Regarding the transition of the island to 100% renewable energy sources, they said that there are a number of different energy systems and solutions that may be implemented based on the local circumstances, development stages, and local availability of different RES. In addition, the author performed research [14] on the process of converting the energy grid on Madeira Island to one that is powered entirely by renewable resources. The authors used the EnergyPLAN modeling program in order to investigate and compare two distinct energy strategies. They noted that due to the absence of international trade, the island must implement a sophisticated energy system and make use of innovative technology in order to increase its level of energy self-sufficiency.

Chatziargiriou and colleagues [15] conducted research on the development of renewable energy sources on the Greek islands. According to the authors, Greece is made up of 124 inhabited islands, any number of which might serve as a prototype for the use of RES in the country's efforts to advance its economy. They also said that the implementation of RES on these islands might provide local inhabitants an extra source of income while offering the chance for the development of novel RES systems that include power storage.

A group of islands in the Nordic region has been formed, and one of its goals is to achieve carbon neutrality [16]. More than sixty (60) different technologies that are based on RES have been identified as having the potential to be implemented on these islands in order to reduce their carbon footprint or bring it down to zero.

Breyer et al. [17] explored the possibility of reaching net-zero CO_2 emissions in Europe by 2050. Building on existing 100% renewable energy models, the authors developed scenarios for achieving carbon neutrality by 2040 and 2050, using a high-resolution, multi-region energy system model covering all energy sectors. The results reveal a cost-neutral pathway toward net zero emissions by 2050, reducing electricity costs. An accelerated transition by 2040 would see a 9% increase in energy costs but a 17% improvement in CO_2 emissions reduction. Solar photovoltaics would dominate the total electricity production, becoming 61–63% of total generation in 2050. The authors suggested a 'power-to-X' economy as a key feature of Europe's energy future by transforming most chemical products into electricity so that solar power can be converted to methanol, kerosene, and hydrogen in order to be stored for seasonal inequality.

1.2. Zero Carbon Crete

The pursuit of a carbon-neutral Crete has been investigated extensively. Vourdoubas [7] has examined the prospect of constructing residential structures on the Greek island of Crete that produce zero net carbon emissions. According to the author, it is possible to meet all of the energy needs of residential structures in Crete using locally sourced renewable energy sources, which would result in such buildings having no net carbon footprint as a result of their usage of energy. Moreover, he claimed that it is possible to reach carbon neutrality with the use of solar thermal energy, solar photovoltaic energy, and high-efficiency heat pumps. Also, he investigated the use of RES in the more rural parts of the island of Crete [18]. According to the author, Crete already makes use of many different types of renewable energy, including solar energy, wind energy, hydro energy, biomass, and low enthalpy geothermal energy. Furthermore, it was noted that renewable energy sources currently contribute over 20% to the island's overall power consumption, with the anticipation of this proportion to increase in the foreseeable future.

1.3. Interconnections of Islands in Greece

Before its completion, Biza et al. [19] conducted research on the subject of the interconnection between Peloponnese and Crete. Per the authors, the connectivity would be completed in two stages. The electric grid of Crete would be coupled with the system of Peloponnese in phase 1 of the project, with a capacity for transmitting power ranging from 150 MW to 180 MW. During phase 2, the electric grid of Crete is expected to be integrated with the system of Attica, and the capacity for transmitting power will be 1000 MW [20].

Tzanes et al. [21] conducted research on the current production of power on the non-interconnected islands in the Aegean Sea, with the exception of the islands of Crete and Rhodos. According to the authors, the yearly contribution of renewable energies in the overall energy mix has remained relatively constant at roughly 15% to 18% due to the current technological limits. They also claimed that the cost of generating energy on tiny islands using thermal power plants is significant, coming in at roughly 600 EUR/MWh. They suggested the creation of electricity storage devices as a means of increasing the proportion of renewable energies that are included in the overall energy mix on these islands.

Tsekeris [22] has focused on the process of energy transition in the islands of Greece that are not connected with each other. According to what was stated in the article, customized transitional policies need to be devised. He also noted that the national objectives for the year 2030 include a decrease in CO_2 emissions of forty (40) percent and a penetration of renewable energy sources of thirty-two (32) percent in the overall energy mix. Regarding the transportation industry, it is anticipated that there will be a complete shift in the kind of fuels that are used, in addition to an increase in the number of electric cars and trains.

1.4. Islandic Electricity Storage

Bertsiou and Baltas [23] conducted research on the topic of combining renewable energy sources with the generation of hydrogen and its usage using fuel cells on a Greek island. The paper discusses the comparison between a single storage system (PHS) and a hybrid storage system (HPHS) for fulfilling energy requirements on islands, with the HPHS showing lower loss of load probability values despite higher cost, conducted on the island of Fournoi Korseon in Greece. The study emphasizes the importance of utilizing wind potential and coupling with storage systems for energy security.

Rodrigues et al. [24] conducted research on the possibility of installing energy storage systems on the island of Terceira, which is located in the Azores and belongs to Portugal. The authors claimed that the installed power on the island in 2012 was 75.1 MW, while the greatest energy demand in 2018 would be approximately 30 MW to 35 MW. This difference was due to the fact that the maximum electricity demand in 2018 would be much lower. They determined that the optimal size of an energy storage system for 2018 would be 8.1 MW, which would be produced by water turbines working in a static waterfall at 340 m with a water flow rate of 3 m^3/s. They also investigated the similarities and differences between electric batteries and hydro pump storage. The cost of a hydro pump storage system with 8.1 MW was predicted to be 15.56 million EUR, whereas the cost of batteries was 28.27 million EUR.

Nikolaou et al. [25] conducted research on the topic of the storage of surplus power generated by wind farms on the Greek island of Crete by using hydro pump storage devices. The writers discussed the "Potamon" dam, which is situated in the prefecture of Rethymno and has the potential to be utilized for the storage of extra power. They indicated that the optimal capacity of power-producing turbines should be 31 MW, and it is projected that the yearly produced energy would account for 2.0–2.2% of the island's annual consumption of electricity.

The introduction of storage methods can also assist in issues arising from RES's increased penetration levels to a microgrid. Evangelos Pompodakis et al. [26] optimized the installation of hybrid power plants through the development of the energy system of the island of Rhodes. The stochastic production of the WT installed on the island presented significant frequency fluctuations, resulting in enforced RES curtailment. The methodology proposed a WT-battery storage coupling, estimating a reduction in frequency fluctuations and a yearly cost reduction above 7.9 million EUR.

1.5. Cretan Power Grid

Electricity generation and fossil fuel consumption in Crete in 2022 has been reported by the Hellenic Electricity Distribution Network Operator (HEDNO) SA [27]. Total electricity generation in Crete during 2022 was 20,784.7 MWh, while most of it was generated in thermal power stations fueled by heating oil and diesel oil. The share of RES in electricity generation was slightly higher than 23%. According to HEnEx data [28], in 2023, the island's annual energy demand was 3.14 TWh, 1.48 TWh of which were produced using the island's conventional power plants, 701.7 GWh from RES production, while the rest were imported through the interconnection. The proposed methodology receives as input the load and means of energy demand hourly values of Crete for 2023. An alternative priority chain of load satisfaction and RES production is proposed, based on which the optimal RES expansion and hydrogen system introduction is determined for an independent, sustainable, and carbon-neutral energy system. The potential of storing power on the island is discussed, and the CO_2 emissions that are caused by the usage of fossil fuels in the production of electricity are estimated. In the following section, the requirements for net zero CO_2 emissions due to electricity use in Crete are stated.

The current operational scheme is dictated by the participation of the island's electricity producers (both RES and power plants) and providers in the HEnEx, with the interconnection facilitating the transmission of purchased quantities. This research explores an alternative model of an independent grid, demonstrating a method to increase the use of RES in energy production. This approach aims to reduce carbon emissions and fossil fuel consumption in Crete, offering a replicable model for other islands. This paper's methodology presents a comprehensive analysis of the economic and environmental aspects of an innovative project design using the most recent data from 2023. The analysis employs a methodological model that utilizes techniques to analyze and simulate large-scale real-world data, including hourly values for an entire year. This approach provides targeted insights into the real-time energy management of a microgrid, capturing the actual dynamics of both energy consumption and production throughout the year. This holistic understanding of microgrid performance allows us to coherently assess its economic and environmental dimensions, paving the way for more informed decision-making.

2. Materials and Methods

2.1. Electricity Generation of Crete

Crete is the fifth biggest island in the Mediterranean and the largest in Greece, with a population of around 620,000 people living on an area of 8303 km^2. During the tourist season, which lasts from May to October, Crete's population and demand for electricity increase. Since November 2021, Crete's energy suppliers and providers have participated in the Hellenic Energy Exchange (HEnEx). The HEnEx's platform provides the island's hourly electricity demand [28]. In Figure 1, the monthly electricity consumption for 2023 is illustrated:

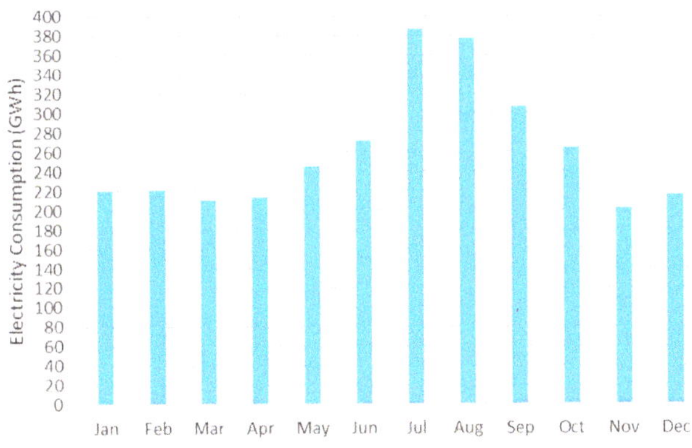

Figure 1. Crete's electricity consumption in 2023.

July and August, characterized by the peak of the tourist season, are the months with the highest energy demand. The significant fluctuations between peak and off-peak energy demand, closely tied to seasonal variations, underscore the necessity for adaptability and highlight the challenge of managing operational costs.

Crete's power system was the biggest independent electrical system in Greece until the first connectivity with the mainland grid happened in 2021, while the second is scheduled to become operational in 2024 [29].

2.2. Conventional Generation of Crete

The majority of Island's energy is generated by three (3) oil-fired thermal power plants, as seen in Figure 2. Three power plants with varied technology, fuels, and response characteristics have a combined capacity of 607 MW.

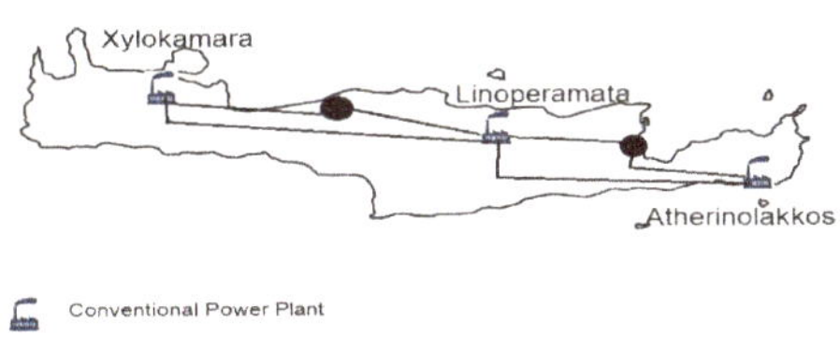

Figure 2. Conventional power plant locations in Crete.

There are, specifically, steam turbines, internal combustion engines, open-cycle gas turbines, and one combined cycle (CC) unit. The fuel analysis for each thermal oil plant is presented in Table 1. The base-load demand is primarily supplied by steam and diesel units. Gas turbines typically provide the daily peak load or the load that cannot be provided by the other units during outages [29].

Table 1. Conventional power plants.

S.L.	Power Plant	Installed Power (MW)	Fuel
1	Power Plant 1 (STs)	105	Diesel
2	Power Plant 2 (ICEs)	47.2	Diesel
3	Power Plant 3 (OCGTs)	118.5	Mazut
4	Power Plant 4 (CC)	132	Mazut
5	Power Plant 5 (STs)	100	Diesel
6	Power Plant 6 (ICEs)	100	Diesel
	TOTAL	606.7	

In Figure 3, the monthly conventional production is depicted, displaying a similar trend to that of Figure 1:

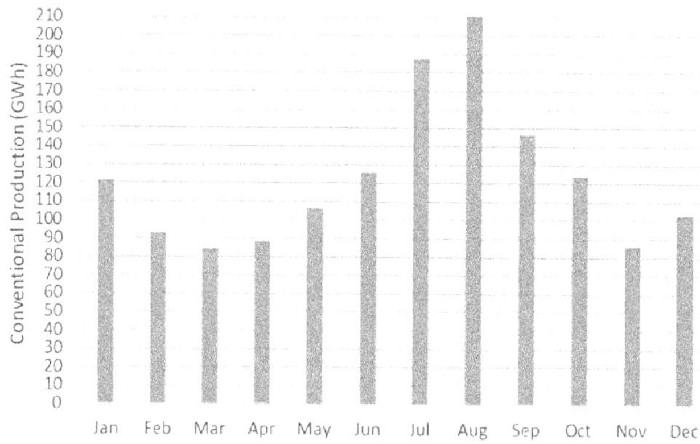

Figure 3. Total conventional production in Crete during 2023.

2.3. Productivity of Existing RES in Crete

Solar and wind speed potential of Cretan Island is one of the greatest in EU, reaching up to 2100 kWh/m^2/year solar radiation and above 8.5 m/s average wind speed in various spots on the island. The percentage of total output from RES contribution was increased for the year 2020 to reach 23.5%. The island is a single administrative region of Greece, which is divided into 4 individual prefectures, which in the context of this study are called areas 1, 2, 3, 4. As can be seen in Table 2, there are a total of 47 wind farms erected around the island, with a combined nominal power output of 215.12 MW [18].

Table 2. Wind power plant allocation.

S.L.	Prefecture of Wind Parks	Plants	Installed Power (MW)
1	Area 1	25	107.12
2	Area 2	8	45.40
3	Area 3	5	24.30
4	Area 4	9	38.30
	TOTAL	47	215.12

In addition, as indicated in Table 3, there are now 1204 small PV plants erected around the island, each with a nominal power output of 80 kW. The majority of these plants are attached to substations with voltages ranging from 0.4 kV to 20 kV, which is the standard practice worldwide [30].

Table 3. Photovoltaic power plant allocation.

S.L.	Prefecture of PV Parks	Plants	Installed Power (MW)
1	Area 1	262	19.90
2	Area 2	501	35.91
3	Area 3	241	18.26
4	Area 4	200	14.75
	TOTAL	1204	88.82

In Figure 4, the Cretan RES production for each month of 2023 is depicted as follows:

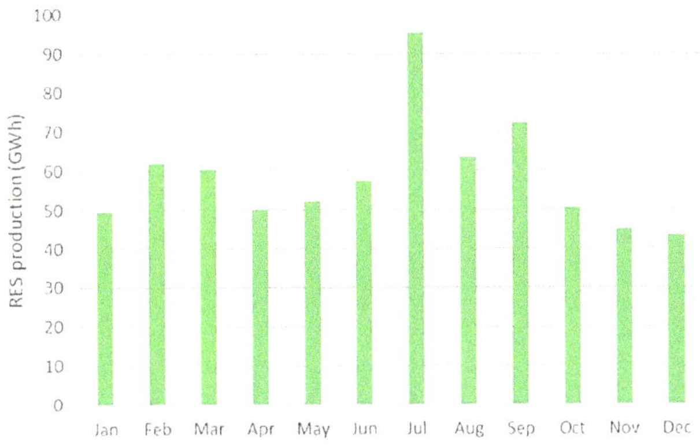

Figure 4. Total RES production in Crete during 2023.

The highest production was noted in July when the PV parks reached their peak due to prolonged sun exposure, accompanied by the production of WT parks. In addition to energy from the sun and the wind, the enormous olive tree farming that takes place on the island of Crete provides the island with a significant amount of solid biomass resources. In Crete, there are insufficient supplies of hydro energy and geothermal energy with a high enthalpy. It is anticipated that in the not-too-distant future, Crete will be host to not just one but many wind farms located offshore in addition to a solar thermal power plant. With the interconnection of Crete's electric grid, it is anticipated that the island would produce more environmentally friendly power that is derived from renewable energy sources. Tables 4 and 5 exhibit Crete's electricity production and carbon emissions from the island's different fossil fuels and energy sources, while Table 4 presents the island's installed capacities for those fuels and sources [31].

Table 4. Crete's carbon emissions and electricity production from various fossil fuel and power sources.

Fuel	Annual Fuel Consumption (Tons)	Annual Electricity Generation (MWh)	%	Annual CO_2 Emissions (Tons)	%
Fuel oil	429,565.50	1,571,182.50	56.52%	1,363,870.46	75.72%
Diesel oil	139,187.10	487,730.20	17.54%	437,325.87	24.28%
Solar Energy	-	164,647.70	5.92%	-	
Wind Energy	-	556,189.50	20.01%	-	
Hydroelectric Energy	-	257.00	0.01%	-	
Fossil Fuels Total	568,752.60	2,058,912.70	74.06%	1,801,196.33	
RES Total		721,094.20	25.94%	-	
TOTAL	568,752.60	2,780,006.90	100%	1,801,196,33	100%

Table 5. Crete's installed power plant capacity and the efficiency of its solar photovoltaic and wind power plants.

Energy Source Technology	Installed Capacity	Annual Electricity Generation (MWh)	Annual Productivity/Capacitor Factor
Solar Photovolatic	88.82 MWp	164,647.70	Annual productivity = 1853 MWh/MWp
Wind plants	215.12 MWel	556,189.50	Capacity factor = 0.291
Hydro plants	0.6 MWel	257.00	
RES Total	304.54 MWel	721,094.20	
Thermo-electric power	606.7 MWel	2,058,912.70	
TOTAL	1121.1 MWel	2,780,006.90	

Based on the information that is now available on the installed power and the output of energy for 2018, which is provided in Table 2, an estimate was made regarding the productivity of the solar-PV systems and the wind parks that are currently operating in Crete. Comparatively, the capacity factor of wind farms is 0.291, whereas the yearly productivity of solar photovoltaic systems is 1412 MWh/MWp.

2.4. Cretan Electric Grid Interconnection

The Greek government's priority is focused on connecting the unconnected islands in order to reduce energy production costs. Notwithstanding the project's technical problems, it was intended, in earlier years, to link the island's electric grid to the mainland's electric grid via two underwater cables. One cable can carry 150–180 MW, while the other can transport 350 MW. The installation of the smaller cable was completed by the end of 2020, whilst the second was anticipated to be completed by 2023. The net position of 2023 imports–exports is shown in Figure 5:

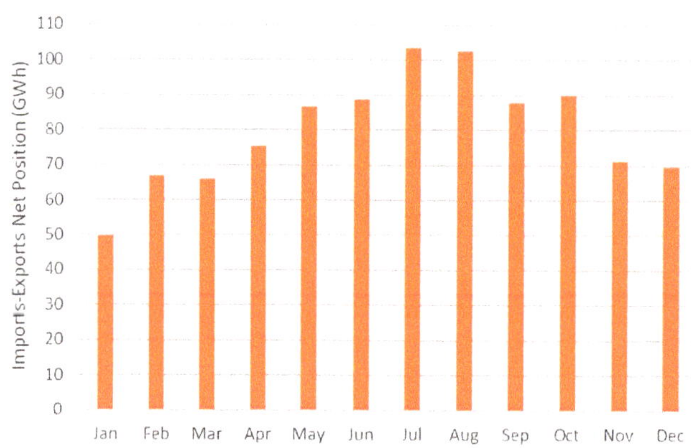

Figure 5. Imports–exports net position in Crete during 2023.

The interconnection is utilized most extensively during the summer season, as demand surges due to sudden increases in population and commercial traffic. In the past, when Crete's electric grid operated on its own, there were restrictions placed on the maximum installed power of intermittent renewable energy systems such as solar-PV and wind-generated electricity. These restrictions were implemented for reasons related to maintaining the stability of the grid. Despite the fact that previous research has shown both their technical and economic viability, hydro pump storage systems for the purpose of power storage have not yet been developed on the island of Crete. With the aforementioned limitations, the contribution of renewable energies to the total mix of local energy sources is now slightly greater than 23%. Due to the operation of undersized and inefficient thermal power facilities, the price of energy in Crete is at present costly, accompanied by high greenhouse emissions. Crete has an abundance of sun and wind, and the technology to turn those resources into power is established and cost-effective. Despite investor enthusiasm for implementing these projects, the island has attracted energy investment in solar-PV systems and wind farms that are much lower than its capacity. Wind farms and solar energy production systems are anticipated to replace the polluting and inefficient thermal power facilities in Crete.

2.5. Crete's Electricity Storage

There are various electrical energy storage technologies, including electrochemical batteries, hydrogen-based storage, and hydropower-pumped storage (HPS) systems. While current developments in their technology allow for the use of huge capacity batteries capable of storing several MWh, electric batteries only have the ability to store relatively tiny amounts of power. It is anticipated that in the not-too-distant future, power will be stored in massive batteries once their prices become competitive. In addition, energy has the capability of being stored in the batteries of electric cars, which is a feature that promotes V2G solutions. Pumped hydro storage systems are more advantageous for the scale of the electric grid that serves Crete, a result of the geomorphology of the island, meaning that various locations may be identified and utilized for the building of energy storage systems. By taking into consideration both technical and economic factors, pumped hydro storage systems are now the greatest alternative for power storage in Crete. Moreover, the deployment of these systems would strengthen the energy security of the island [32]. Because of both economic and technological considerations, the use of H_2 for the storage of power is not yet advised. Yet, it continues to be an appealing choice for Crete's economic future, particularly in light of the impending requirement of the switch to a hydrogen-based economy.

2.6. Crete's Carbon Emissions via Electricity Generation

It is anticipated that fossil fuel power plants will release 1,801,196 metric tons of carbon dioxide into the atmosphere in 2021 (Table 4). Taking into consideration the fact that there are a total of 620,000 permanent residents on the island, the amount of CO_2 emissions that are caused by the production of electricity has been determined to be 2.9 tn CO_2 per capita. According to research conducted by the World Bank [31], Greece's CO_2 emissions in 2019 were 5.6 tn CO_2 per capita. As a result, the CO_2 emissions that are caused by the production of electricity in Crete account for more than 50% of the total CO_2 emissions that are produced on the island.

2.7. Crete's Carbon Emissions Due to Electricity Generation

The interconnection of the electric networks of Crete and mainland Greece enables power to be transferred through the two underwater cables. Since low-cost energy can be transmitted to the island, the electricity generation cost in Crete using the existing thermal power plants is significantly reduced. The electric lines enable power production from the plentiful RES in Crete, ideally solar and wind energy, contributing to the decarbonization of power generation. When there is a surplus of environmentally friendly power produced in Crete, it will be sent to the Greek mainland. Conversely, in scenarios where electricity generation on Crete falls short of demand, electrical power will be transferred via the subsea cables. As a result, the interconnection of the island's several electrical grids makes the decarbonization of the island a lot easier. To achieve net zero carbon emissions from power production in Crete, the following requirements must be met:

Limiting the reliance on fossil fuels in the island's electricity production is crucial, with a simultaneous push for the promotion of renewable energy sources. While fossil fuel-generated electricity from mainland Greece will be transmitted to Crete through the subsea interconnector, an annual equivalent of "green electricity" generated by Crete's renewable sources should be transmitted back. Any excess carbon emissions that are the result of the use of fossil fuels for the production of electricity in Crete must be offset yearly by the carbon sequestration that ought to come from tree plantations that are required to be grown in Crete.

Renewable energy technologies that are currently used or could be used in the future for electricity generation in Crete are presented in Table 6, including utilization of solar energy, wind energy, hydroelectric energy, and biomass.

Table 6. Sustainable Energy Technologies can be applied in Crete for electricity generation.

Energy Source	Technology	Current Use in Crete
Solar energy	Photovoltaic cells	Yes
Solar energy	Solar thermal power	Planned
Wind power (on-shore)	Wind turbines	Yes
Wind power (off-shore)	Wind turbines	Planned
Biogas	Burning	Yes
Solid Biomass	Burning	No
Hydro energy	Hydro turbines	Yes
Electrolytic hydrogen	Fuel cells	No

2.8. Hydrogen Production

The overall energy efficiency of a hydrogen production system, whether connected or not, is closely linked to the performance of its key components, namely the electrolyzers and hydrogen storage tanks, which fundamentally impact the final quantity of produced hydrogen as well as the production cost [33]. Su et al. demonstrate the close relationship between electrolyzer performance and the energy efficiency of the hydrogen production process [34]. Improving electrolyzer efficiency leads to a reduction in energy losses and an increase in the final hydrogen output, offering significant energy savings.

On the other hand, Sharma et al.'s study [35] emphasizes the importance of enhancing hydrogen storage tanks in reducing production costs. Enhancing the efficiency of these tanks can minimize energy losses and improve the economic efficiency of the hydrogen production process. Hence, the combined improvement of electrolyzers and hydrogen storage tanks holds paramount importance for the integrated energy and economic performance of the hydrogen production system.

The typical performance in hydrogen production and storage varies depending on the technologies used and operational conditions. Electrolyzers, the primary subsystems for hydrogen production, often exhibit efficiencies ranging between 60% and 80% [36]. This percentage is a crucial parameter for process efficiency, as higher efficiencies lead to reduced energy losses and increased hydrogen output.

However, hydrogen storage tanks also impact the overall system efficiency. Typically, the performance of these tanks hovers around 60%, resulting in losses during hydrogen storage [37]. Nevertheless, technological advancements in recent years have allowed improvements in these figures, potentially increasing the efficiency of the storage tanks.

2.9. Methodology

Minimizing CO_2 emissions on an electrically interconnected island represents an integral and challenging campaign to enhance the sustainability and efficiency of the energy system. With the coexistence of conventional power plants and renewable energy sources on the island, flexibility and optimal use of the energy produced is vital. An innovative approach involves the use of a hydrogen tank to store excess energy produced from RES sources. Additionally, the sustainability and independence of the island's grid are also pursued, utilizing the interconnection only for power export to the mainland for financial gain. Therefore, the methodology's pursuits are the achievement of the following grid properties:

- Independence;
- Sustainability;
- Carbon neutrality.

Hydrogen storage is a central element in the mechanism for storing the energy produced, allowing it to be converted into electricity during periods of peak demand. This enhances the reliability of the energy grid, allowing excess energy to be sold either as electricity to other regions through the interconnection or as compressed hydrogen gas for industrial use.

The combination of conventional and renewable energy sources with the use of hydrogen storage allows for flexible and more efficient management of energy demand and supply. With the help of algorithms and computing systems solutions, it is feasible to determine the optimal combination of RES, conventional production units, and interconnection usage in collaboration with hydrogen production and storage in order to achieve a minimum environmental impact from CO_2 emissions.

- Creating a precise mathematical model to compute the parameters outlined in this study is intricate and heavily reliant on the unique conditions and attributes of the energy grid of the specific island under consideration. Based on aforementioned data, a simplified model has been devised to estimate the optimal performance of a hydrogen energy storage system, which follows these steps:
- Simulation of Crete's network as proposed.
- Calculation of annual values based on key parameters.
- Estimation of environmental and financial metrics based on real data.
- Development of an optimization model to reach the desired goals under certain constraints.

The power dispatch strategy prioritizes grid stability and minimizes reliance on conventional fuels by employing a tiered approach. The priority chain commences by activating only the minimum essential load coverage from conventional power plants, ensuring stable operation while conserving conventional fuel resources. Subsequently, available energy from locally produced RES is directly utilized to meet demand, maximizing the use of clean and sustainable energy. If real-time RES production proves insufficient, the strategy incorporates stored energy, potentially captured from prior periods of high-RES production through electrolysis. This stored energy can be converted for use through fuel cells. In scenarios where demand surpasses both real-time RES production and stored energy capacity, conventional generation is proportionally increased up to their technical limits or until the hourly load is satisfied. As a final resort, if all previous stages are inadequate, a controlled import of energy from the external grid can be implemented, though this option is used sparingly due to potential dependence on external energy sources. The specified priority chain is summarized in Figure 6:

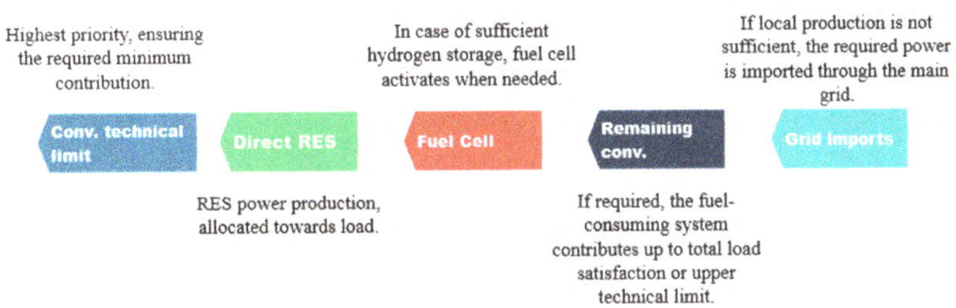

Figure 6. Priority axis of load satisfaction.

The proposed approach for managing excess RES production exceeding the total hourly load and minimum conventional unit operation requirements employs a multi-step strategy. Firstly, the system prioritizes conversion of any surplus into hydrogen for storage in available tanks through electrolysis. The electrolysis process efficiency is estimated to be around 80% [37]. This approach capitalizes on hydrogen as a viable storage solution, enabling the use of excess RES during low-demand periods. Following hydrogen storage prioritization, any remaining excess beyond the technical minimum and storage capacity is exported to the main grid. This export is limited to a pre-determined value. The export of surplus electric energy through the interconnector is subject to a pre-established nominal value. This value is contingent upon either the interconnector's maximum load capacity or negotiated power purchase agreements with the relevant electricity provider.

Finally, if a surplus persists after storage and export, the strategy suggests the additional option of hydrogen production for sale, primarily targeting the industrial sector up to the full capacity of the electrolysis system. Any remaining excess production beyond this point is, thus, rejected by the system. The specified process is summarized in Figure 7:

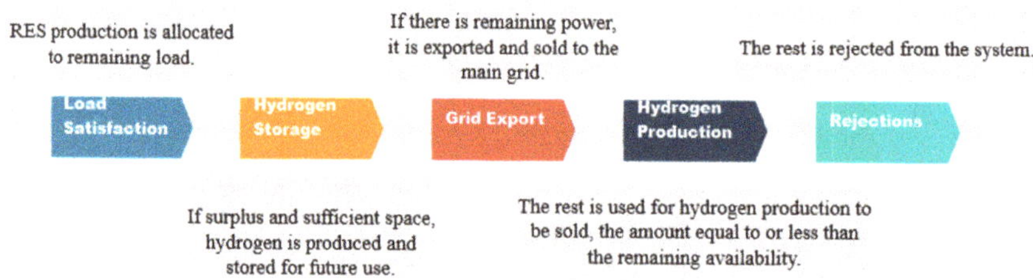

Figure 7. RES production allocated following the proposed priority.

On an hourly basis, the simulation calculates the hourly RES production. This includes the RES production recorded for each hour T of the year 2023, CRETE RES[T] (sourced from HEnEx data) and the additional production from the RES system expansion. The production from the expansion is determined by multiplying the installed capacities of PV and WT systems (denoted as PV_{cap} and WT_{cap}) by the normalized output of a photovoltaic and wind park located in Crete, using Equation (1):

$$RES[T] = CRETE\ RES[T] + PV_{cap} \times PV_{norm} + WT_{cap} \times WT_{norm} \quad (1)$$

Direct RES refers to the renewable energy production that directly meets demand, as calculated in Figure 8.

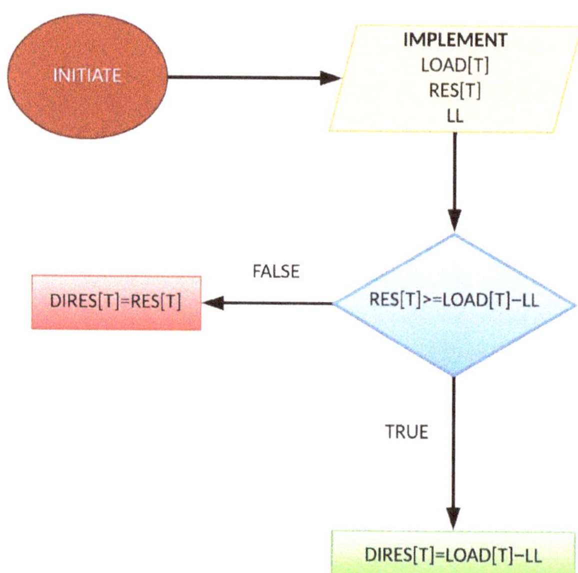

Figure 8. Computation of RES production successfully allocated for demand satisfaction.

The determination of the hydrogen storage system's state of charge is delineated across two figures: Figure 9, illustrating the charging instances, and Figure 10, depicting the discharging instances. The hydrogen system discharges hydrogen to contribute to load through the fuel cell system until satisfaction or total release. The tank is filled with green hydrogen produced through electrolysis by an amount attributed to the surplus of RES energy production or until full capacity.

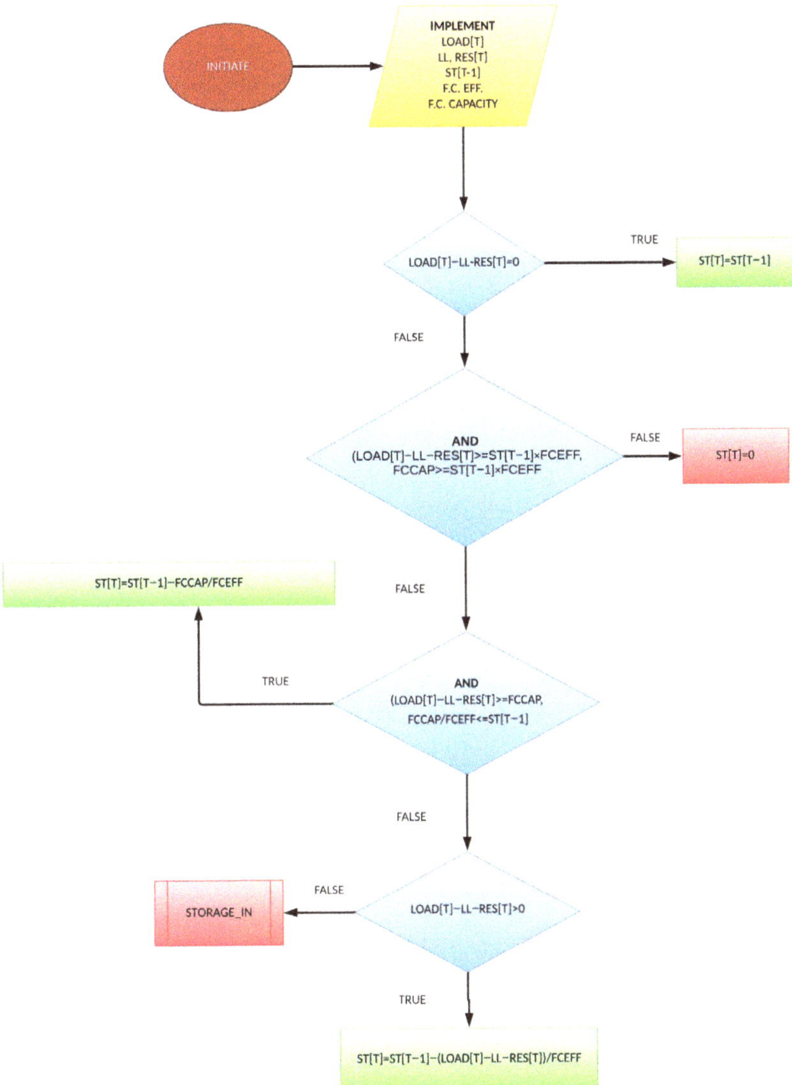

Figure 9. Hydrogen storage's state of charge in case of required discharge.

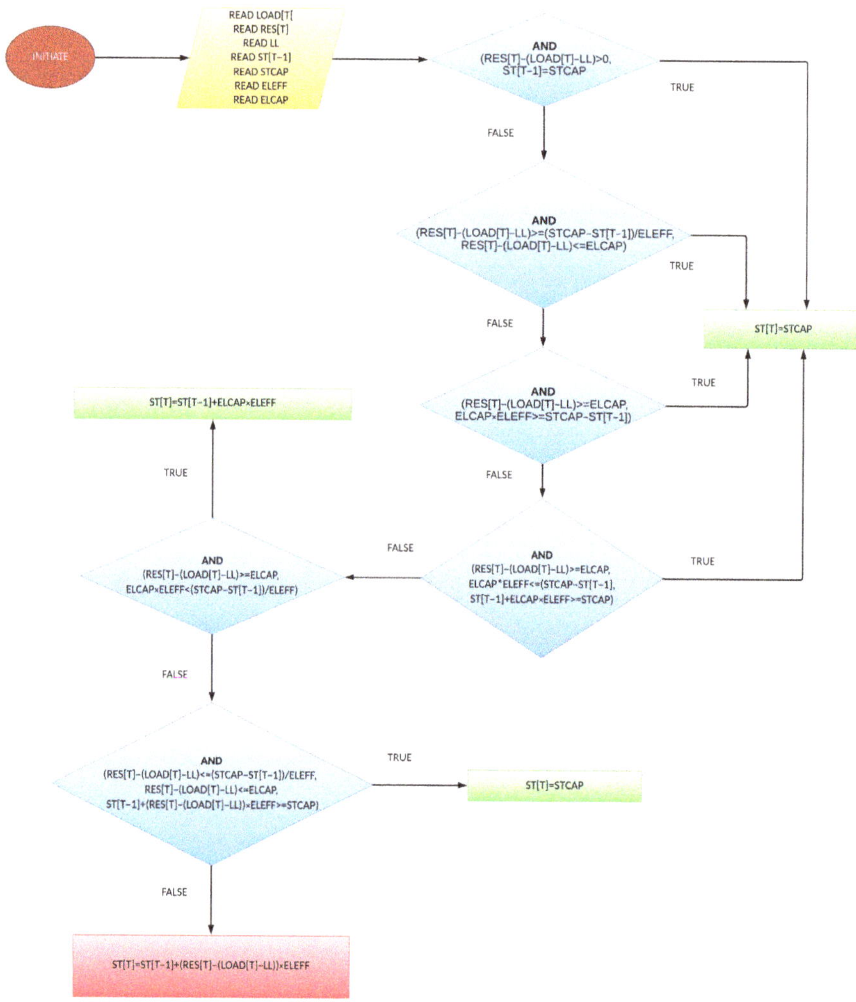

Figure 10. Hydrogen storage's state of charge in case of electrolysis process after load satisfaction.

If hydrogen storage is ample, Figure 11 computes the contribution of the fuel cell system to the hourly demand. The fuel cell system's priority is higher than the power plants', which is so in order to decrease the carbon footprint associated with the emissions originating from the plants' fuel consumption. The contribution of the fuel cell system is countified through the monitoring of the hydrogen tank's state of charge for each hour.

Figure 12 calculates the amount of surplus energy effectively stored in the hydrogen tank for each hour.

In order to estimate the current annual CO_2 emissions derived from diesel and mazut, the data were used in [38] and presented in Table 7. Since the data concern the year 2021, the assumption being made is that the energy mix has not deviated significantly in the span of 2 years.

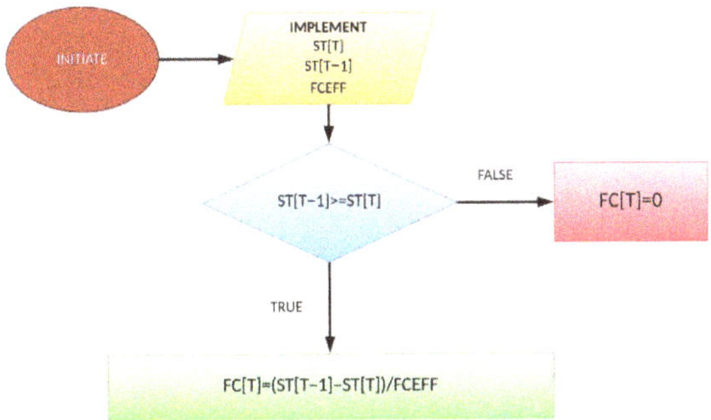

Figure 11. Calculation of hourly fuel cell system's contribution.

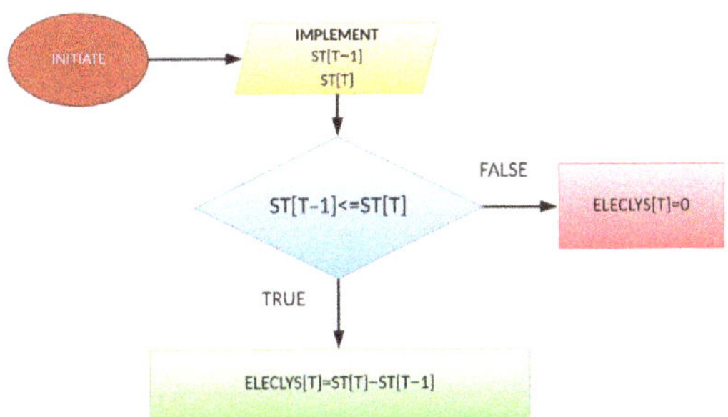

Figure 12. Hourly green hydrogen production through successfully stored electrolysis.

Table 7. Mixed energy sources of Crete in 2021.

Energy Source	Technology
RES	24.1%
Interconnection	11.0%
Diesel	17.3%
Mazut	47.6%

Using Table 7, the shares of the two fuels of the total conventional production are estimated. This is achieved by estimating the efficiency of the two fuels, which is calculated using Table 8 and [39]. Similarly to Table 7, the contents of Table 8, which concern the year 2020, are assumed to remain relatively the same in 2023.

Table 8. Electricity generation and consumption for the two fuels for the Cretan system in 2020.

Fuel Type	Production (MWh)	Consumption
Diesel	622,574.2	193,545.6 (klt)
Mazut	1,503,131.8	390,405.8 (m. ton.)

Figure 13 computes the combined hourly contribution of the diesel and Mazut power plants for each hour of the simulation.

To estimate the annual carbon emissions generated during the simulation, the emission factors of diesel and Mazut were used ([40,41]) to calculate the annual CO_2 production, as presented in Table 9.

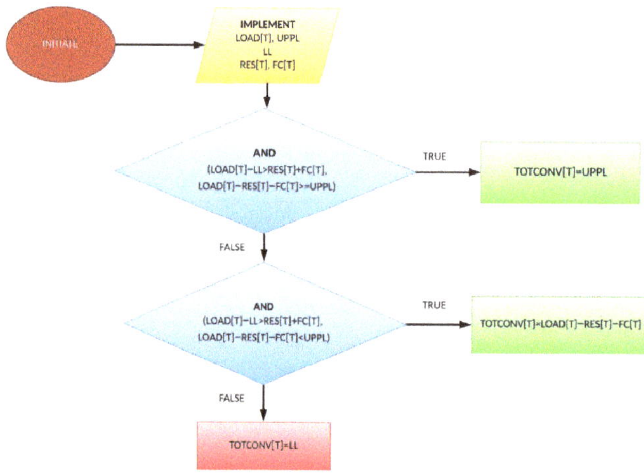

Figure 13. Power plant system's total hourly contribution.

Table 9. Emission factors for the two fuels.

Fuel Type	Emission Factor (tCO_2/MWh)
Diesel	0.306
Mazut	0.279

By referring to Figure 13 and Tables 7 and 9, the annual emissions can be calculated as the percentage of each fuel's contribution to the total power plant production multiplied by their respective emission factor under the assumption of identical fuel participation in 2023. Equation (2) represents this relationship.

$$\text{Emissions} = \sum_{\text{i}=1}^{8760} (0.08157 + 0.37988) \times \text{TOTAL CONV}[i] \tag{2}$$

The system's emissions are linearly correlated to the annual energy produced by the conventional system, which can be decreased down to the total sum of the hourly lower limit, a crucial aspect to be commented on in the appropriate section.

The calculation of the energy import needed to fulfill the remaining demand is depicted in Figure 14. To ensure the simulation's hourly power balance, the imported energy is equal to the amount needed if all other measures do not suffice. The imports are attributed with the least priority, opting for their removal after the optimization process.

In contrast to imports, exported energy is constrained by the capacity of the interconnection, which imposes an upper limit. The constraint is especially useful in case of extended renewable energy production, particularly from PV and WT systems, that could substantially increase during periods of intense sunlight and wind. Such limitation would turn out to be crucial, as surplus energy can be directed towards hydrogen production or rejected energy, which cannot be utilized by the system. From a financial perspective, this may also be advantageous, especially if revenue from selling green hydrogen exceeds that from exporting electricity via the interconnection, which often fluctuates in price according to the country's energy exchange.

The calculation of surplus energy sold to the main grid is illustrated in both Figures 15 and 16. The algorithm has been divided into two images to improve readability.

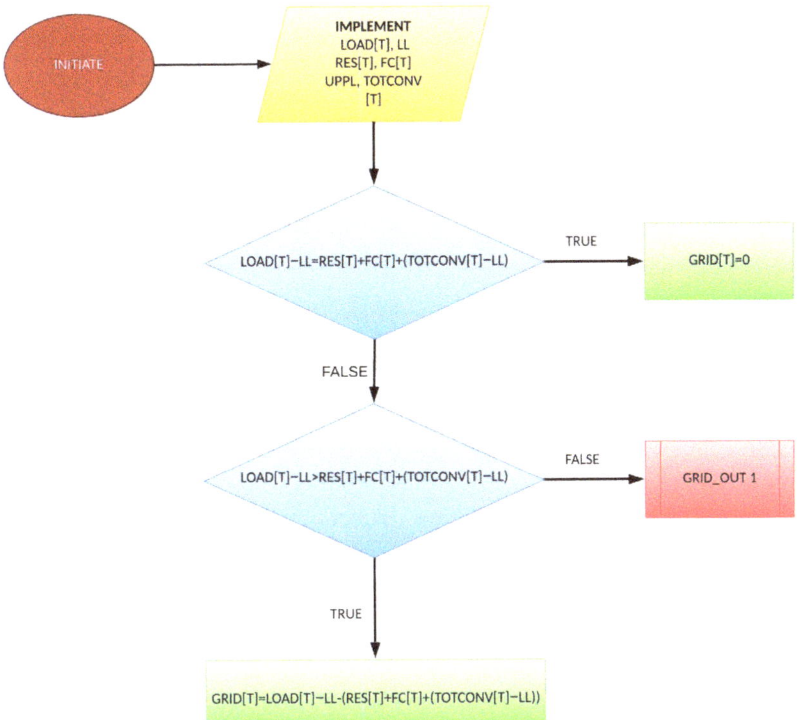

Figure 14. Imports from the Peloponnese–Crete interconnection.

If the electrolyzed energy has not reached its nominal value, the remaining energy is utilized for hydrogen production intended for industrial purposes, a crucial part of the business scheme, as it is the main contributor to the system's financial gain. The quantity is determined as illustrated in Figure 17.

The calculation of the mass of hydrogen produced (in kilograms) requires the use of the lower heating value of hydrogen (LHV H_2) set at 33.33 kWh/kg. The resultant mass of hydrogen to be compressed for sale is then determined using Equation (3):

$$\text{Hydrogen}_{\text{mass}}[\text{T}] = \frac{\text{IND}\cdot\text{H2}[\text{T}]}{\text{LHV H}_2} \tag{3}$$

Losses are attributed solely to the hydrogen system. Losses associated with electricity production and transmission from RES and the conventional system are considered negligible due to their minimal impact on the overall efficiency. Additionally, the transmission losses through the interconnection are regarded as practically nonexistent, given the high efficiency involved in the interconnection infrastructure. This assumption simplifies the analysis by focusing on the most significant source of energy loss, the hydrogen system.

Figure 18 provides a quantification of the losses linked to the operation of the electrolyzer and/or fuel cell system.

The limited hydrogen storage, export capacity of the interconnection, and maximum hourly capacity of the electrolyzer system imply the possibility of rejection of RES production in case of surplus. The annual rejections of the system are considered an important metric of the design's efficiency. High rates of rejections imply RES installation that neither contributes to the island's electricity demand nor is exported through the interconnection, therefore failing to contribute financially. Combined with their high cost of installation and maintenance, they could potentially burden the expenses of the investment needlessly.

The flowchart in Figure 19 details the methodology used to estimate the system's rejected power per operating hour, ensuring power balance for all hours of the simulation:

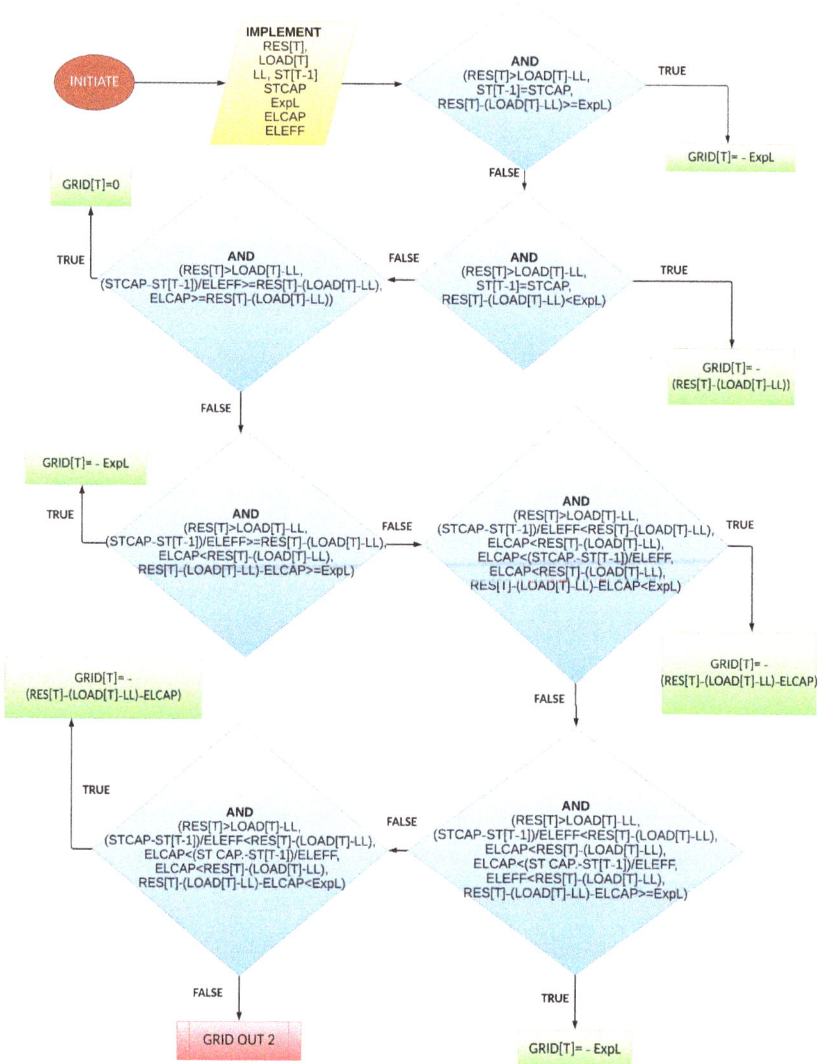

Figure 15. Power exported to the mainland through the interconnection (1st half).

After computing the hourly values to simulate the system's operation, the sustainability metrics utilized encompass the assessment of capital expenditure and net present value. This evaluation leverages the prices associated with the necessary components for investment and consumable costs, as detailed in Table 10. Using as reference [42,43], Table 10 lists inverters necessary for converting DC current generated by PV, WT, and the fuel cell system into AC current, as well as a compressor for compressing electrolysis-produced green hydrogen and storing it in suitable hydrogen tanks.

Table 10. Investment-related costs.

Component/Consumable	Price
Hydrogen Price:	3.50 EUR/kg
Hydrogen Tank Cost:	6 EUR/kWh
PV Cost:	600 EUR/kW
WT Cost:	1000 EUR/kW
Electrolyzer Cost:	1500 EUR/kW
Fuel Cell Cost:	1500 EUR/kW
Compressor Cost:	50 EUR/kW
Inverter Cost:	100 EUR/kW
Diesel Price:	2.00 EUR/lt
Mazut price:	595.35 EUR/lt
RES O&M cost rate:	2%
Total O&M cost rate:	5%
Annual Discount Rate:	3%

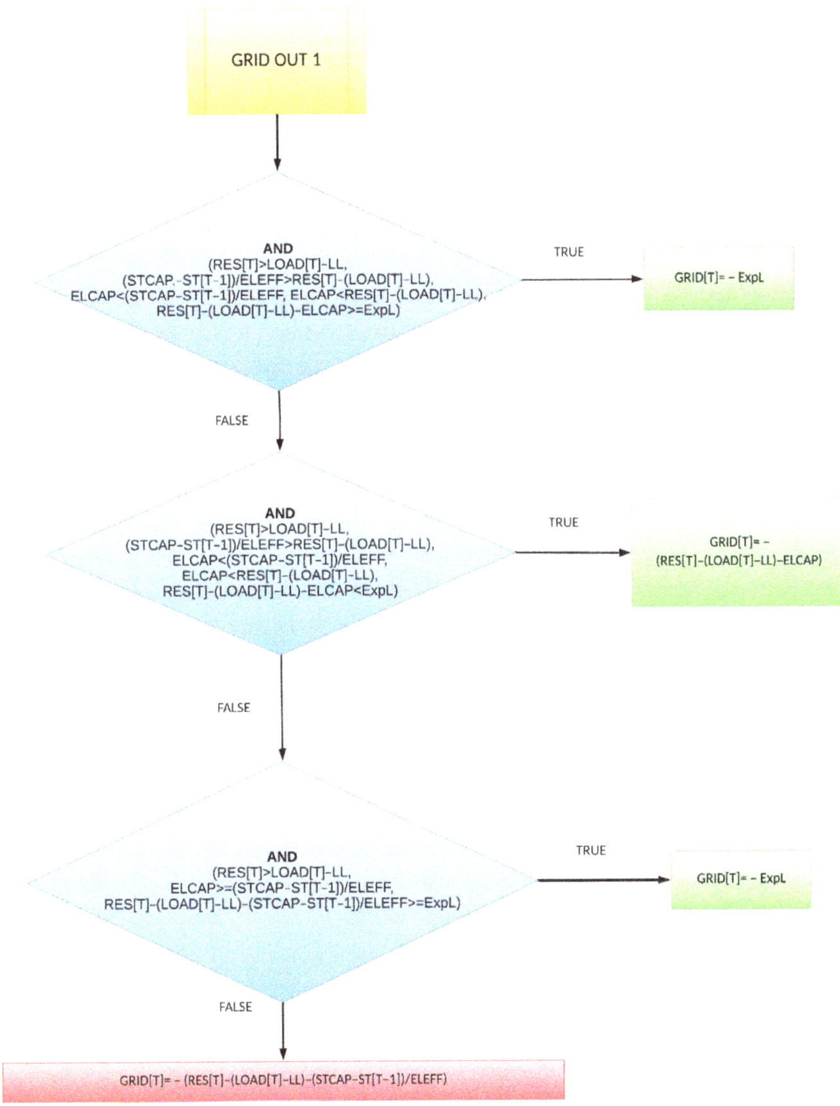

Figure 16. Power exported to the mainland through the interconnection (2nd half).

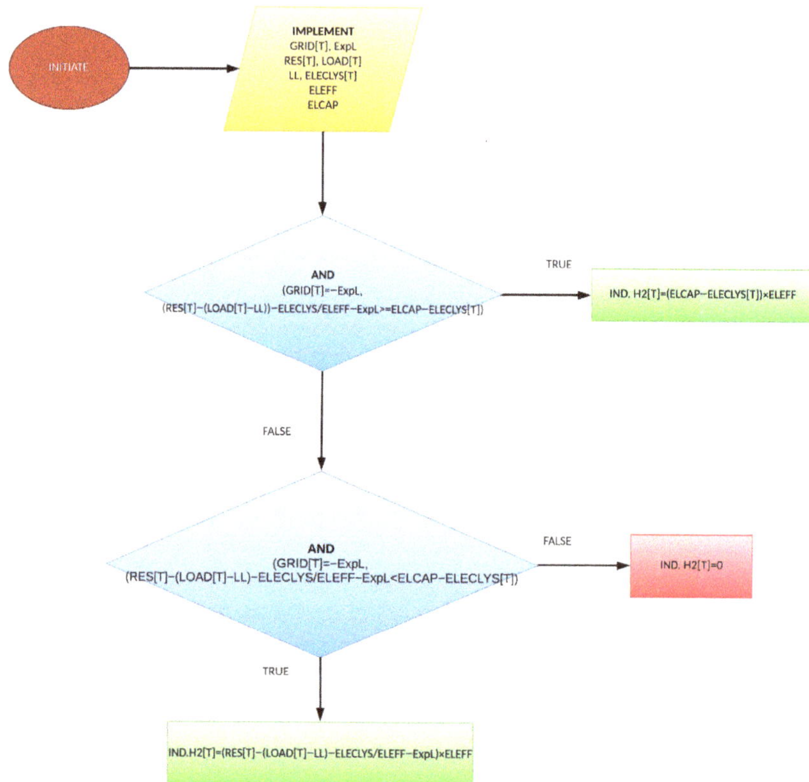

Figure 17. Hydrogen produced and sold for industrial purposes, calculated in energy units.

The interconnection price is determined to be equivalent to the HEnEx Market Clearing Price (MCP) published in the HEnEx platform for each hour of the year, denoted in EUR/MWh, to which Crete, like all market participants, is subject.

The expected lifespans of the fuel cell and electrolyzer system are estimated to be 10 years, whereas those of PV and WT are projected to be 20 years. Using Table 10, the NPV is calculated using Equation (4):

$$\mathrm{NPV} = \sum_{i=1}^{t} \frac{\mathrm{Cashflows(i)}}{(1+\mathrm{r})^{i}} - \mathrm{CAPEX} \tag{4}$$

The CAPEX comprises the costs associated with the PV and WT installation, coupled with inverters and the hydrogen system, which includes the electrolysis system, fuel cell system, hydrogen tank, inverter connected to the fuel cell, and compressor.

Cash inflows comprise revenue generated from exported energy, calculated by multiplying the exported energy amount by the interconnection price, along with income from hydrogen sales, determined by multiplying the quantity of hydrogen sold (in metric tons) by its price. Conversely, cash outflows encompass expenses such as grid import costs, fuel expenses for conventional system operation, total operational and maintenance costs of the RES system (estimated as an annual percentage of the installation cost), as well as operational and maintenance (O&M) costs of the hydrogen system.

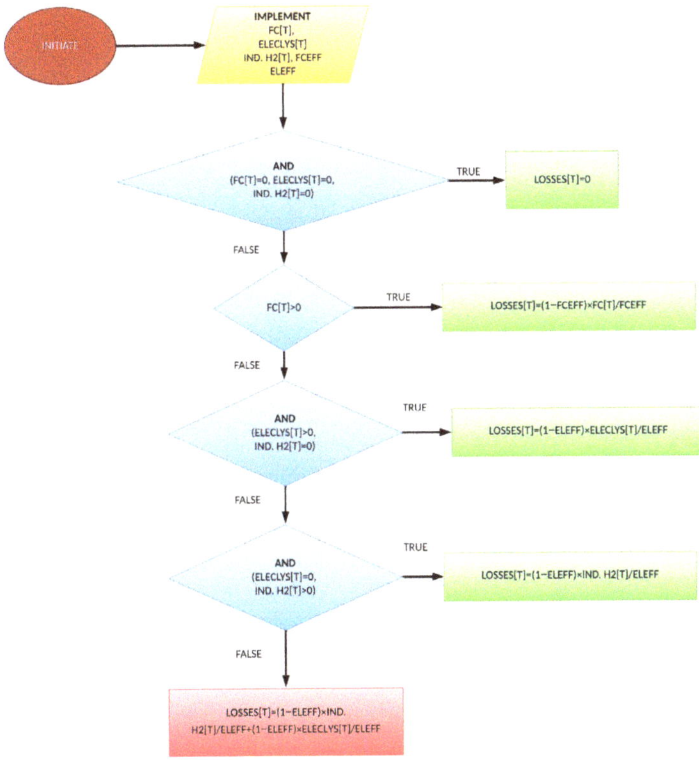

Figure 18. Hourly energy system's losses attributed to electrolysis or fuel cell use.

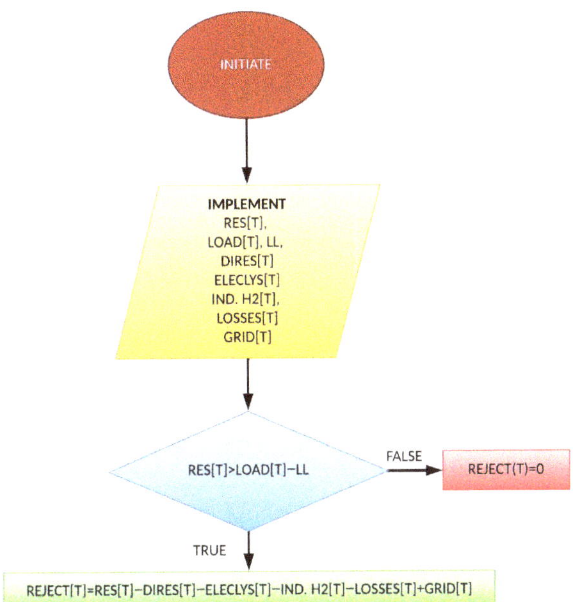

Figure 19. Total RES production rejected from the system after each hour.

The carbon neutrality metrics are calculated through the estimation of carbon emissions produced by energy production, transport and residential heating, and carbon absorption through the island's primary cultivation. Along with the system's emissions, the island absorbs the carbon emissions at a certain rate through the natural process of carbon sequestration. According to [44], olive tree is the dominant cultivated tree in Crete, occupying an area of 142,900 ha. The same reference presents the annual absorption of CO_2 from olive groves and annual emission for heating purposes in residences, while [45] presents the annual generation of CO_2 for transportation within the island's road grid. The data are compiled in Table 11.

Table 11. Annual carbon emission and absorption in Crete for 2018.

Source of Emission/Sink	Annual Carbon Balance (tCO_2)
Existing olive groves in Crete	−1,310,393
Use of heating oil for heating buildings in Crete	95,619
Transport	1,030,188

Table 11 outlines the overall emissions attributed to electricity production for achieving net zero carbon in Crete. In the proposed simulation without hydrogen storage, emissions are estimated to amount to 920,138.9 tCO_2. The objective is to align the emissions from all three activities (electricity generation, residential heating, and transportation) with the annual absorption rate from olive groves.

To achieve that goal, an optimization model is formulated, using as the model's decision variables the PV installed capacity (PV_{cap}), WT capacity (WT_{cap}) to be installed in order to increase the annual RES production, the capacity of the electrolyzer system (ELCAP), and the storage capacity (in energy units) of the hydrogen tank (STCAP), serving as storage method for the compressed hydrogen gas stemming from electrolysis. The installed capacity of the fuel cell is selected to be equal to the capacity of the electrolyzer.

The objective is to decrease CO_2 emissions from the energy system to match the disparity between emissions from heating and transportation and carbon absorption. Consequently, the objective function represents the annual CO_2 emissions generated by electricity production, computed using Equation (2). The constraints encompass the positivity of decision variables and zero NPV to determine the CAPEX required for the long-term sustainability of the system. In essence, the optimization problem is structured as follows:

$$\text{min. Emissions}(PV_{cap}, WT_{cap}, \text{ELCAP}, \text{STCAP}) \tag{5}$$

Subject to the constraints:

$$PV_{cap}, WT_{cap}, \text{ELCAP}, \text{STCAP} \geq 0 \tag{6}$$

$$\text{NPV} = 0 \tag{7}$$

The emissions of the energy system, determined by Equation (2), rely on the values of TOTAL CONV[T]. The conditional computation of these values introduces a non-linear attribute to the objective function.

The simulation is conducted within a specific scenario where the technical lower and upper limits of total conventional production are set to the minimum and maximum values observed in the time series depicted in Figure 3. Additionally, the maximum hourly power import or export through the interconnection is set at 150 MW, and the initial storage of the tank equals its capacity. The chosen fuel cell technology is the PEM fuel cell, with an overall efficiency of 60% and electrolysis efficiency of 80%.

3. Results

Prior to commencing the optimization process, the initial solution values are outlined in Table 12.

Table 12. Initial solution.

Variable	Initial Value
PV_{cap}:	6 MW
WT_{cap}:	7 MW
ELCAP:	5 MW
STCAP:	8 MWh

The implementation of the methodology produces the values of the decision variables and the system's metrics, as illustrated in Table 13.

Table 13. Optimal solution and results.

Variable	Value
PV_{cap}:	2.13 GW
WT_{cap}:	3.55 GW
ELCAP:	278.83 MW
F.C. CAP.:	278.83 MW
STCAP:	69.20 MWh
CAPEX:	6,516,043,003.31 EUR
NPV:	0.49 EUR
Net carbon emissions:	94,772.22 tCO_2

The optimal energy system's annual emissions are estimated at 279,358.2 tCO_2. For the specific solution, the island's energy mix is illustrated in Figure 20:

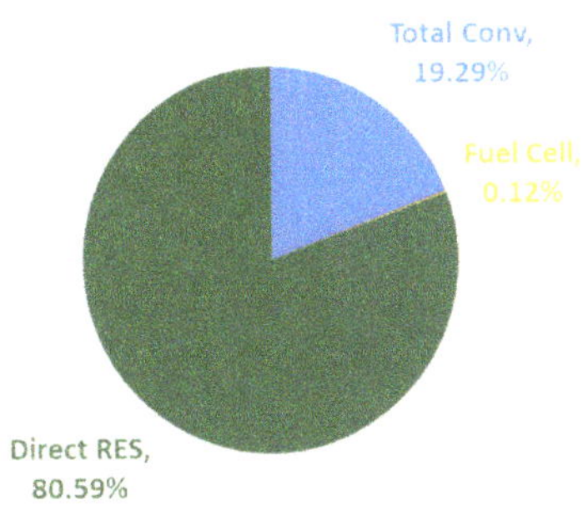

Figure 20. Optimal energy mix of Crete.

The methodology has effectively eradicated imports through the interconnection, rendering the island self-sufficient. The extensive expansion of RES ensures that the majority of the load is met through Direct RES production, with conventional sources minimized to an average hourly contribution equal to 69.1 MW, slightly above their hourly lower technical limits. The utilization of fuel cells is rare, occurring only for a few hours annually when RES production proves insufficient due to weather conditions.

On a monthly basis, Figure 21 illustrates the energy mix for the load time series of 2023.

The main contribution stems from Direct RES generation, while total conventional production operates mostly at its lower technical limit. This implies that additional conventional reduction would necessitate the shutdown of power plants, facilitating further penetration of RES and leading to a subsequent reduction in CO_2 emissions.

The annual RES production is shown in Figure 22 as percentage shares:

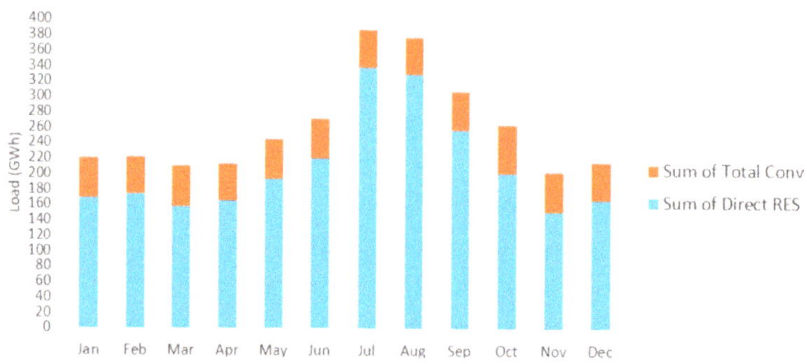

Figure 21. Optimal monthly energy mix for 2023.

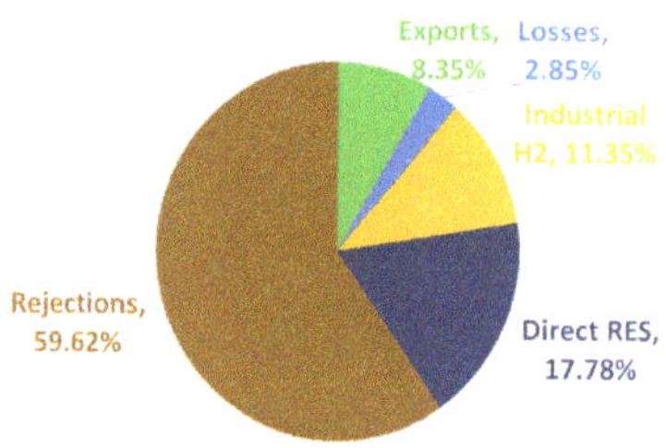

Figure 22. Optimal RES allocation for 2023.

In a scenario where RES production is exceptionally high, there is almost zero electrolysis leading to storage (0.05%). Instead, electrolysis serves almost exclusivelyfor conversion to industrial hydrogen, with the majority of the produced energy being rejected, rendering the system inefficient. This underscores a mismatch between energy supply and demand, highlighting the need for achieving the desired goals with means other than the expansion of the installed capacity of RES to this extent.

On a monthly basis, Figure 23 shows the monthly production, mostly comprised of rejections of the system.

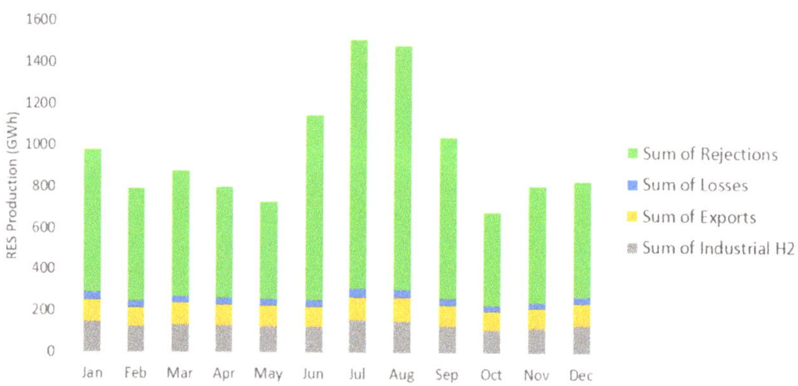

Figure 23. Optimal monthly RES allocation for 2023.

Despite the substantial profits estimated, the optimized system fails to manage the surplus energy produced in a financially viable manner, as production significantly exceeds the island's total consumption.

The initial capital expenditure for the optimal solution is presented in Figure 24, demonstrating the ratio of RES system expenditure to the hydrogen system cost.

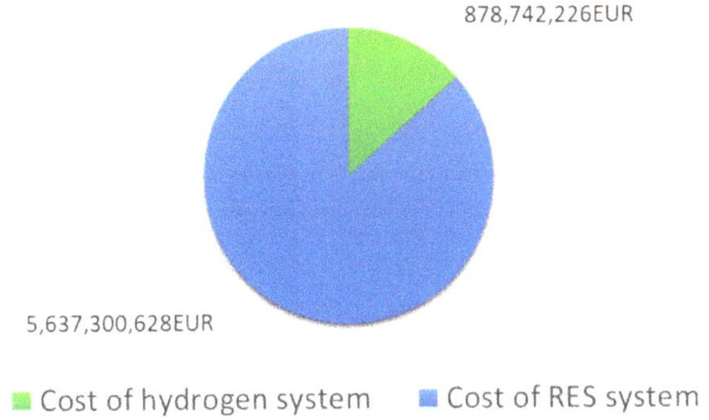

Figure 24. Capital expenditure for the RES system's expansion and hydrogen system's installation.

The majority of the costs stem from the expansion of RES, primarily due to the high installed capacities and elevated costs in comparison to the components of the hydrogen system.

The annual cashflows for the entirety of the investment's lifespan are presented in Figure 25, presenting the most impactful contributors to the business model's operation:

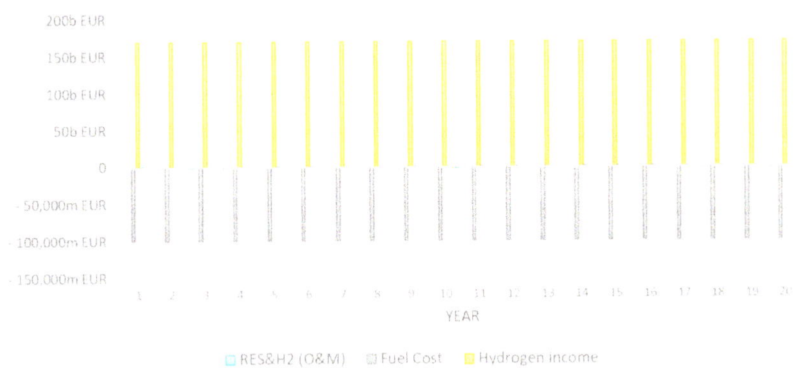

Figure 25. Estimated annual cashflows for 20 years of operation.

In a scenario where the income from hydrogen sales surpasses any maintenance or consumable costs, assuming an infinite demand for hydrogen mass, the system appears financially robust. Interestingly, the interconnection plays a negligible role, neither utilized for imports nor serving as a significant source of income. Throughout each year, inflows and outflows remain relatively constant, excluding the tenth year when replacement of the electrolyzer and fuel cell system becomes necessary. Despite the need for replacement, the overall investment presents a profitable outlook, albeit within a strictly defined context.

4. Discussion

The methodology emphasized the island's autonomy from the mainland grid and aimed to optimize specified targets within specific constraints, resulting in independence from the mainland grid, pursuing, at the same time, sustainability and smooth operation over the project's 20-year lifespan. However, the calculated CAPEX relied on assumptions such as constant MCP values, stable fuel and hydrogen prices, and unlimited demand for industrial hydrogen. Any deviation or flexibility in these assumptions could result in a negative NPV. Moreover, despite the failure to achieve carbon-neutral operation, the proposed system could only meet one of the three requirements since the estimated CAPEX exceeds the total funding for projects related to RES and hydrogen systems in Crete ([46,47]). Therefore, the proposed methodology is infeasible under the specified conditions. However, the methodology presented in this paper for achieving net zero carbon emissions in a microgrid transcends the specific case study and offers a versatile framework applicable to a broader spectrum of microgrid configurations. Its strength lies in its ability to integrate existing energy sources, like RES, hydrogen, and conventional units, with a customized fuel mix of various emission factors. The implementation also extends to future advancements in diversified energy generation and the emerging hydrogen market. This adaptability acknowledges the dynamic nature of the energy landscape, where new, environmentally friendly sources are constantly developed. By incorporating these potential additions, the methodology can be continuously optimized to maintain zero carbon emissions in microgrids with diverse energy mixes (e.g., biomass and geothermal energy) and fluctuating fuel prices.

Future research will focus on pursuing the three specified requirements by introducing a hybrid storage system comprising lithium batteries and green hydrogen storage. This approach would allow for increased penetration of RES without the high costs associated with PV and WT installations or significant amounts of rejected energy. This is also crucial for the investigation of any financial feasibility to map the current market state of hydrogen supply and demand, as well as current supply routes and methods. Additionally, a more detailed estimation of carbon absorption on the island would lead to more precise environmental targets and evaluation metrics. Moreover, carbon capture and storage (CCS) technologies should be explored as more reliable alternatives to carbon sequestration for achieving a net-zero carbon footprint on islands, aiming to enhance the logistical sustainability of the system.

5. Conclusions

This paper examines the feasibility of a standalone energy system for Crete, relying on hydrogen-based storage infrastructure. The suggested energy system prioritizes consumption coverage through renewable energy production, local fuel cell production using green hydrogen, and conventional production from locally produced Mazut and diesel power plants, supplemented by imports via the mainland grid interconnection. Surplus energy generation is earmarked for storage as green hydrogen is produced through electrolysis, either for future use or potential sale. The system's optimal sizing is determined using financial and environmental metrics, such as NPV and carbon neutrality. Carbon neutrality is pursued by balancing emissions from transportation, residential heating, and energy production with absorption through carbon sequestration in olive groves. This study highlights the significant research idea of enhancing renewable energy integration in islands, particularly focusing on Crete, to reduce carbon emissions and fossil fuel consumption. The research achievement lies in proposing a method to boost the use of renewable energy sources, contributing to the European goal of achieving carbon neutrality by 2050. The methodology fails to introduce a feasible sizing for the proposed installation that meets all objectives. Yet, its high level of replicability justifies the study of the installation under these conditions. This study underscores the pivotal role of islands in pioneering renewable energy systems and emphasizes the importance of transitioning towards sustainable energy solutions to combat climate change effectively.

Author Contributions: Conceptualization, E.K. and I.M.; methodology, I.M. and M.N.; validation, E.K., I.M., M.N., and A.T.; formal analysis, I.M. and M.N.; investigation, I.M. and M.N.; resources, I.M. and M.N.; writing—original draft preparation, I.M. and M.N.; writing—review and editing, I.M., M.N., and E.K.; visualization, M.N.; supervision, E.K.; project administration, E.K. All authors have read and agreed to the published version of the manuscript.

Funding: This research received no external funding.

Institutional Review Board Statement: Not applicable.

Informed Consent Statement: Not applicable.

Data Availability Statement: The original data regarding Crete's participation in the HEnEx presented in the study are openly available at https://www.enexgroup.gr/web/guest/markets-publications-el-day-ahead-market (accessed on 5 May 2024). The data regarding fuel prices are publicly available at http://oil.gge.gov.gr/. The recorded data of photovoltaic and wind park production are not publicly accessible as they were shared in confidence by the relevant stakeholders.

Conflicts of Interest: The authors declare no conflicts of interest.

Nomenclature

Symbol	Description	Unit
CAPEX	Capital Expenditure	EUR
Cashflows(i)	Sum of the investment's total expenses and income in year i	EUR
CC	Combined Cycle	-
CCS	Carbon Capture and Storage	-
CRETE RES[T]	Cretan RES production recorded in hour T	MWh
DIRES[T]	RES production on load in hour T	MW
ELCAP	Electrolyzer System's hourly maximum H2 production	MWh
ELECLYS[T]	Electrolyzer System's hourly H2 production in hour T	MWh
ELEFF	Electrolyzer System's overall efficiency	-
ExpL	Maximum hourly energy exported	MWh
GRID[T]	Total energy imported/exported in hour T	MWh
FC[T]	Fuel Cell System's electricity production in hour T	MWh
F.C. CAP.	Fuel Cell System's hourly maximum electricity production	MWh
F.C. EFF.	Fuel Cell System's overall efficiency	-
HEnEx	Hellenic Energy Exchange	-
HPS	Hydropower Pumped Storage	-
$Hydrogen_{mass}[T]$	Produced H2 for sale in mass units	kg
IND. H2[T]	Produced H2 for sale in energy units	MWh
LHV	Lower Heating Value	-
LL	Conventional System's Technical Lower Limit	MW
LOAD[T]	System's Demand in hour T	MW
LOSSES[T]	Hydrogen System's total losses in hour T	MWh
MCP	Market Clearing Price	-
NPV	Net Present Value	EUR
PEM	Proton Exchange Membrane	-
PV	Photovoltaic	-
PV_{cap}	PV system's installed capacity	MW
r	Annual Discount Rate	-
REJECT[T]	RES production's rejected energy	MWh
RES	Renewable Energy Sources	-
RES[T]	RES electricity production in hour T	MWh
RES O&M cost	Operation and Maintenance cost of RES system	EUR
SG	Smart Grid	-
ST[T]	Hydrogen Tank's State of Charge in hour T	MWh
STCAP	Hydrogen Tank's storage capacity	MWh
Total O&M cost	RES and hydrogen systems' operation and maintenance cost	EUR
TOTCONV[T]	Conventional System's total production in hour T	MWh
UPLL	Conventional System's Technical Upper Limit	MW
V2G	Vehicle to Grid	-
WT	Wind Turbine	-
WT_{cap}	Wind turbine system's total installed capacity	MW

References

1. Sun, J.; Sun, H.; Jiang, J. An Improved Modulation Method for Low Common-Mode Current Non-Isolated Series Simultaneous Power Supply Dual-Input Inverters for New Energy Generation Applications. *Electr. Eng.* **2024**. [CrossRef]
2. Jha, A.V.; Ghazali, A.N.; Appasani, B.; Ravariu, C.; Srinivasulu, A. Reliability analysis of smart grid networks Incorporating hardware failures and packet loss. *Rev. Roum. Des. Sci. Tech.* **2021**, *65*, 245–252.
3. Allen, M.R.; Dube, O.P.; Solecki, W.; Aragón-Durand, F.; Cramer, W.; Humphreys, S.; Kainuma, M.; Kala, J.; Mahowald, N.; Mulugetta, Y.; et al. *IPCC Special Report on Global Warming of 1.5 °C*; Cambridge University Press: Cambridge, UK; New York, NY, USA, 2018; pp. 49–92. [CrossRef]
4. Yaqoob, S.J.; Motahhir, S.; Agyekum, E.B. A New Model for a Photovoltaic Panel Using Proteus Software Tool under Arbitrary Environmental Conditions. *J. Clean. Prod.* **2022**, *333*, 130074. [CrossRef]
5. Wei, Y.; Sun, H.; Zhang, T.; Jiang, J.; Su, X.; Zeng, N. Study of Inductively Coupled Fuel Cell DMPPT Converters. *Electr. Eng.* **2024**. [CrossRef]
6. Zhang, C.; Cao, C.; Chen, R.; Jiang, J. Three-Leg Quasi-Z-Source Inverter with Input Ripple Suppression for Renewable Energy Application. *Energies* **2023**, *16*, 4393. [CrossRef]
7. Vourdoubas, J. Creation of Net Zero Carbon Emissions Residential Buildings due to Energy use in the Mediterranean Region: Are they Feasible? *Civ. Eng. Res. J.* **2020**, *10*, 555777.
8. Directorate-General for Climate Action. *Going Climate-Neutral by 2050: A Strategic Long Term Vision for a Prosperous, Modern, Competitive and Climate Neutral EU Economy*; European Commission: Brussels, Belgium, 2019. [CrossRef]
9. Juswanto, W.; Ali, Z. Renewable energy and sustainable development in Pacific Island countries. In Proceedings of the Pacific Update Conference: Inclusive Growth for Enhanced Resilience, Suva, Fiji, 18–19 July 2016; Available online: https://www.adb.org/sites/default/files/publication/215216/adbi-pb2016-5.pdf (accessed on 5 February 2021).
10. Meschede, H.; Bertheau, P.; Khalili, S.; Breyer, C. A Review of 100% Renewable Energy Scenarios on Islands. *WIREs Energy Environ.* **2022**, *11*, e450. [CrossRef]
11. International Renewable Energy Agency (IRENA). A Path to Prosperity: Renewable Energy for Islands. In Proceedings of the Framework Convention on Climate Change, Marrakesh, Morocco, 7–18 November 2016. Available online: https://www.irena.org/-/media/Files/IRENA/Agency/Publication/2016/IRENA_Path_to_Prosperity_Islands_2016.pdf (accessed on 5 February 2021).
12. Bunger, J. Renewable Energy Islands in Europe. In Proceedings of the European Seminar on Renewable Energy Islands, Samso Island, Denmark, 29–30 June 1998; Available online: https://www.osti.gov/etdeweb/servlets/purl/317117 (accessed on 5 February 2021).
13. Marczinkowski, H.; Alberg Østergaard, P.; Roth Djørup, S. Transitioning Island Energy Systems—Local Conditions, Development Phases, and Renewable Energy Integration. *Energies* **2019**, *12*, 3484. [CrossRef]
14. Marczinkowski, H.M.; Barros, L. Technical Approaches and Institutional Alignment to 100% Renewable Energy System Transition of Madeira Island—Electrification, Smart Energy and the Required Flexible Market Conditions. *Energies* **2020**, *13*, 4434. [CrossRef]
15. Chatziargiriou, N.; Margaris, I.; Dimeas, A. *Renewable Energy Developments in Greek Islands*; Friedrich Ebert Stiftung: Athens, Greece, 2016; Available online: https://library.fes.de/pdf-files/bueros/athen/13050.pdf (accessed on 9 March 2023).
16. Refsgaard, K.; Kull, M.; Bogason, Á.; Salolammi, P.; Slätmo, E.; Zelenkauskė, M. *Carbon Neutral Islands*; Nordregio Projects: Stockholm, Sweden, 2021; Available online: https://nordregioprojects.org/carbon-neutral-islands/ (accessed on 9 March 2023).
17. Breyer, C.; Bogdanov, D.; Ram, M.; Khalili, S.; Vartiainen, E.; Moser, D.; Román Medina, E.; Masson, G.; Aghahosseini, A.; Mensah, T.N.O.; et al. Reflecting the Energy Transition from a European Perspective and in the Global Context—Relevance of Solar Photovoltaics Benchmarking Two Ambitious Scenarios. *Prog. Photovolt.* **2023**, *31*, 1369–1395. [CrossRef]
18. Vourdoubas, J. Use of renewable energy sources for energy generation in rural areas in the island of Crete, Greece. *Eur. J. Environ. Earth Sci.* **2020**, *1*, 1–7. [CrossRef]
19. Biza, S.; Piromalis, D.; Barkas, D.; Psomopoulos, C.S.; Tsirekis, C.D. Crete—Peloponnese 150kV AC Interconnection. Simulation Results for Transient Phenomena in Main Switches. *Energy Procedia* **2019**, *157*, 1366–1376. [CrossRef]
20. Home | Ariadne Interconnection. Available online: http://www.ariadne-interconnection.gr/en (accessed on 16 April 2024).
21. Tzanes, G.; Zafeiraki, E.; Papapostolou, C.; Zafirakis, D.; Konstantinos, M.; Kavadias, K.; Chalvatzis, K.; Kaldellis, J.K. Assessing the Status of Electricity Generation in the Non-Interconnected Islands of the Aegean Sea Region. *Energy Procedia* **2019**, *159*, 424–429. [CrossRef]
22. Tsekeris, D.; The Energy Transition of the Greek Non-Interconnected Islands. Greek Minist. *Environ. Energy 2019.* Available online: https://hybridpowersystems.org/wp-content/uploads/sites/13/2019/06/1_1_HYB19_088_presentation_Tsekeris_Dimitris.pdf/ (accessed on 17 April 2024).
23. Bertsiou, M.M.; Baltas, E. Power to Hydrogen and Power to Water Using Wind Energy. *Wind* **2022**, *2*, 305–324. [CrossRef]
24. Rodrigues, A.; Machado, D.; Dentinho, T. Electrical Energy Storage Systems Feasibility; the Case of Terceira Island. *Sustainability* **2017**, *9*, 1276. [CrossRef]
25. Nikolaou, T.; Stavrakakis, G.S.; Tsamoudalis, K. Modeling and Optimal Dimensioning of a Pumped Hydro Energy Storage System for the Exploitation of the Rejected Wind Energy in the Non-Interconnected Electrical Power System of the Crete Island, Greece. *Energies* **2020**, *13*, 2705. [CrossRef]

26. Pompodakis, E.E.; Kryonidis, G.C.; Karapidakis, E.S. Optimizing the Installation of Hybrid Power Plants in Non-Interconnected Islands. *J. Energy Storage* **2023**, *74*, 109511. [CrossRef]
27. Hellenic Electricity Distribution Network Operator (HEDNO) SA, Annual Financial Report 2020. Available online: https://deddie.gr/media/36008/annual-financial-report-2020-%CE%B1%CE%B3%CE%B3%CE%BB%CE%B9%CE%BA%CE%B1.pdf (accessed on 15 April 2024).
28. Day-Ahead Market—EnExGroup. Available online: https://www.enexgroup.gr/web/guest/markets-publications-el-day-ahead-market (accessed on 5 May 2024).
29. IPTO (Independent Power Transmittion Operator). Available online: https://www.admie.gr/erga/erga-diasyndeseis/diasyndesi-tis-kritis-me-tin-peloponniso (accessed on 15 April 2024).
30. Al-Shetwi, A.Q.; Hannan, M.A.; Jern, K.P.; Alkahtani, A.A.; Pg Abas, A.E. Power Quality Assessment of Grid-Connected PV System in Compliance with the Recent Integration Requirements. *Electronics* **2020**, *9*, 366. [CrossRef]
31. CO2 Emissions (Metric Tons per Capita)—Greece. Available online: https://data.worldbank.org/indicator/EN.ATM.CO2E.PC?locations=GR (accessed on 3 April 2024).
32. Katsaprakakis, D.; Antonakakis, I.; Dakanali, I.; Christakis, D. Turning Crete into an energy independent island. In Proceedings of the 4th International Hybrid Power Systems Workshop, Agia Pelagia Crete, Greece, 22–23 May 2019; Available online: https://www.researchgate.net/publication/333701601_Turning_Crete_into_an_energy_independent_island (accessed on 3 April 2024).
33. Takatsu, N.; Farzaneh, H. Techno-Economic Analysis of a Novel Hydrogen-Based Hybrid Renewable Energy System for Both Grid-Tied and Off-Grid Power Supply in Japan: The Case of Fukushima Prefecture. *Appl. Sci.* **2020**, *10*, 4061. [CrossRef]
34. Su, W.; Zheng, W.; Li, Q.; Yu, Z.; Han, Y.; Bai, Z. Capacity Configuration Optimization for Green Hydrogen Generation Driven by Solar-Wind Hybrid Power Based on Comprehensive Performance Criteria. *Front. Energy Res.* **2023**, *11*, 1256463. [CrossRef]
35. Sharma, S.; Agarwal, S.; Jain, A. Significance of Hydrogen as Economic and Environmentally Friendly Fuel. *Energies* **2021**, *14*, 7389. [CrossRef]
36. Jadhawar, P.; Saeed, M. Optimizing the Operational Efficiency of the Underground Hydrogen Storage Scheme in a Deep North Sea Aquifer through Compositional Simulations. *J. Energy Storage* **2023**, *73*, 108832. [CrossRef]
37. Shiva Kumar, S.; Himabindu, V. Hydrogen Production by PEM Water Electrolysis—A Review. *Mater. Sci. Energy Technol.* **2019**, *2*, 442–454. [CrossRef]
38. Karapidakis, E.; Mozakis, I.; Iliadis, I. Ultra-High Share of Renewable Energy Sources in Interconnected Island Systems. In Proceedings of the International Conference on Electronics, Engineering Physics and Earth Science (EEPES'23), Kavala, Greece, 21–23 June 2023; p. 15.
39. Available online: https://www.admie.gr/en/market/market-statistics/detail-data?process[0]=529&data_type[0]=518&since=&until=&op=Submit&_wrapper_format=html&page=4 (accessed on 14 May 2024).
40. Brigitte, K.; Alessandro, C.; Marlene, D.; Andreea, I.; Albana, K.; Greet, J.-M. CoM Default Emission Factors for the Member States of the European Union—Version 2017, European Commission, Joint Research Centre (JRC) [Dataset]. 2017. Available online: http://data.europa.eu/89h/jrc-com-ef-comw-ef-2017 (accessed on 15 April 2024).
41. Klimenta, D.; Mihajlović, M.; Ristić, I.; Andriukaitis, D. Possible Scenarios for Reduction of Carbon Dioxide Emissions in Serbia by Generating Electricity from Natural Gas. *Energies* **2022**, *15*, 4792. [CrossRef]
42. Available online: http://oil.gge.gov.gr/ (accessed on 2 May 2024).
43. Ahmed, A.; Pompodakis, E.E.; Katsigiannis, Y.; Karapidakis, E.S. Optimizing the Installation of a Centralized Green Hydrogen Production Facility in the Island of Crete, Greece. *Energies* **2024**, *17*, 1924. [CrossRef]
44. Vourdoubas, J. Estimation of Carbon Sequestration from Olive Tree Groves in the Island of Crete, Greece. *Int. J. Agric. Environ. Res.* **2018**, *6*, 560. [CrossRef]
45. Vourdoubas, J. Studies on the Electrification of the Transport Sector in the Island of Crete, Greece. *OJEE* **2018**, *7*, 19–32. [CrossRef]
46. Available online: https://cordis.europa.eu/project/id/101112169 (accessed on 14 May 2024).
47. Available online: https://energypress.eu/nsrf-support-worth-over-e40m-announced-for-energy-communities/ (accessed on 14 May 2024).

applied sciences

MDPI

Article

Energy Harvester Based on a Rotational Pendulum Supported with FEM

Grzegorz Litak [1,*], Mirosław Kondratiuk [2], Piotr Wolszczak [1], Bartłomiej Ambrożkiewicz [1] and Abhijeet M. Giri [1]

[1] Faculty of Mechanical Engineering, Lublin University of Technology, Nadbystrzycka 36, 20-618 Lublin, Poland; p.wolszczak@pollub.pl (P.W.); b.ambrozkiewicz@pollub.pl (B.A.); a.giri@pollub.pl (A.M.G.)

[2] Faculty of Mechanical Engineering, Białystok University of Technology, Wiejska 45C, 15-351 Białystok, Poland; m.kondratiuk@pb.edu.pl

* Correspondence: g.litak@pollub.pl

Abstract: The proposed energy harvesting system is based on a rotational pendulum-like electromagnetic device. Pendulum energy harvesting systems can be used to generate power for wearable devices such as smart watches and fitness trackers, by harnessing the energy from the human body motion. These systems can also be used to power low-energy-consuming sensors and monitoring devices in industrial settings where consistent ambient vibrations are present, enabling continuous operation without any need for frequent battery replacements. The pendulum-based energy harvester presented in this work was equipped with additional adjustable permanent magnets placed inside the induction coils, governing the movement of the pendulum. This research pioneers a novel electromagnetic energy harvester design that offers customizable potential configurations. Such a design was realized using the 3D printing method for enhanced precision, and analyzed using the finite element method (FEM). The reduced dynamic model was derived for a real-size device and FEM-based simulations were carried out to estimate the distribution and interaction of the magnetic field. Dynamic simulations were performed for the selected magnet configurations of the system. Power output analyses are presented for systems with and without the additional magnets inside the coils. The primary outcome of this research demonstrates the importance of optimization of geometric configuration. Such an optimization was exercised here by strategically choosing the size and positioning of the magnets, which significantly enhanced energy harvesting performance by facilitating easier passage of the pendulum through magnetic barriers.

Keywords: vibration energy harvesting; FEM in electromagnetism; pendulum system

1. Introduction

Small- and micro-scale energy harvesting (EH) usually refers to the conversion of ambient energy from different sources, including mechanical vibration energy, into electrical energy, which can be further accumulated or used for powering up devices demanding low amounts of energy [1,2]. One of the main advantages of such energy harvesting systems is the capability of onsite energy transduction and power supply for the integrated operation of small- and microelectronic devices. EH systems can also be used for remote and inaccessible locations or maintenance-free applications [3,4]. Another advantage is the improvement in the efficiency of the parent process, which acts as an ambient energy source, due to recovery of energy which could be otherwise lost [x1]. These energy harvesting systems can readily contribute to the current industrial revolution called Industry 4.0 or the internet of things (IoT) [5,6], which demand independent sources of energy for powering up electronic units and sensors. In general, such EH systems are classified as per the source of ambient energy, i.e., mechanical vibrations, thermal, solar, wind, radio waves (RF), fluid motion, etc. (see Figure 1). Many devices have been developed by researchers based on the abovementioned energy sources. In this research work, we focused on energy conversion from ambient mechanical vibrations using a pendulum-like rotating body with modifications possible in the system's potential [7].

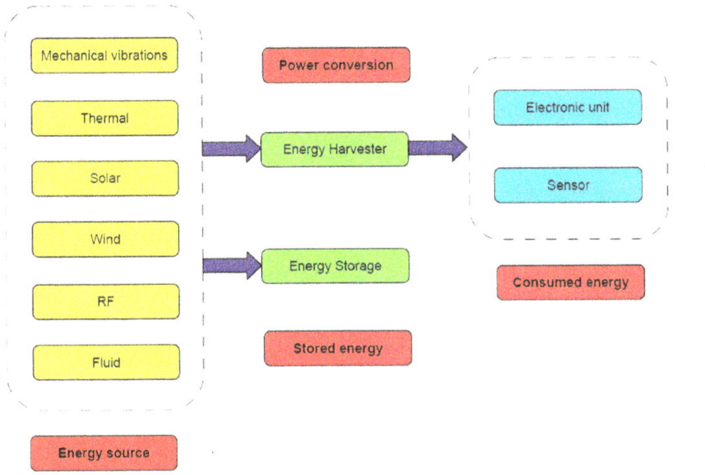

Figure 1. The energy conversion process in the energy harvesting systems.

It is worth mentioning that energy harvesting systems that are based on ambient mechanical vibrations are generally divided into electromagnetic [8,9], piezoelectric [10,11], and electrostatic [12,13] categories. In many scientific articles, they are called vibration energy harvesters (VEH) [14,15]. VEH systems have gained popularity because they provide the simplest transduction mechanism and highest efficiency of energy transduction [16,17]. Practical applications of VEH systems span diverse sectors. In the automotive industry, these systems can capture energy from vehicle vibrations to power tire pressure monitoring systems or onboard sensors [18]. Similarly, in structural health monitoring, they can be employed to generate power for sensors embedded in bridges or buildings while facilitating continuous monitoring. This eliminates the need for frequent battery replacements as well [19]. Furthermore, wearable devices [20] can utilize VEH system to extend battery life or even eliminate the need for batteries altogether, enhancing user convenience and sustainability.

Being a linear dynamic system, a simple VEH responds well only when excited near its resonance frequency. This is a major disadvantage because the harvested power from the device drops when excited at other non-resonant operational frequencies resulting from uncertain ambient conditions [21]. However, to overcome this problem, many current research works are focusing on the broadband frequency-response effect to build a flexible system adjusting to the wider range of operational frequencies [22–24]. Compared to a linear system, the broadband frequency effect causes a reduction in the mechanical quality factor (Q-factor). The highest power output obtained under resonant conditions also reduces because a nonlinear system may exhibit multiple low- and high-amplitude solutions [25–27]. However, as the nonlinear systems exhibit better performance at non-resonant conditions, overall harvesting efficiency is improved due to the broadband response. Such nonlinear behavior has been obtained by using nonlinear mechanical and magnetic springs with softening or hardening effects [28]. Some of the interesting proposals include energy harvesters with multi-stable mechanical or magnetic potential [29–32]. A few other approaches involve active and autonomous tuning of system parameters to sustain the resonance conditions [33,34].

Until now, many VEH systems have been explored based on their applications, source of excitation, and transduction methods, with the primary objective of improving efficiency and broadband response. One of the recent trends is to improve performance by introducing different types of nonlinearities into the system. Concurrently, studies to support experiments with mathematical modeling and numerical simulations are also being undertaken. Numerical simulations using finite element analysis (FEA) show potential for nonlinear modeling and are capable of revealing the future directions in this field.

2. The Concept of Pendulum-Based Electromagnetic Energy Harvesting Systems

In recent times, energy harvesting systems using rotational movement have been receiving attention to scavenge energy from the environment [34–37]. Very often, pendulums are used in electromagnetic energy harvesting systems because they allow versatility in their configuration. Electromagnetic transduction of vibration energy in the standard magnet–coil generators is based on the electromagnetic induction phenomena. In this, electromotive force is induced in an induction coil, which cuts the magnetic flux of a core during its relative displacement. For improving the electromagnetic coupling with the strongest flux density, permanent magnets made from an alloy of neodymium, iron, and boron (NdFeB) are most suitable. These magnets are also available in various configurations based on the direction of polarization, shapes, sizes, grades of neodymium, and thermal resistance.

Historically, many concepts of electromagnetic energy harvesting systems with pendulums have been proposed among the other types of low-frequency energy harvesting systems based on cantilevers, mechanical springs, magnetic springs, and spiral springs [38]. Electromagnetic pendulum-based harvesters can depict both oscillations and full rotation of the pendulum depending on the character of excitation. In real applications, a particular design faces challenges such as sustaining rotational motion under random excitations, and sizing of the system based on constraints or power supply requirements [39]. In order to overcome such challenges and optimize the system for a particular application, several modifications were introduced. These modifications were based on the research on mechanical couplings [40,41], using sub-harmonic resonance for the broadband power output [42], and using nonlinearities as an additional source of energy [43].

One of the analysis approaches used to study the dynamics of EH systems is comparing the experimental results with the results of mathematical modeling. Simplified mechanical models based on spring-mass systems have provided fairly good results in many studies [44–46]. Another approach is using the FEA for analyzing the electromagnetic interactions in such systems [47–49]. Currently, several commercial FEA software packages are available for analyzing electromagnetism problems including 2D axis-symmetric problems and advanced 3D problems. Some of the popular FEA programs are NEC (Numerical Electromagnetics Code), FEKO (Altair Engineering), COMSOL Multiphysics (COMSOL Inc.), and Momentum (Keysight EEsof EDA).

Referring to the study on electromagnetic energy harvesters using rotational movement [7], we propose a rotational pendulum concept with an improved approximation. The originality of this system is the variable potential introduced by the pairs of additional adjustable magnets whose strength can be changed. The dynamic response of this system, depending on the system's configuration, provides full rotation or oscillations of the pendulum, which in turn influences the output power. In our analysis, we studied the magnetic interactions and the dynamics of a system using the FEA and mathematical models. In our previous work [7], the effect on performance was analyzed through magnetic interactions with a simplified FEA model. In this study, the effect of magnet positioning on potential modulation and on attaining complete rotational response is explored using numerical simulations. To avoid magnetic interaction with the housing, the system was realized using the 3D printing technique. The originality of the concept lies in the incorporation of adjustable additional magnets to modulate the magnetic potential, allowing a customizable system configuration that improves both pendulum dynamics and output power.

This article is organized in the following way. In the next section (Section 3), the proposed concept of the rotational VEH system with the pendulum is described. Mathematical models of the electromagnetic EH system, with inputs from the FEA, are introduced and described. Further, details about the experiment and measuring circuit are given. In Section 4, dynamics of the pendulum caused by different configurations of additional magnets are discussed and visualized with the results of numerical simulations. Lastly, Section 5 summarizes and provides conclusions of this analysis.

3. Design and Modeling of an Energy Harvesting System with a Rotational Pendulum

In electromagnetic energy harvesting systems, the voltage is induced in coils according to Faraday's law by the changing of the magnetic flux in the coils. In this pendulum-based electromagnetic EH system, modulation of magnetic potential using the alignment of additional magnets is proposed. In the analyzed prototype, the adjustable potential is produced by two symmetrically mounted magnetic pairs (see Figure 2), changing the character of the system. This EH system was designed entirely with non-magnetic or weak magnetic materials using 3D printing (ABS—acrylonitrile butadiene styrene), plastic (nylon), and aluminum; the motivation was to avoid undesirable magnetic interaction with the system's other parts. Optionally, external pendulums can also be attached to the system on its sides to increase the moment of inertia and to overcome potential barriers by full rotations. The magnetic-electrical circuit consisted of 12 magnetic coils connected in series and two types of permanent magnets, one in the form of the tip mass mounted at the end of the pendulum and two pairs of adjustable magnets placed inside the plastic housing tunnels (Figure 3).

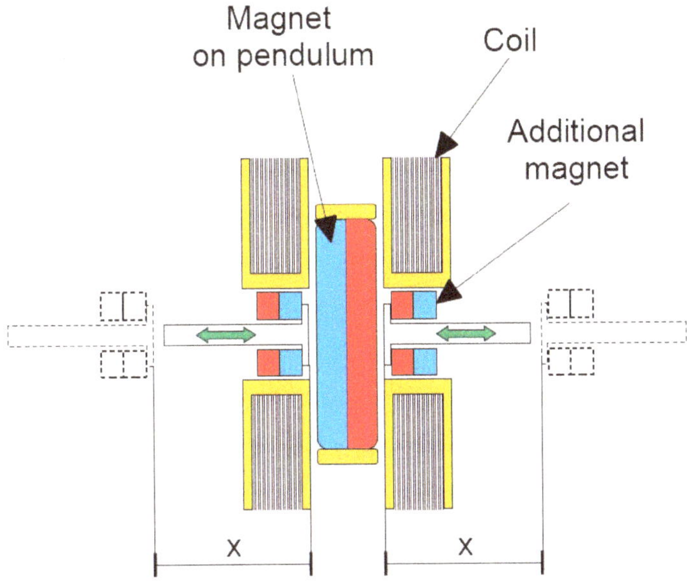

Figure 2. Magnetic interaction in the energy harvester. The additional magnets are movable axially, which changes the magnetic force between permanent magnets. The maximum distance by which the additional magnet can be moved axially is x = 13 mm.

A compact EH system was designed here by limiting the system's dimensions to 10 cm in any direction from the center (see Figure 3a). The design allowed for versatile coil connections; for example, we connected six coils on a single wall of the system. Within these coils, we strategically placed four additional adjustable permanent magnets—two on each side of the system (Figure 3b). These magnets played a crucial role in electromagnetic induction within the system and influenced the movement of the pendulum by creating a double potential well.

(a) 
(b) 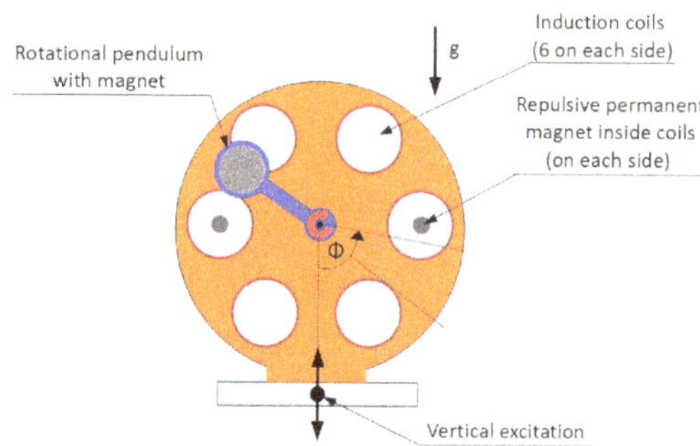

Figure 3. The physical concept of the electromagnetic energy harvesting system with rotational pendulum: (**a**) physical prototype view; (**b**) the scheme of the system (cross-section view).

Mechanical and electrical properties of this EH system are listed in Table 1. More information on the technical issues of such EH systems can be found in [49], in which experiments and the simplified FEA (2D planar) are described. As this study will combine two-stage modeling, i.e., a 3D FEM electromagnetic model and simulating the mathematical model using MATLAB (v2022b) software, the actual physical quantities of the system are specified in Table 1. Mechanical and electrical features were used for preparation of the 3D model in COMSOL (v5.3) Multiphysics software.

Table 1. Mechanical and electrical properties of the electromagnetic EH system.

Mechanical Part	
Mass of the system (g)	~900
Pendulum magnet (mm)	⌀ 20 (O.D) × 5 (thick) (cylinder)
Side magnets (mm)	⌀ 8 (O.D) × ⌀ 6 (I.D) × 3.5 (thick) (ring)
Diameter of rotation (mm)	⌀ 66.5
Limiting dimensions of the system (mm)	>100
Electrical Part	
Number of coils	12 (connected in series)
Coil induction (mH)	4.545
Number of turns	253
Coil length (mm)	8
Wire type	28 AWG
Resistance (Ω)	30.3
Type of magnets	NdFeB 37

3.1. Electromagnetic FEM Model

In comparison to the previous FEA work in [7], the current system was analyzed in 3D space keeping the existing coil layout and the geometry of the energy harvester (Figure 4). The motivation behind achieving precise calculation using 3D FEA was to obtain the information about the magnetic force between magnets during rotation of the pendulum and its influence on the dynamics of the pendulum with magnet. Once this was obtained, the magnetic torque vector was utilized in the mathematical model. Such magnetic interactions strongly affected the dynamics and changed the energetic potential of the energy harvester [49]. In the finite element (FE) model, the influence of housing can be neglected as the entire housing is made of non-magnetic materials and there are no undesirable external magnetic interactions.

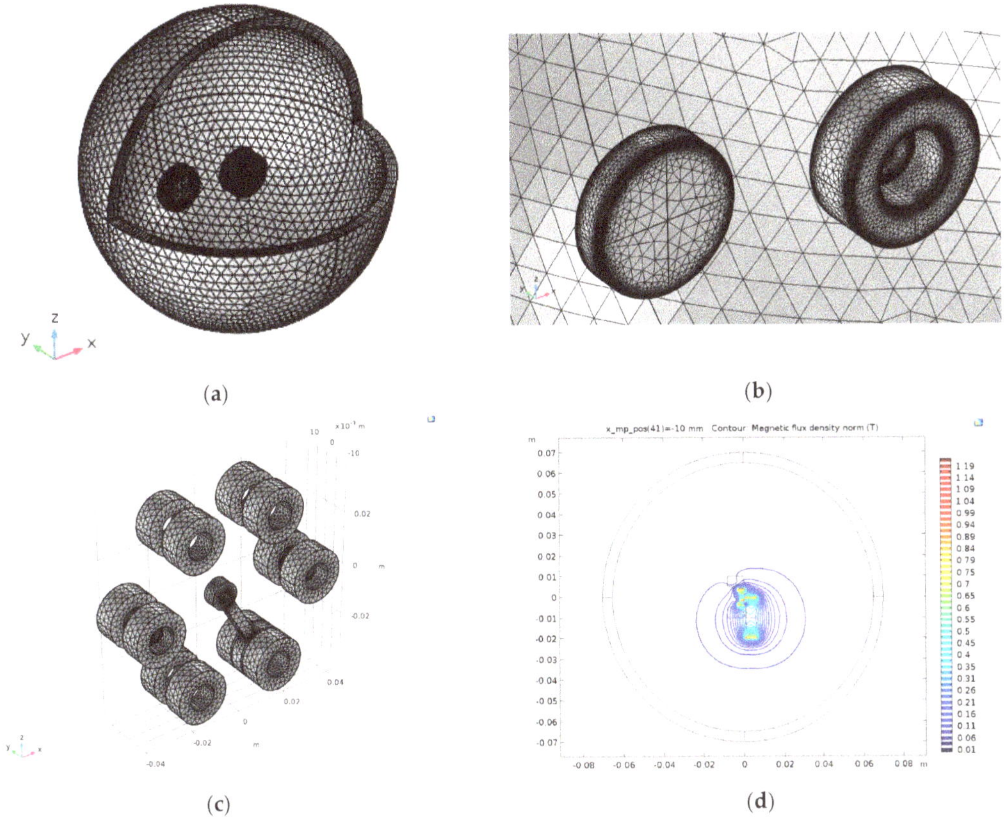

Figure 4. The physical concept of the electromagnetic energy harvesting system with rotational pendulum modeled in COMSOL Multiphysics. (**a**) The entire domain considered for the analysis of magnetic force interaction between pendulum magnet and fixed additional magnet in meshed state. (**b**) Enlarged meshed view of the two interacting magnets mentioned above. (**c**) Meshed geometry considered for the MBD analysis of the system, and (**d**) the distribution of the magnetic field during interactions between the magnets shown in the top planer view of (**b**) (directly taken from the COMSOL results report).

The current problem is a typical multi-body dynamics (MBD) problem dealing with the combination of the dynamics and electromagnetism. Additionally, an electrical circuit was added to analyze the power output of the system. Moreover, couplings between the magnet–magnet and magnet–coil pairs were investigated in the form of acting forces and torques. Thus, the obtained torque vector using FEA was crucial for the analysis of the pendulum's dynamics. As the FEA model was linear, the system could be analyzed for one rotation of the pendulum. Hence, the results were obtained for one rotation per second of the pendulum.

In order to obtain magnetic force interaction between the pendulum magnet and the additional fixed magnet using FEA, a stationary model was considered, as shown by a meshed view in Figure 4a. In this view, the relative size and positions of the magnets as well as the surrounding air space considered for the magnetic interaction is shown. The linear position of the pendulum magnet relative to the additional magnet–coil pair was changed along the x-axis (see Figure 4b), from which the magnet force in terms of x-, y-, and z-axis components was computed. A detailed description of this procedure is mentioned in the attached Supplementary Materials to this article. Further, using these calculations, resultant magnet forces and torques for magnet–magnet and magnet–coil pairs were estimated. These force and torque quantities along with the electrical variables were then considered for the MBD analyses of the system shown Figure 4c. A top view of Figure 4b showing magnetic flux interaction between the magnets is shown in Figure 4d. Readers can find the geometric dimensions and other physical variables used for these FEM simulations in the Supplementary Materials attached with this article.

3.2. Reduced Mathematical Model

In Figure 5, the scheme of the mathematical model for the energy harvester with terms applied in the equation of motion is presented. Vertical excitation coming from the shaker initiated the movement of the rotational pendulum around the pivot in the form of oscillations or full rotations, depending on the applied strength of excitation and the strength of magnetic interactions in the system. The mentioned terms had significant influence on the system's response, which is discussed in the next section. In Table 2, the values of the terms used in the mathematical model are listed.

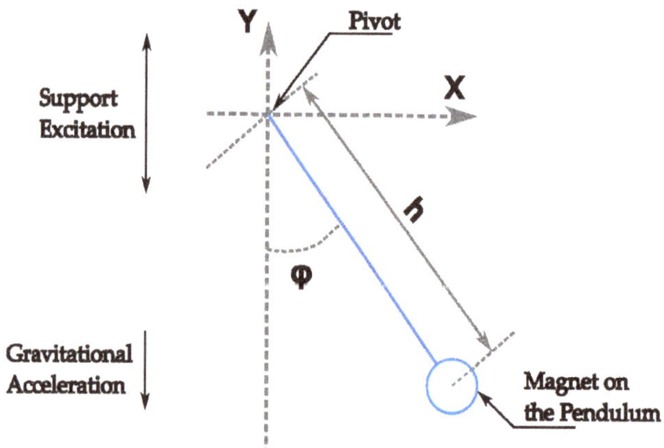

Figure 5. Variables in the mathematical model are shown in the simplified configuration of the energy harvester with magnetic pendulum.

Table 2. Parameters used in the mathematical model (see Equation (1)).

Term	Value
I [kg-m^2]	0.000377
m [kg]	0.0186
g [m/s^2]	9.81
h [m]	0.03325
α, β	0 and 1
λ	0.048
h [m]	0.034
ω [rad/s]	15
A [m]	0.2

In comparison with the previously derived mathematical model of the rotational pendulum [7], additional terms were added, which consider magnet–magnet and magnet–coil couplings. Relative displacement between the pendulum magnet and permanent magnet pair in its proximity created strong magnetic interactions in the form of two opposite potential wells, affecting the acting magnetic forces in the system, as can be observed in Figure 6a. During the relative displacement between magnets, the peak-to-peak amplitude of the pendulum's torque changed drastically, creating a direct influence on its dynamics. The coupling between the pendulum's magnet and the coils had a periodic response (Figure 6b) in each of the 12 coils, showing a relatively smaller impact on its dynamics. Thus, the governing equation of motion (Equation (1)) for the pendulum considering these additional forces can be expressed in the following form:

$$\ddot{\varphi} + \frac{1}{I}(c\dot{\varphi} - \alpha F_{mc}(\varphi)\dot{\varphi} + mgh\sin(\varphi) - \beta F_{mm}(\varphi) - m\omega^2 Ah\,sin(\varphi)\cos(\omega t)) = 0, \quad (1)$$

where I—moment of inertia of the pendulum, m—mass of pendulum and additional inertial masses, g—gravitational acceleration, h—distance from center of gravity to pivot axis, F_{mc}—torque between magnet–coil couplings obtained with FEA, ω—angular velocity of the pendulum, A—amplitude of the vertical excitation, F_{mm}—torque between magnet–magnet couplings in the system obtained with FEA, c—coefficient of mechanical damping (in the calculations we assumed $c = 0$ for simplicity), α, β = 0 or 1—are the coefficients of magnetic force that are dependent on the distances between magnets and coils. We have neglected the macroscopic eddy currents in the magnets.

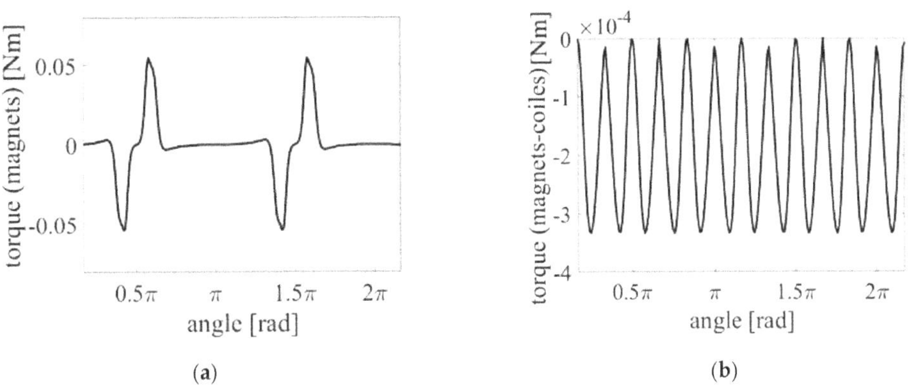

Figure 6. (**a**) The torque from magnet–magnet coupling F_{mm} [Nm] (see Equation (1)) calculated with the FEM model. (**b**) The torque from magnet–coil F_{mc} coupling [Nm] (see Equation (1)) during the clockwise rotation of the pendulum with a constant rotational velocity (1 revolution per second). The initial angle in FEM calculations was 0.25π.

The output power of the system in time domain t is calculated using the following formula (Equation (2)):

$$P = \frac{1}{R_l}\frac{1}{t}\int_0^t U^2(t)dt = (1-\lambda)\alpha F_{mc}(\varphi)\dot{\varphi}^2, \quad (2)$$

where U is the voltage output estimated on the resistor and $\lambda = R_c/(R_l + R_c)$ is the coil resistivity part of the electrical circuit (R_c and R_l denote coil and load resistances, respectively).

4. Numerical Simulations

The analysis of the influence of magnetic forces on the dynamics of the rotational pendulum started with the FEM computations conducted in the COMSOL (v5.3) Software. In the system, there were two unstable equilibrium points where the axes of moving and additional magnets coincided. The amplitude and the distribution of the magnetic force with the angular position of the rotating pendulum depended on the geometry of the magnets and the distance of their alignment, which were considered in this research. It was observed in the prior work [7] during experiments that the system's frame needed excitation in the specific range of frequencies, i.e., f = 10–15 Hz, in order to obtain a completely rotational solution of the pendulum to attain the highest voltage/power performance. Outside the mentioned range of frequencies, the pendulum became stuck at the equilibrium point and did not move. Additionally, it was observed that changing the distance between additional magnets mounted on the sides influenced the potential barrier at the equilibrium positions. In this work, the proposed model considers the influence of varying magnetic forces with the alignment distance on the magnetic pendulum. We also focused on solutions with complete rotation of the pendulum.

In the equation of motion, the magnet–magnet and magnet–coil interactions are important terms influencing significantly the dynamics of the pendulum. To estimate these interactions, the 3D model of the energy harvester was meshed for the FEA (Figure 4) and a series of torques was obtained, which are depicted in Figure 6. The FEA results showed the influence of the magnet–magnet interaction was significantly stronger than that of the coil–magnet coupling on the dynamics of the pendulum. Regarding the interactions between the magnet–magnet in the system (Figure 6a), mutually opposite areas with the strongest magnetic interactions were found. Such a repulsive configuration of magnets helps the pendulum to escape from this area if the potential energy is enough to go through the potential barrier created by the magnetic coupling. This is shown by the red dotted lines in Figure 7. For comparison, gravitational potential is also shown by black dotted lines in Figure 7. It had an almost 10 times weaker influence on the pendulum. The magnet–coil couplings (Figure 6b) had smaller amplitude than the magnet–magnet couplings; however, their influence on the pendulum was also considered in this study.

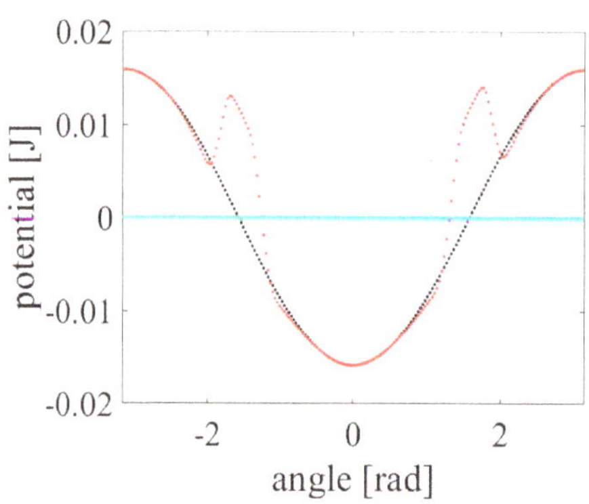

Figure 7. The potential energy of the magnetic force interaction was obtained for one revolution of the pendulum and is shown by the red-dotted line. The potential of the gravity force is shown by a black-dotted line.

We considered the following distinct configurations of the harvester for revealing additional details about the dynamics:

(a) System without coils and without magnets, only pendulum with magnet (only gravitational acceleration): $\alpha = 0$, $\beta = 0$.
(b) System with coils and pendulum with magnet (only magnet–coil coupling): $\alpha = 0$, $\beta = 1$,
(c) System with all magnets and coils.

In Figure 8, the dynamic response of the system is illustrated with the change of angular position Φ and angular velocity ω for the abovementioned different configurations. In all considered cases, the transient state can be observed in the beginning of the system's operation. The reason for this is that the pendulum must first synchronize with the external excitation, and then the steady state operation is observed. For energy harvesting studies, the steady state response is considered from which the highest output voltage/power is observed. In the first case, the system without additional magnets and without coils tended to maintain constant rotations after the synchronization of the pendulum and the vertical excitation. The variations in the angular velocity between positive and negative values depict the pendulum's oscillations around the equilibrium point. Further, the pendulum was able to pass this equilibrium point after around 50 s. Adding induction coils (Figure 8b) established the coil–magnet coupling. The pendulum at first started oscillating, but later when its energy was dissipated by coils, the angular velocity was reduced to zero. Mounting additional magnets strongly influenced the dynamics of the pendulum, which can be seen in Figure 8c. The magnetic force was adjusted by adjusting the axial distance between one of the pair of magnets. Initially, the magnetic potential barrier could not be passed by the pendulum to perform a full rotation. However, after increasing the magnetic force, the repulsive force assisted in easier passing of the magnetic barrier and the pendulum oscillated between the two positions.

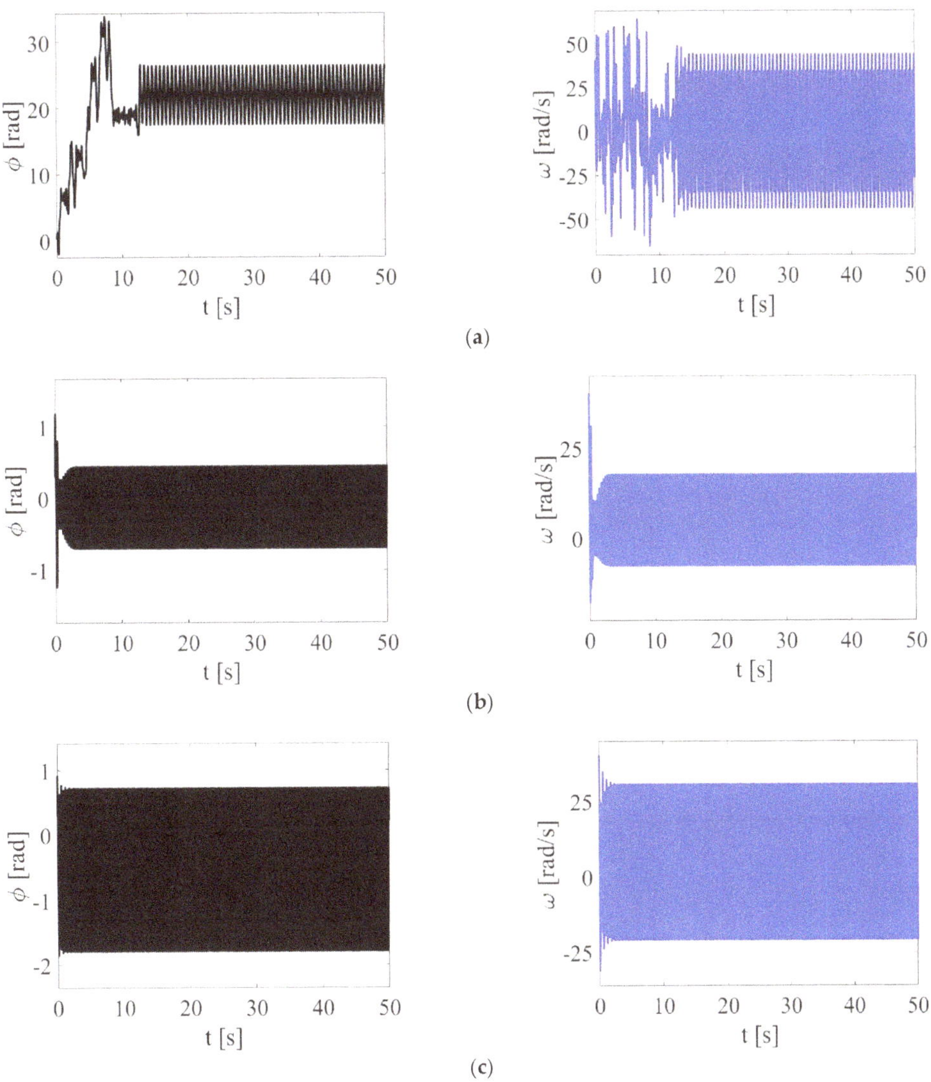

Figure 8. Selected solutions of the dynamic system in terms of the instantaneous rotation angle (left panel—black plots) and angular velocity (right panel—blue plots) of pendulum are shown here. Solutions for (**a**) a system without magnets and coils, (**b**) a system with coils but without additional magnets, and (**c**) a system with additional magnets and coils are presented.

In cases b–c, the voltage in the coils was induced due to the electromagnetic transduction during the parallel passing of the magnet-pendulum around the vicinity of the coils. In Figure 9, the power generated by the pendulum's movement is illustrated. The level of generated power increased as the magnetic force interaction became stronger. The mean value of power was estimated by the moving mean algorithm (see Figure 9—black line), whose value was in the range 0.01–0.02 [W] for case (b). For case (c), the mean harvester power was around 0.5 [W] because of larger and magnetic force-assisted oscillations.

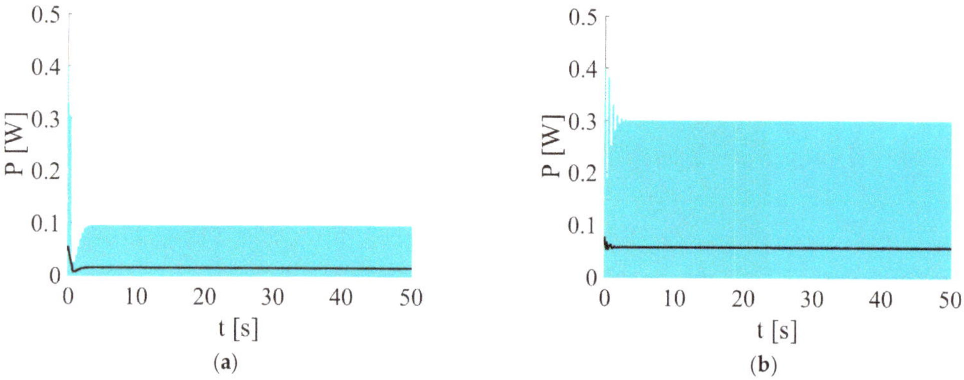

Figure 9. Power output plots corresponding to the solutions of cases (**a**,**b**) (shown in Figure 8b,c, respectively) are shown here. Note, there is no power output in the case shown in Figure 8a, because no electromagnetic transduction occurred in the absence of coils.

The phase portraits of the considered three cases are shown in Figure 10, which illustrates the trajectories drawn by the excited pendulum. Excitation of pendulum without any source of damping caused free rotation under the influence of gravitational potential only (see Figure 10a). After adding coils into the system, its damping caused small amplitude oscillations around the central equilibrium point, as shown in Figure 10b. This was caused by the transduction of energy between the pendulum magnet and coils. Installation of additional magnets led to inter-well oscillations around the offset equilibrium point, where magnetic interactions had maximum impact (see Figure 10c). The phase portrait took a square-based shape and depicted the smaller potential well formed at the specific angular position. It was observed that, as the strength of magnets increased, it became easier to pass through the magnetic barrier at the offset pendulum. Figure 10c shows oscillation of the pendulum between the central and offset equilibrium positions.

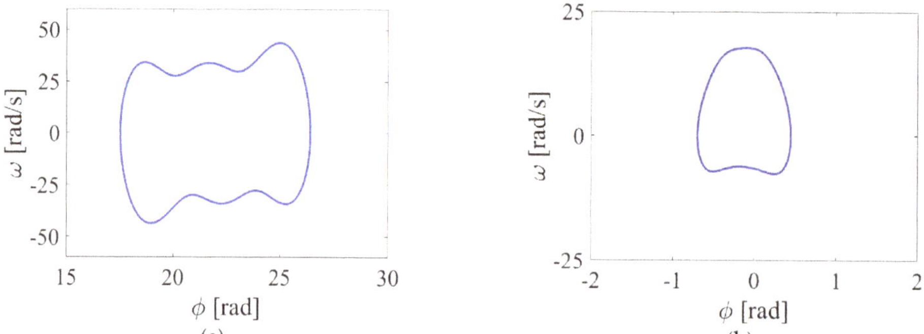

Figure 10. *Cont.*

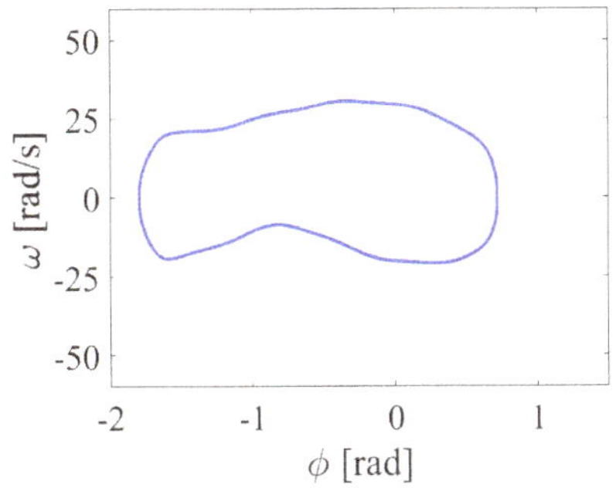

(c)

Figure 10. Phase portraits of the considered cases (**a–c**) for the solutions shown in Figure 8a–c.

5. Conclusions

This paper investigates the dynamic response of an electromagnetic energy harvester with a rotational pendulum with variable magnetic-force interactions. Based on the concept proposed in [7] with a simplified model, in this work a more rigorous two-stage mathematical model is proposed that includes FEM analysis of magnet–magnet and magnet–coil couplings. The governing equation of motion is presented, which includes the magnetic interaction terms from FEA from the first stage. Based on a physically realized model built using the 3D printing method, physical quantities and terms in the mathematical model were obtained. In this investigation, three majorly different cases were tested, which were configurations with and without coils and magnet couplings, and with adjustable magnet positions. The findings from the constant amplitude excitation experiments highlighted the significant impact of magnetic force on the pendulum's traversal through the magnetic barriers. When larger magnets were employed in a repulsive setup, the pendulum more easily overcame the initial barrier. Furthermore, as the pendulum exited the magnetic field of additional magnets, it experienced a substantial acceleration, which contributed to the increased kinetic energy. This heightened energy level enabled the pendulum to navigate through the subsequent magnetic barriers more effectively.

The results suggest that optimization of the system should involve strategically adjusting the position and strength of repulsive or attractive magnets for a fine tuning of the potential landscape. By doing so, nonlinearities will dominate the dynamics of the pendulum and enhance energy harvesting efficiency. Additionally, a comprehensive understanding of the system's behavior requires investigations on various initial conditions and excitation frequencies. Future works involving such analyses will reveal a spectrum of potential operating regimes, each offering advantages in terms of energy harvesting capabilities. By delving into these aspects, novel strategies for improving the rotational pendulum-based energy harvesting systems can be explored, thereby broadening the practical applications. In our next step, a comprehensive investigation of the model under different angular and axial positions of the side magnets is to be carried out.

Supplementary Materials: The following supporting information can be downloaded at: https://www.mdpi.com/article/10.3390/app14083265/s1, Figure S1: (**a**) Global mesh model of a stationary pendulum and magnet-coil pair. (**b**) Local enlarged mesh view of a stationary pendulum and a magnet-coil pair. (**c**) Contours of magnetic flux density [T] at a distance of −27.5 mm along x-axis; Figure S2: Variation of components of magnet force in x, y, and z axes with the distance between the magnets along x-axis are shown in (**a–c**), respectively; Figure S3: Under the application of 1A current through coil, variation of components of magnet force in x, y, and z axes with the distance between the magnets along x-axis are shown in (**a–c**), respectively; Figure S4: (**a**) Local mesh of a quasi-dynamic model for a pendulum-magnet and a coil with induced voltage. (**b**) Induced voltage in coil for different speeds of the pendulum. (**c**) Reference characteristics of induced voltage obtained after normalizing with speed; Table S1: List of variables used for the FEA simulation of magnetic force interactions.

Author Contributions: Conceptualization, G.L., M.K., P.W., B.A. and A.M.G.; methodology, G.L., M.K., P.W., B.A. and A.M.G.; software, G.L., M.K. and B.A.; validation, G.L. and M.K.; formal analysis, G.L.; investigation, G.L. and B.A.; resources, B.A.; data curation, G.L., M.K., P.W. and B.A; writing—original draft preparation, G.L., B.A. and A.M.G.; writing—review and editing, G.L., B.A. and A.M.G.; visualization, G.L., M.K., P.W., B.A. and A.M.G.; supervision, G.L.; project administration, P.W. and G.L.; funding acquisition, G.L. All authors have read and agreed to the published version of the manuscript.

Funding: Supported by the National Science Centre, Poland under the project SHENG-2 (Grant No. 2021/40/Q/ST8/00362).

Institutional Review Board Statement: Not applicable.

Informed Consent Statement: Not applicable.

Data Availability Statement: The data presented in this study are available on request from the corresponding author.

Acknowledgments: The authors would like to thank Prof. Shengxi Zhou for useful discussions.

Conflicts of Interest: The authors declare no conflict of interest.

References

1. Kaźmierski, T.J.; Beeby, S. *Energy Harvesting Systems, Principles, Modelling and Applications*; Springer: New York, NY, USA, 2011.
2. Priya, S.; Inman, D.J. *Energy Harvesting Technologies*; Springer: Blacksburg, VA, USA, 2009.
3. Elvin, N.; Erturk, A. *Advances in Energy Harvesting Methods*; Springer: New York, NY, USA, 2013.
4. Yang, T.; Zhou, S.; Litak, G.; Jing, X. Recent advances in correlation and integration between vibration control, energy harvesting and monitoring. *Nonlinear Dyn.* **2023**, *111*, 20525–20562. [CrossRef]
5. Alegret, R.N.; Aragones, R.; Oliver, J.; Ferrer, C. Exploring IIoT and Energy Harvesting Boundaries. In Proceedings of the IECON 2019—45th Annual Conference of the IEEE Industrial Electronics Society, Lisbon, Portugal, 14–17 October 2019; pp. 6732–6736. [CrossRef]
6. Zeadally, S.; Shaikh, F.K.; Talpur, A.; Sheng, Q.Z. Design architectures for energy harvesting in the Internet of Thing. *Renew. Sustain. Energy Rev.* **2020**, *128*, 109901. [CrossRef]
7. Ambrożkiewicz, B.; Litak, G.; Wolszczak, P. Modelling of electromagnetic energy harvester with rotational pendulum using mechanical vibrations to scavenge electrical energy. *Appl. Sci.* **2020**, *10*, 671. [CrossRef]
8. He, J.; Wen, T.; Qian, S.; Zhang, Z.; Tian, Z.; Zhu, J.; Mu, J.; Hou, X.; Geng, W.; Cho, J.; et al. Triboelectric-piezoelectric-electromagnetic hybrid nanogenerator for high-efficient vibration energy harvesting and self-powered wireless monitoring system. *Nano Energy* **2018**, *43*, 326–339. [CrossRef]
9. Zhang, B.; Zhang, Q.; Wang, W.; Han, J.; Tang, X.; Gu, F.; Ball, A.D. Dynamic modelling and structural optimization of a bistable electromagnetic vibration energy harvester. *Energies* **2019**, *12*, 2410. [CrossRef]
10. Cottone, F.; Vocca, H.; Gammaitoni, L. Nonlinear energy harvesting. *Phys. Rev. Lett.* **2009**, *102*, 080601. [CrossRef] [PubMed]
11. Erturk, A.; Inman, D.J. An experimentally validated bimorph cantilever model for piezoelectric energy harvesting from base excitations. *Smart Mater. Struct.* **2009**, *18*, 025009. [CrossRef]
12. Tvedt, L.G.W.; Nguyen, D.S.; Halvorsen, E. Nonlinear behaviour of an electrostatic energy harvester under wide- and narrowband excitation. *J. Microelectromech. Syst.* **2010**, *19*, 305–316. [CrossRef]
13. Zhang, Y.; Wang, T.; Luo, A.; Hu, Y.; Li, X.; Wang, F. Micro electrostatic energy harvester with both broad bandwidth and high normalized power density. *Appl. Energy* **2018**, *212*, 362–371. [CrossRef]
14. Rubes, O.; Brablc, M.; Hadas, Z. Nonlinear vibration energy harvester: Design and oscillating stability analyses. *Mech. Syst. Signal Process.* **2019**, *125*, 170–184. [CrossRef]
15. Aravindan, M.; Ali, S.F. Exploring 1:3 internal resonance for broadband piezoelectric energy harvesting. *Mech. Syst. Signal Process.* **2021**, *153*, 107493. [CrossRef]
16. Haroun, A.; Yamada, I.; Warisawa, S. Study of electromagnetic vibration energy harvesting with free/impact motion for low frequency operation. *J. Sound Vib.* **2015**, *349*, 389–402. [CrossRef]
17. Phan, N.P.; Bader, S.; Oelmann, B. Performance of an electromagnetic energy harvester with linear and nonlinear springs under real vibrations. *Sensors* **2020**, *20*, 5456. [CrossRef] [PubMed]
18. Pepe, G.; Dorla, A.; Roveri, N.; Carcaterra, A. Vibration energy harvesting for cars: Semi-active piezo controllers. *Arch. Appl. Mech.* **2023**, *93*, 663–685. [CrossRef]
19. Cahill, P.; Hazra, B.; Karoumi, R.; Mathewson, A.; Pakrashi, V. Vibration energy harvesting based monitoring of an operational bridge undergoing forced vibration and train passage. *Mech. Syst. Signal Process.* **2018**, *106*, 265–283. [CrossRef]

20. Fan, W.; Zhang, C.; Liu, Y.; Wang, S.; Dong, K.; Li, Y.; Wu, F.; Liang, J.; Wang, C.; Zhang, Y. An ultra-thin piezoelectric nanogenerator with breathable, superhydrophobic, and antibacterial properties for human motion monitoring. *Nano Res.* **2023**, *16*, 11612–11620. [CrossRef]
21. Huang, D.; Zhou, S.; Han, Q.; Litak, G. Response analysis of the nonlinear vibration energy harvester with an uncertain parameter. *Proc. Inst. Mech. Eng. Part K J. Multi-Body Dyn.* **2020**, *234*, 393–407. [CrossRef]
22. Malaji, P.V.; Ali, S.F. Broadband energy harvesting with mechanically coupled harvesters. *Sens. Actuators A* **2017**, *255*, 1–9. [CrossRef]
23. Kullah, H.; Najafi, K. Energy Scavenging from Low-Frequency Vibrations by Using Frequency Up-Conversion for Wireless Sensor Applications. *IEEE Sens. J.* **2008**, *8*, 261–268. [CrossRef]
24. Jia, Y. Review of nonlinear vibration energy harvesting: Duffing, bistability, parametric, stochastic and others. *J. Intell. Mater. Syst. Struct.* **2020**, *31*, 921–944. [CrossRef]
25. Giri, A.M.; Ali, S.F.; Arockiarajan, A. Characterizing harmonic and subharmonic solutions of the bi-stable piezoelectric harvester with a modified harmonic balance approach. *Mech. Syst. Signal Process.* **2023**, *198*, 110437. [CrossRef]
26. Ferrari, M.; Ferrari, V.; Guizzetti, M.; Ando, B.; Baglio, S.; Trigona, C. Improved energy harvesting from wideband vibrations by nonlinear piezoelectric converters. *Sens. Actuators A Phys.* **2010**, *162*, 425–431. [CrossRef]
27. Huguet, T.; Badel, A.; Lallart, M. Exploiting bistable oscillator subharmonics for magnified broadband vibration energy harvesting. *Appl. Phys. Lett.* **2017**, *111*, 173905. [CrossRef]
28. Pellegrini, S.P.; Tolou, N.; Schenk, M.; Herder, J.L. Bistable vibration energy harvesters: A review. *J. Intell. Mater. Syst. Struct.* **2013**, *24*, 1303–1312. [CrossRef]
29. Giri, A.M.; Ali, S.F.; Arockiarajan, A. Influence of asymmetric potential on multiple solutions of the bi-stable piezoelectric harvester. *Eur. Phys. J. Spec. Top.* **2022**, *231*, 1443–1464. [CrossRef]
30. Dong, L.; Closson, A.B.; Jin, C.; Trase, I.; Chen, Z.; Zhang, J.X.J. Vibration-Energy-Harvesting System: Transduction Mechanics, Frequency Tuning Techniques and Biomechanical Applications. *Adv. Mater. Technol.* **2019**, *4*, 1900177. [CrossRef] [PubMed]
31. Bouhedma, S.; Rao, Y.; Schutz, A.; Yuan, C.; Hu, S.; Lange, F.; Bechtold, T.; Hohfeld, D. System-Level Model and Simulation of a Frequency-Tunable Vibration Energy Harvester. *Micromachines* **2020**, *11*, 91. [CrossRef] [PubMed]
32. Huguet, T.; Lallart, M.; Badel, A. Orbit jump in bistable energy harvesters through buckling level modification. *Mech. Syst. Signal Process.* **2019**, *128*, 202–215. [CrossRef]
33. Lee, S.; Lee, Y.; Dongseob, K.; Yan, Y.; Lin, L.; Lin, Z.H.; Hwang, W.; Wang, Z.L. Triboelectric nanogenerator for harvesting pendulum oscillation energy. *Nano Energy* **2013**, *2*, 1113–1120. [CrossRef]
34. Xu, J.; Tang, J. Multi-directional energy harvesting by piezoelectric cantilever-pendulum with internal resonance. *Appl. Phys. Lett.* **2015**, *107*, 213902. [CrossRef]
35. Wu, Y.; Qiu, J.; Shengpeng, Z.; Ji, H.; Chen, Y.; Li, S. A piezoelectric spring pendulum oscillator used for multi-directional and ultra-low frequency vibration energy harvesting. *Appl. Energy* **2018**, *231*, 600–614. [CrossRef]
36. Wang, Y.J.; Chen, C.D.; Sung, C.K. Design of a frequency-adjusting device for harvesting energy from a rotating wheel. *Sens. Actuators A* **2010**, *159*, 196–203. [CrossRef]
37. Naifar, S.; Bradai, S.; Viehweger, C.; Kanoun, O. Survey of electromagnetic and magnetoelectric vibration energy harvesters for low frequency excitation. *Measurement* **2017**, *106*, 251–263. [CrossRef]
38. Yurchenko, D.; Alevras, P. Parametric pendulum based wave energy converter. *Mech. Syst. Signal Process.* **2018**, *99*, 504–515. [CrossRef]
39. Malaji, P.V.; Ali, S.F. Analysis of energy harvesting from multiple pendulums with and without mechanical coupling. *Eur. Phys. J. Spec. Top.* **2015**, *224*, 2823–2838. [CrossRef]
40. Malaji, P.V.; Doddi, S.; Friswell, M.I.; Adhikari, S. Analysis of pendulums coupled by torsional springs for energy harvesting. *MATEC Web Conf.* **2018**, *211*, 05008. [CrossRef]
41. Kuang, Y.; Hide, R.; Zhu, M. Broadband energy harvesting by nonlinear magnetic rolling pendulum with sub-harmonic resonance. *Appl. Energy* **2019**, *255*, 113822. [CrossRef]
42. Dai, X. An vibration energy harvester with broadband and frequency-doubling characteristics based on rotary pendulums. *Sens. Actuators A Phys.* **2016**, *241*, 161–168. [CrossRef]
43. Koszewnik, A. Analytical Modeling and Experimental validation of an energy harvesting system for the smart plate with and integrated piezo-harvester. *Sensors* **2019**, *19*, 812. [CrossRef]
44. Borowiec, M.; Syta, A.; Litak, G. Energy harvesting optimizing with a magnetostrictive cantilever beam system. *Int. J. Struct. Stab. Dyn.* **2019**, *19*, 1941002. [CrossRef]
45. Smilek, J.; Hadas, Z.; Vetiska, J.; Beeby, S. Rolling mass energy harvester for very low frequency of input vibrations. *Mech. Syst. Signal Process.* **2019**, *125*, 215–228. [CrossRef]
46. Toyabur, R.M.; Salauddin, M.; Cho, H.; Park, J.Y. A multimodal hybrid energy harvester based on piezoelectric-electromagnetic mechanisms for low-frequency ambient vibrations. *Energy Convers. Manag.* **2018**, *168*, 454–466. [CrossRef]
47. Li, Z.; Yan, Z.; Luo, J.; Yang, Z. Performance comparison of electromagnetic energy harvesters based on magnet arrays of alternating polarity and configuration. *Energy Convers. Manag.* **2019**, *179*, 132–140. [CrossRef]

48. Cui, X.; Hu, J. An infrasonic vibration energy harvester using pendulum impact. *Int. J. Appl. Electromagn. Mech.* **2015**, *47*, 467–474. [CrossRef]
49. Avila Bernal, A.G.; Linares Garcia, L.E. The modelling of an electromagnetic energy harvester architecture. *Appl. Math. Model.* **2012**, *36*, 4728–4741. [CrossRef]

applied sciences

MDPI

Review

Energy Storage in Carbon Fiber-Based Batteries: Trends and Future Perspectives

Antonio del Bosque [1], Diego Vergara [1,*], Georgios Lampropoulos [2,3,*] and Pablo Fernández-Arias [1]

1 Technology, Instruction and Design in Engineering and Education Research Group (TiDEE.rg), Catholic University of Ávila, C/Canteros s/n, 05005 Ávila, Spain; antonio.bosque@ucavila.es (A.d.B.); pablo.fernandezarias@ucavila.es (P.F.-A.)
2 Department of Applied Informatics, University of Macedonia, 54636 Thessaloniki, Greece
3 Department of Education, University of Nicosia, 2417 Nicosia, Cyprus
* Correspondence: diego.vergara@ucavila.es (D.V.); glampropoulos@uom.edu.gr (G.L.)

Abstract: Carbon fiber-based batteries, integrating energy storage with structural functionality, are emerging as a key innovation in the transition toward energy sustainability. Offering significant potential for lighter and more efficient designs, these advanced battery systems are increasingly gaining ground. Through a bibliometric analysis of scientific literature, the study identifies three primary research areas: (i) the development of anodes for lithium-ion batteries, tackling challenges such as dendrite formation and performance degradation; (ii) the creation of new carbon fiber-based cathodes with coatings of $LiFePO_4$, $LiCoO_2$, or other nanoparticles, alongside efforts to develop cobalt-free alternatives; and (iii) the advancement of solid electrolytes that achieve a balance between ionic conductivity and mechanical strength. These advancements position carbon fiber-based batteries as promising solutions for seamless integration into various structural applications. The analysis of publication trends, citation patterns, and collaboration networks provides critical insights into the ongoing technological developments, current research challenges, and emerging trends in this field. Moreover, the study highlights potential research directions, underscoring the importance of continuous innovation to fully realize the potential of carbon fiber-based energy storage technologies.

Keywords: batteries; carbon-fiber; structural composites; bibliometric analysis; review

1. Introduction

Nowadays, there is an increasing concern about global sustainability and the growing energy demand. A significant focus has been placed on reducing CO_2 emissions, which are known to be highly detrimental to human health and the integrity of various ecosystems. Thus, the development of advanced energy storage technologies has become a critical area of research [1–3]. Battery systems are now being tailored to meet specific needs: low-current, long-duration applications like clock circuits; medium-current uses such as mobile devices; and high-current demands like electric vehicles and drones. In this regard, battery electric vehicles and drones are now attracting the interest of industry as they significantly reduce urban air pollution [4,5]. Among the materials being investigated for energy storage applications, carbon fibre stands out as a particularly promising candidate [6–8].

Carbon fiber, traditionally utilized in the aerospace, automotive, and sports equipment industries, possesses unique structural characteristics that enable the development of multifunctional materials. These materials can simultaneously serve as both the structural component and the energy storage medium [9–11]. As a result, conventional heavy batteries can be either replaced by or integrated into carbon fiber-based batteries, allowing them to fulfill both structural and energy storage roles. For instance, they could be incorporated into the trunk of a car or the fuselage of an airplane, leading to significantly lighter and more efficient designs [7,12–14]. The ability of carbon fibers to be coated into various forms and integrated with other active materials further enhances their versatility in battery design.

The general architecture of carbon fiber-based batteries is illustrated in Figure 1. It consists of a carbon fiber-reinforced polymer composite, where the carbon fibers serve as both the anode (negative electrode) and the cathode (positive electrode) [15,16]. A separate interlayer or fiberglass functions as the separator, ensuring that the electrodes do not short-circuit [17]. The entire structure is embedded within a solid electrolyte, which could be a polyelectrolyte, polymer-based electrolyte, or ceramic electrolyte, depending on the specific design and performance requirements [18,19]. Here, solid electrolytes can function effectively as separators. These systems leverage the unique properties of solid electrolytes, which not only facilitate ion transport between the anode and cathode but also serve as a physical barrier. The dual functionality of solid electrolytes enhances the overall safety and performance of the battery, making them an attractive option in the development of advanced energy storage solutions [20].

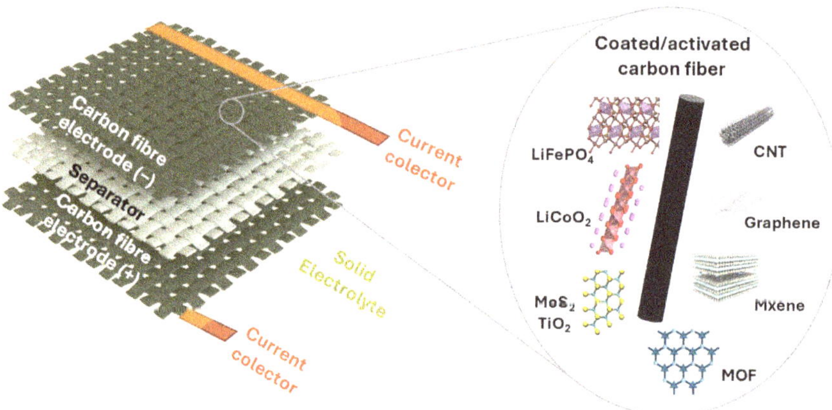

Figure 1. Schematic representation of the innovative carbon fibre-based battery architecture. The carbon fibres, integrated into a reinforced polymer composite, serve dual roles as both anode and cathode.

Electrode polarization is a key factor in the performance of carbon fiber-based batteries, affecting charge/discharge rates, energy efficiency, and lifespan [21,22]. It is influenced by internal resistance, electrode surface characteristics, and electrolyte composition. In carbon fiber systems, factors like fiber structure, porosity, and coatings impact ion transfer and reaction kinetics. Additionally, the choice of electrolyte, especially solid-state or polymer-based, can alter polarization levels. Mitigating polarization involves optimizing electrode design and choosing electrolytes that enhance ionic conductivity, leading to more efficient energy storage and longer battery life. This understanding is crucial for advancing carbon fiber-based batteries.

The scientific community has been particularly focused on leveraging the high surface area, tunable porosity, and excellent electrochemical stability of coated carbon fibers to improve the performance of batteries. One of the most significant advantages of carbon fiber-based electrodes is their potential to enhance charge/discharge rates, increase energy density, and extend the lifespan of batteries [23]. Moreover, the inherent flexibility and durability of carbon fibers open new possibilities for the development of flexible and wearable energy storage devices, which are increasingly in demand in consumer electronics and biomedical applications. For this purpose, carbon fibers are activated as cathodes or coated as anodes/cathodes with $LiFePO_4$ [24–26], $LiCoO_2$ [27,28], MoS_2 [29,30], TiO_2 [31,32], metal organic frameworks (MOFs) [33,34], Mxenes [35,36], graphene [8,37] or carbon nanotubes (CNTs) [9,38,39].

However, despite these advantages, the integration of carbon fiber into battery systems presents several challenges. Key issues include the optimization of the fiber-matrix interface to ensure efficient electron and ion transport, the prevention of degradation mechanisms that could compromise battery performance, and the scalability of manufacturing processes for large-scale production. Addressing these challenges requires a deep understanding of the interplay among the microstructure of carbon fibers, their electrochemical behavior, and the overall battery architecture.

Various authors have made scientific reviews on structural batteries [40–42] or multifunctional composites for energy storage [43–46]. However, there are no existing studies that provide a bibliometric analysis focusing on trends and future perspectives in carbon fiber-based batteries. Thus, this study aims to provide a detailed bibliometric analysis, tracing the evolution of this research field over the past decade while highlighting key trends, emerging technologies, and critical challenges. Additionally, it explores the potential of carbon fiber to revolutionize energy storage by discussing future directions and opportunities for innovation in this rapidly growing area of research. The insights gained from this study could pave the way for significant breakthroughs in the design and application of multifunctional materials in energy storage, ultimately supporting the global transition to a cleaner and more sustainable energy future.

2. Methods

Due to the nature and breadth of the topic, the Preferred Reporting Items for Systematic reviews and Meta-Analyses (PRISMA) statement [47] was selected as a suitable approach to search for and identify relevant documents. Additionally, a bibliometric analysis and scientific mapping approach was followed to analyze the documents which is regarded as an appropriate and effective approach to examine such topics based on the existing literature [48]. Scopus and Web of Science (WoS) were selected as the most suitable databases due to their containing highly impactful documents and sources related to the topic, and their ability to generate data that can be used by "Bibliometrix" which was the tool used to analyze the documents. Specifically, "Bibliometrix" is an open-source R package developed to aid in the conduct of bibliometric analysis and scientific mapping studies [49].

The following query was used: (TITLE-ABS-KEY (["carbon fibre" OR "carbon fibres"]) AND TITLE-ABS-KEY (["battery" OR "batteries"]) AND TITLE-ABS-KEY (["energy" OR "storage"]) AND NOT TITLE-ABS-KEY (["supercapacitor" OR "supercapacitors" OR "fuel cell"])). All types of documents were assessed. However, only documents published in English and during 2013–2023 were considered. The specific duration was selected to explore the advancement of the field in the last decade. Data from 2024 was not included as 2024 was still ongoing when this study took place. For a study to be deemed as eligible, the following inclusion criterion was set: The study should focus on carbon fiber-based batteries and their use as a means to store energy.

When using the aforementioned query in August 2024, a total of 1014 documents (Scopus: 940 and WoS: 74) were identified. Out of these documents, 44 were duplicates and were removed. Hence, 970 documents were assessed for eligibility. After manually processing the documents, 130 documents were removed. Specifically, 17 documents were removed as they were proceedings books, 1 was removed because it was a book, 2 were removed as they were short communications and notes, 1 was removed because it was an erratum, and 109 documents were removed since they did not meet the inclusion criterion set. As a result, the document collection examined in this study consisted of 840 documents. Figure 2 presents the document processing.

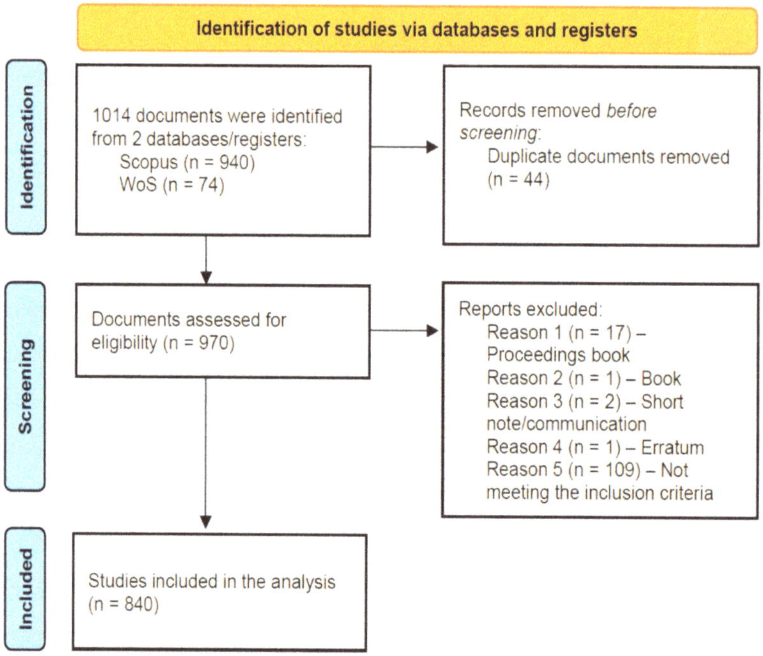

Figure 2. Comprehensive Workflow for Document Processing.

3. Results and Discussion

This section presents the information of the document collection examined and goes over different aspects of the documents. Specifically, the number of published documents and the number of citations received are examined, the sources used are explored, and the authors' affiliations and countries are looked into. Moreover, the documents are further analyzed using keywords to identify the trend topics, the conceptual structure map, the thematic map, and the evolution of the topic.

3.1. Document Collection

The information of the document collection is summarized in Table 1. Specifically, the document collection consists of 840 documents that were published during the last decade (2013–2023). The documents were published in 248 different sources and were written by 2183 authors. Within the document collection, only five documents were single-authored which further highlights the large number of co-authors per document (6.49). However, the international co-authorship rate is extremely low (1.43%) as most collaborations materialized in inter-country settings. Furthermore, the documents of the collection have an average age of 4.12 years and received 36.81 citations on average. The relatively high annual growth rate (24.08%) is in line with the increasing interest in the topic and the need to further explore it. The significant majority of documents were published in journals and in comparison, a limited number of documents were published as conference/proceedings papers. Within the documents, 4516 unique keywords plus and 2089 unique author's keywords were used to characterize and encapsulate the contents and topic of each document.

Table 1. Document collection information (2013–2023).

Description	Results	Description	Results
Main information about data		**Document types**	
Timespan	2013:2023	article	756
Sources (Journals, Books, etc)	248	book chapter	2
Documents	840	conference/proceedings paper	64
Annual Growth Rate %	24.08	review	18
Document Average Age	4.12	**Authors**	
Average Citations per Document	36.81	Authors	2183
		Authors of single-authored docs	4
Document contents		**Authors collaboration**	
Keywords Plus (ID)	4516	Single-authored docs	5
Author's Keywords (DE)	2089	Co-Authors per Doc	6.49
		International co-authorships %	1.43

3.2. *Annual Published Documents and Citations*

When taking into account the high annual growth rate (24.08%) and the average document age (4.12) during the 2013–2023 time period as well as the fact that most documents were published in the last five years (2019–2023) (freq. = 597 and perc. = 71.1%), the significance and recency of the topic become evident. Moreover, it is worth mentioning that a significant portion of the documents were published in the last two years (2022–2023) (freq. = 291 and perc. = 34.6%) which indicates the high interest in this topic which is expected to further increase in the future. Therefore, it comes with no surprise that most documents were published in 2023 (17.5%), followed closely by those published in 2022 (17.1%) which is a significant increase compared to the documents published in 2014 (1.9%) and 2013 (2.0%) respectively. The annual scientific production of the document collection examined is presented in Figure 3. However, when taking into account the mean total citations per document, it becomes evident that the documents published in 2013 (MeanTCperDoc = 109.47) and 2014 (MeanTCperDoc = 95.31) laid the foundations for this field of study to mature and further advance. Other impactful years were 2018 (MeanTCperDoc = 68.18) and 2016 (MeanTCperDoc = 60.89). However, it should be noted that the total number of documents published in each year and the citable years strongly influence these outcomes. Hence, it is expected that these results will change in the future. To provide a more in-depth look into the most impactful documents, the documents are further examined in document analysis section. The annual scientific production and citations are presented in detail in Table 2.

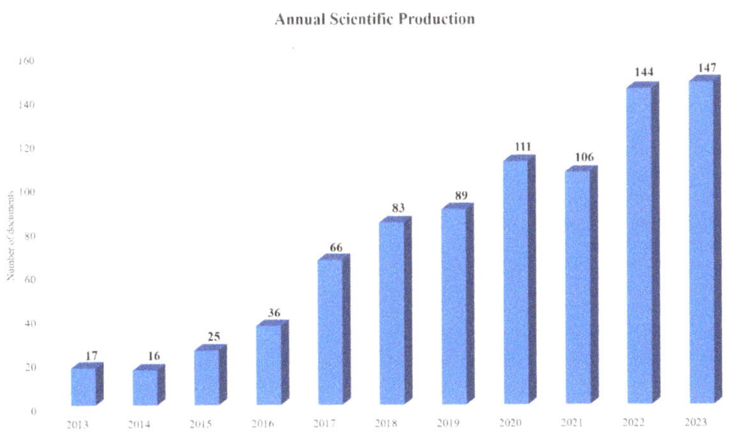

Figure 3. Annual published documents (2013–2023).

Table 2. Annual scientific production and citations (2013–2023).

Year	MeanTCperDoc	N	MeanTCperYear	CitableYears
2013	109.47	17	9.12	12
2014	95.31	16	8.66	11
2015	33.08	25	3.31	10
2016	60.89	36	6.77	9
2017	50.92	66	6.36	8
2018	68.18	83	9.74	7
2019	54.25	89	9.04	6
2020	40.95	111	8.19	5
2021	24.4	106	6.1	4
2022	16.73	144	5.58	3
2023	7.69	147	3.85	2

3.3. Sources

Of the 840 documents examined, the vast majority were published as journal articles (90.0%). To a significantly less extent, documents were published as conference/proceedings paper (7.6%) and review articles (2.1%). Only two documents (0.2%) were published as book chapters. Additionally, "ACS Applied Materials & Interfaces" (h-index: 25 and total citations: 2208), "Journal of Materials Chemistry A" (h-index: 23 and total citations: 1810), "Electrochimica Acta" (h-index: 22 and total citations: 1874), "Journal of Power Sources" (h-index: 21 and total citations: 1406), and "Electrochimica Acta" (h-index: 19 and total citations: 997) emerged as the top-5 sources when considering the h-index and total citations received. The detailed list of the most impactful sources is depicted in Table 3.

Table 3. Most impactful sources based on h-index and total citations.

Sources	h-Index	g-Index	m-Index	TC	NP	PY_Start
ACS Applied Materials & Interfaces	25	44	2.083	2208	44	2013
Journal of Materials Chemistry A	23	30	1.917	1810	30	2013
Electrochimica Acta	22	26	2.75	1874	26	2017
Journal of Power Sources	21	27	1.909	1406	27	2014
Chemical Engineering Journal	17	29	2.125	997	29	2017
Small	15	19	1.875	1012	19	2017
Advanced Functional Materials	14	15	1.273	1381	15	2014
Carbon	14	19	1.167	705	19	2013
Nano Energy	14	15	1.167	1480	15	2013

Using Bradford's law, the sources were further examined. The documents were further analyzed using clustering by coupling. Specifically, documents were set as the analysis unit, references were used as the coupling measurement and citation score was used as the impact measure. Moreover, the clustering algorithm used was Walktrap, the number of units were set to 250, and the min cluster frequency was set to 5 which are the default values suggested by the tool used in this study. Specifically, the sources were grouped into three clusters. Cluster 1 has the most impactful sources, followed by cluster 2 and cluster 3. Based on the outcomes, cluster 1 had 11 (4.4%) sources, cluster 2 had 42 (16.9%) sources, and cluster 3 had 195 (78.6%) sources. Moreover, cluster 1 had a total of 292 (34.8%) published documents, cluster 2 had 274 (32.6%) published documents, and cluster 3 also had 274 (32.6%) published documents. Table 4 presents the top-10 sources according to Bradford's law. Out of all sources, "ACS Applied Materials & Interfaces", "Electrochimica Acta", "Journal of Materials Chemistry A", "Chemical Engineering Journal", and "Journal of Alloys and Compounds" arose as the top-5 most impactful ones following Bradford's law.

Table 4. Most impactful sources based on Bradford's law.

Source	Rank	Freq	cumFreq	Cluster
ACS Applied Materials & Interfaces	1	44	44	Cluster 1
Electrochimica Acta	2	35	79	Cluster 1
Journal of Materials Chemistry A	3	30	109	Cluster 1
Chemical Engineering Journal	4	29	138	Cluster 1
Journal of Alloys and Compounds	5	28	166	Cluster 1
Journal of Power Sources	6	27	193	Cluster 1
Energy Storage Materials	7	26	219	Cluster 1
ACS Applied Energy Materials Journal	8	19	238	Cluster 1
Carbon	9	19	257	Cluster 1
Small	10	19	276	Cluster 1
Composites Science and Technology	11	16	292	Cluster 1

3.4. Affiliations

The top-10 affiliation based on the number of published documents is presented in Figure 4. More specifically, KTH Royal Institute of Technology (Sweden) (52 documents), Central South University (China) (45 documents), Hunan University (China) (44 documents), Chalmers University of Technology (Sweden) (33 documents), and Zhejiang University (China) (32 documents) arose as the top-5 affiliations that have contributed the largest number of documents to this topic. The affiliation collaboration network (Figure 5) highlights the high number of intra-country collaborations (SCP) and simultaneously, the lack of inter-country collaborations (MCP) as it becomes more evident when analyzing the country specific information in the next section.

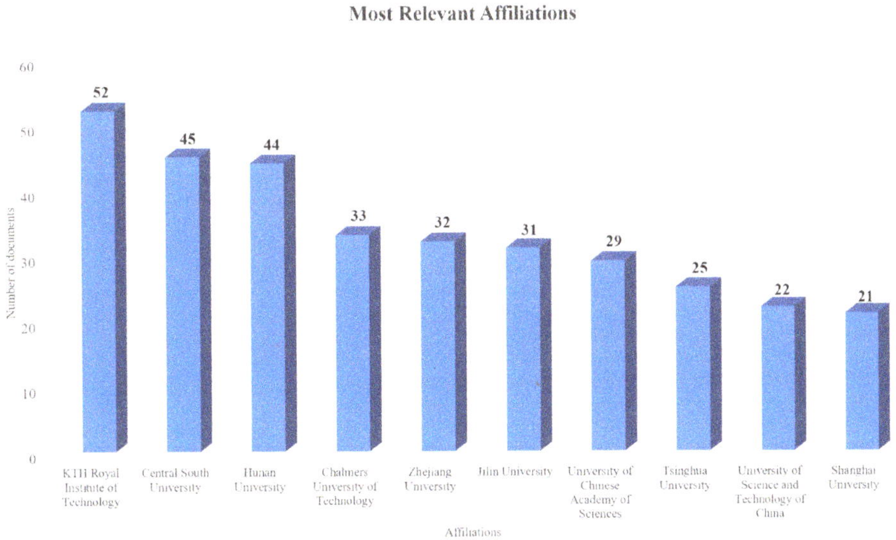

Figure 4. Top affiliations based on the number of documents published.

Figure 5. Global collaboration network among affiliations.

3.5. Countries

The documents of the document collection examined were written by 2183 authors from 32 countries. For each document, the country of the corresponding author was used. In the case that there was no corresponding author indicated in the document, the country of the first author was used. The top-10 countries according to the number of documents published are presented in Table 5. Based on the results, China (517 documents), the United States (70 documents), Sweden (52 documents), South Korea (37 documents), and Australia (31 documents) emerged as the top-5 countries that have contributed the largest number of relevant to the topic documents. When examining the SCP and MCP, once again, China (SCP = 515), the United States (SCP = 70), Sweden (SCP = 51), South Korea (SCP = 37), and Australia (SCP = 29) were the countries with the highest number of intra-country collaborations. When considering the low levels of inter-country collaborations and taking into account the extremely low rate of international collaborations (1.43%), the drastic need for more international collaboration to materialize to advance this field of study becomes evident. This fact becomes more obvious when examining the country collaboration network (Figure 6) and collaboration map (Figure 7) in which the limited number of international collaborations can be seen.

Table 5. Countries that published the most over time.

Country	Articles	SCP	MCP	Freq	MCP_Ratio
China	517	515	2	0.615	0.004
United States	70	70	0	0.083	0
Sweden	52	51	1	0.062	0.019
South Korea	37	37	0	0.044	0
Australia	31	29	2	0.037	0.065
India	19	19	0	0.023	0
Germany	18	17	1	0.021	0.056
Hong Kong	14	14	0	0.017	0
Singapore	12	12	0	0.014	0
Japan	10	10	0	0.012	0

Furthermore, Table 6 presents the top countries based on the total citations received. According to the findings, China (20,033 citations), the United States (2963 citations), Sweden (1694 citations), Australia (1408 citations), and South Korea (934 citations) were the top-5 countries that have received the most citations. Among the top-10 countries in terms of total citations received presented in Table 7, Georgia (ACR = 90), Czech Republic (ACR = 88), Singapore (ACR = 74.8), Hong Kong (ACR = 55.6), and Australia (ACR = 45.4) were the countries with the largest number of ACR per document.

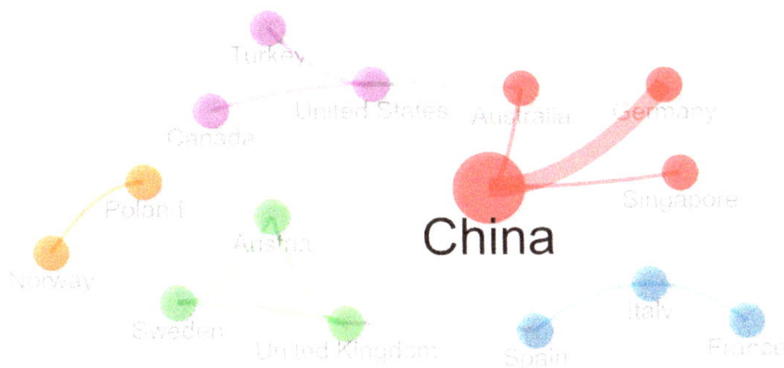

Figure 6. Country collaboration network.

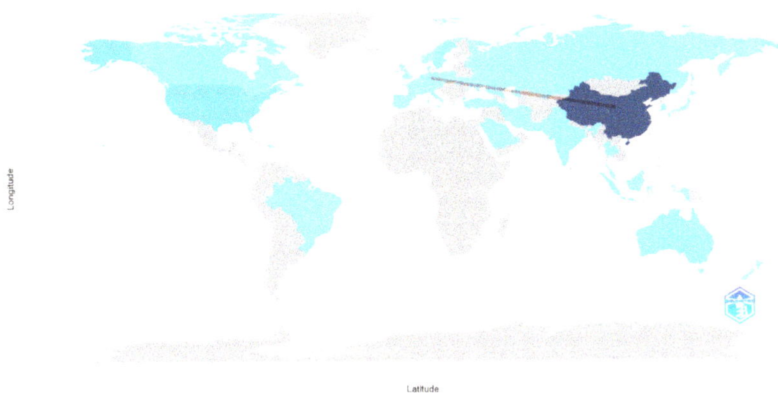

Figure 7. Country collaboration map.

Table 6. Countries that received the most citations.

Country	TC	Average Document Citations
China	20,033	38.7
United States	2963	42.3
Sweden	1694	32.6
Australia	1408	45.4
South Korea	934	25.2
Singapore	897	74.8
Hong Kong	779	55.6
Germany	396	22
India	376	19.8
Georgia	270	90

The number of total citations received was also examined to identify the most impactful countries in terms of the total number of citations the documents written by authors affiliated with an institute in this specific country received. Particularly, the United States (1752 total citations), Australia (1124 total citations), Israel (705 total citations), and China (572 total citations) were the top-4 countries which received the most citations. The list of the top-10 countries in terms of the total citations received is presented in Table 6. Additionally, when looking at the average article citations for each country, the documents written by authors in Israel received a significantly higher number of average citations per document (352.5). Documents coming from Australia (62.4) and Denmark (43.2) also showcased a high level of average citations received (ACR) per document.

Table 7. Countries that with the highest average citations received per document.

Country	TC	Average Document Citations
Georgia	270	90
Czech Republic	88	88
Singapore	897	74.8
Hong Kong	779	55.6
Australia	1408	45.4
United States	2963	42.3
China	20,033	38.7
Switzerland	37	37
Turkey	37	37
Thailand	173	34.6

3.6. Document Analysis

Table 8 presents the most impactful documents based on the global citations they have received. Based on the results, the studies of R. Zhang et al. (2018), Zhu et al. (2013), Fu et al. (2014), W. Zhang et al. (2018), and Zhong et al. (2018) emerged as the top-5 most impactful ones. Taking these outcomes into account, it can be inferred that the focus is on developing high-performance, durable electrochemical anodes by exploring various coatings on carbon fiber that incorporate nanoparticles and lithium. This is because the practical use of non-coated lithium metal anodes has been limited by challenges such as low cycling efficiency, short lifespan, and safety concerns.

Table 8. Most impactful documents based on the total number of citations received.

Document	Total Citations	Total Citations per Year	Normalized Total Citations
R. Zhang et al. (2018) [50]	622	88.86	9.12
Zhu et al. (2013) [51]	598	49.83	5.46
Fu et al. (2014) [52]	536	48.73	5.62
W. Zhang et al. (2018) [53]	462	66	6.78
Zhong et al. (2018) [54]	360	51.43	5.28
Liu et al. (2017) [55]	331	41.38	6.5
Pan et al. (2017) [56]	317	39.63	6.22
Shi et al. (2019) [57]	300	50	5.53
Wang et al. (2013) [58]	295	24.58	2.69
Huang et al. (2020) [59]	266	53.2	6.49

To better understand the scope and breadth of the topic and how it has been shaped throughout the years, keywords plus were used to further analyze the documents. Specifically, keywords plus result in more accurate and representative outcomes when compared to author's keywords (Zhang et al., 2016) and therefore, have been used in the analysis. Among the keywords plus used, "carbon fibers" (n = 560), "anodes" (n = 317), "lithium-ion batteries" (n = 288), "electrodes" (n = 179), and "lithium compounds" (n = 177) emerged as the top-5 most commonly used. Figure 8 summarizes the top-10 most frequently used keywords plus.

Based on the outcomes, the scientific community has focused on three main areas. Firstly, the most investigated is the development of anodes for lithium-ion batteries, as this area faces high global demand and presents greater challenges compared to cathodes. Anodes are subject to issues such as dendrite formation during charging [60], significant volume changes of lithium during charge and discharge, leading to crack formation [61], and potential side reactions with the electrolyte, which generate products that can degrade battery performance and reduce capacity [62]. Secondly, the development of a new generation of carbon fiber-based cathodes involves activating them with coatings of $LiFePO_4$, $LiCoO_2$, or their derivatives. Additionally, recent efforts have focused on developing cobalt-free cathodes due to the high cost and difficulty of extraction. These cobalt-free alternatives offer more sustainable solutions for large-scale production while maintaining comparable performance metrics [63,64]. The final area of focus is the development of solid electrolytes, which must simultaneously achieve high ionic conductivity and elevated mechanical properties to guarantee the multifunctionality of the composite. However, it has been demonstrated that these two properties are often in conflict [18,65].

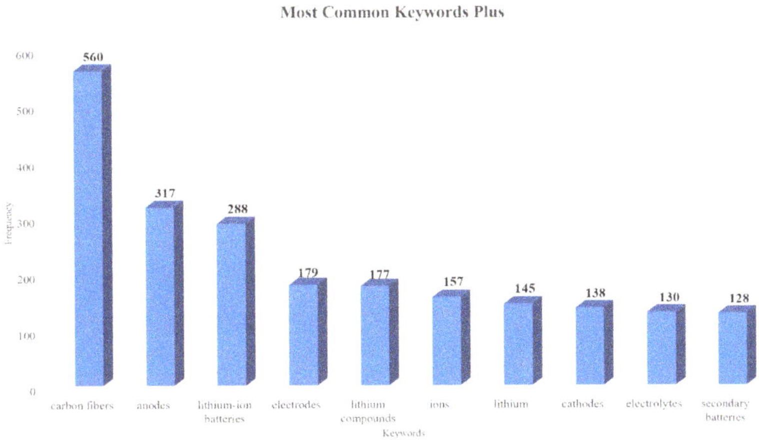

Figure 8. Most frequent keywords plus.

The keyword co-occurrence network presented in Figure 9 revealed three main areas. Specifically, the central and most prominent cluster is focused on 'carbon fibers', which is heavily linked to "lithium-ion batteries" and "anodes". This central cluster indicates a strong relationship between these terms within the research or context provided. Surrounding this main cluster are two other significant areas: one that includes terms like "vanadium", "flow batteries", and "energy storage" suggesting a focus on alternative battery technologies and energy storage solutions. Here, vanadium-based energy storage, particularly in vanadium redox flow batteries, has shown promise for large-scale applications due to its high energy capacity and long cycle life. Flow batteries, in general, provide a scalable solution for renewable energy integration, as they allow for independent scaling of power and energy capacities, making them ideal for grid-level storage. The other area includes terms such as "solid electrolytes", "redox reactions", and "electrochemical performance", pointing towards research in advanced battery materials and their performance. Figure 10 presents the relationships among the top countries, keywords, and sources.

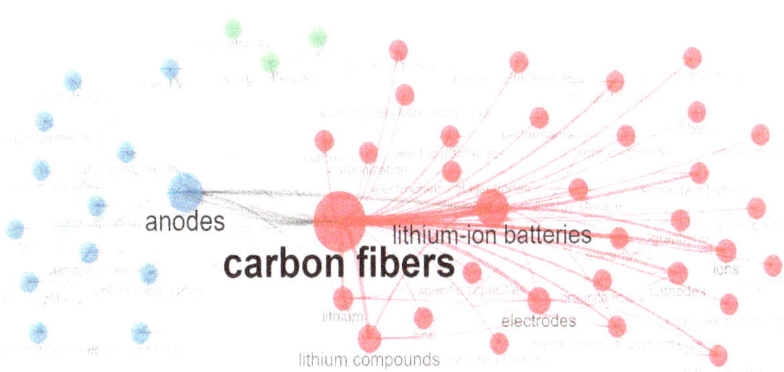

Figure 9. Keywords plus co-occurrence network.

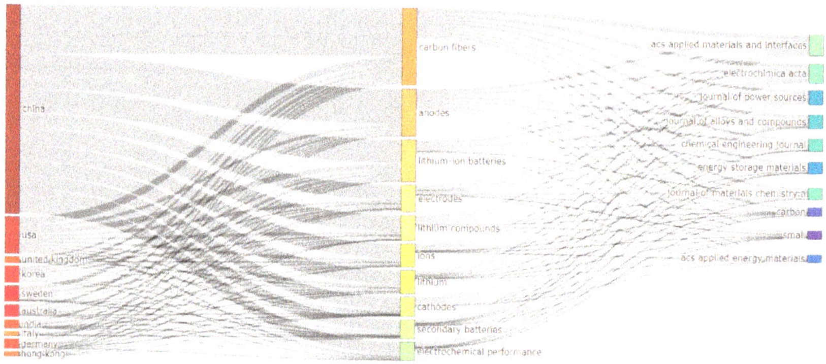

Figure 10. Countries, keywords, and sources relationship.

Furthermore, Figure 11 displays the most popular topics and their progression over the years based on keywords plus. According to the outcomes, the current focus in the development of carbon fiber-based batteries is on creating low-cost batteries utilizing sodium ions, with an optimized structural design. Moreover, we can see the importance of scanning electron microscopy in the development of these devices, as it is a critical machine to master their operation and optimization.

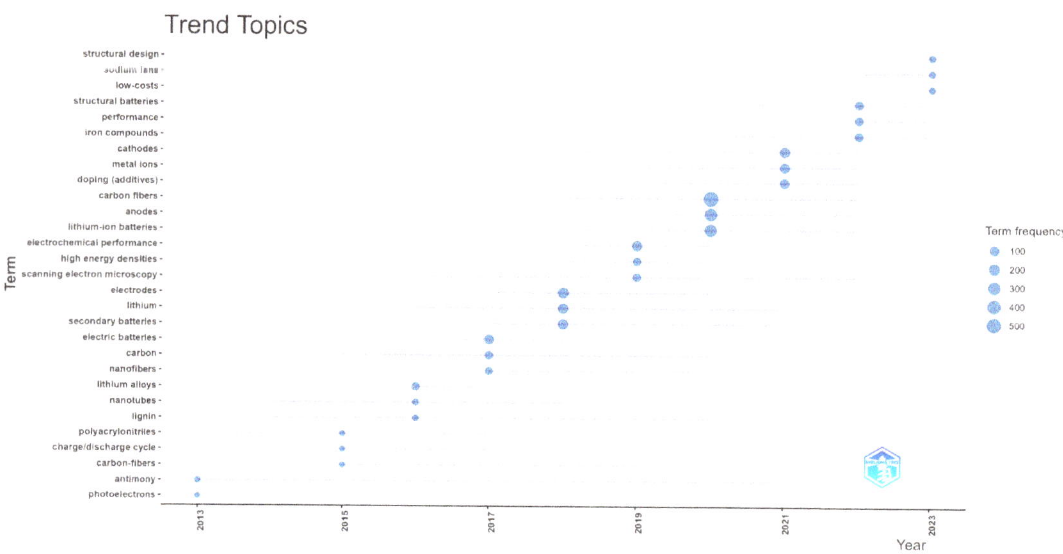

Figure 11. Trend topics based on keywords plus.

When examining the conceptual structure map of the topic, four main dimensions arose as it can be seen in Figure 12. According to the results, the dimensions were related to: (i) carbon fibers, anodes, and lithium, (ii) carbon fibers, anodes, and metal ions, (iii) carbon fibers, lithium-ion batteries, and anodes, and (iv) carbon fibers, lithium-ion batteries, and electrodes. Based on the dimensions that emerged, it can be inferred that carbon fibers play a central role in the development of advanced battery technologies. The repeated association of carbon fibers with anodes, lithium, and lithium-ion batteries highlights their importance in enhancing the performance and efficiency of these components. Additionally, the inclusion of metal ions and electrodes in the dimensions suggests a focus on optimizing the materials and structures used in batteries to improve their overall functionality and longevity.

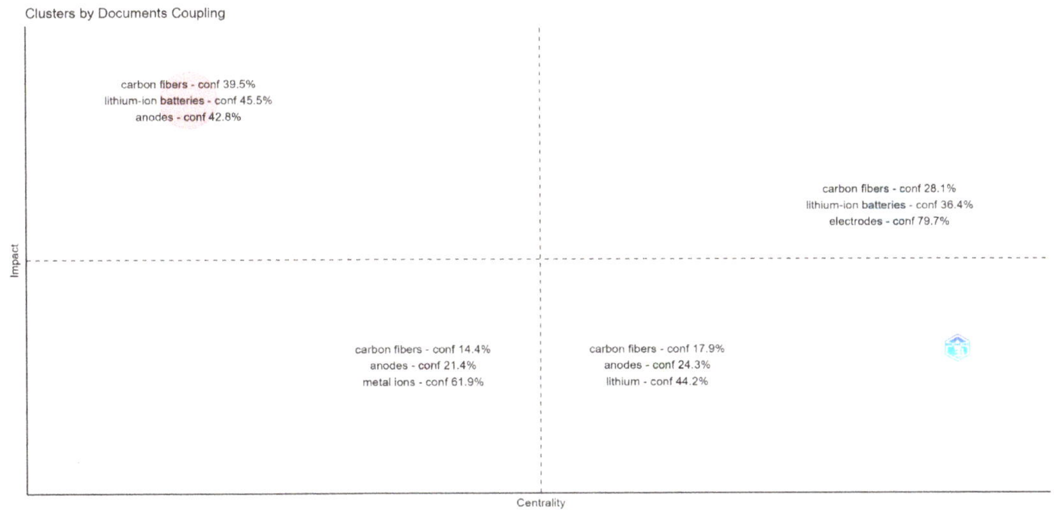

Figure 12. Conceptual structure map.

The thematic map of the topic which is displayed in Figure 13 presents the six themes that emerged. The themes are separated into basic themes, emerging or declining themes, niche themes, and motor themes. Two basic themes arose that were related to: (i) lithium compounds, electrodes, and lithium and (ii) electrochemical electrodes, redox reactions, and cyclic voltammetry. One emerging or declining theme arose which was associated with (i) energy efficiency, flow batteries, and vanadium. Two niche themes were identified that were related to (i) structural batteries, multifunctional composites, and polyelectrolytes and (ii) zinc, oxygen, and electrolytic reduction. The motor theme that arose was associated with (i) carbon fibers, anodes, and lithium-ion batteries. Based on these outcomes, it can be noted that the development of carbon fiber-based batteries is a central focus, particularly in enhancing the performance and efficiency of lithium-ion batteries. The emphasis on electrochemical electrodes and redox reactions highlights the importance of understanding and optimizing the fundamental processes within these batteries. Additionally, the exploration of structural batteries and multifunctional composites indicates a trend toward integrating multiple functionalities into battery materials –structural and energy storage functions especially–leading to more versatile and efficient energy storage solutions.

The thematic evolution of the topic is presented in Figure 14 in five time periods. The first time period (2013–2015) reveals that the focus was primarily on "anode", "carbon fibres", "energy storage", and "mechanical properties". These themes indicate an early emphasis on understanding and improving the fundamental components and properties of energy storage systems. As we move to the second time period (2016–2017), the themes evolve to include "cathodes", "carbon fibres" and "energy efficiency", reflecting a shift towards optimizing both the anode and cathode materials and improving the overall efficiency of energy storage devices. In the third time period (2018–2019), the focus expands to "carbon fibres", "anodes", and "composite materials" suggesting advancements in integrating carbon fibres into composite materials to enhance the performance and durability of batteries. The fourth time period (2020–2021) shows a continued emphasis on "carbon fibres", "anodes", and "energy storage" indicating ongoing research into improving these critical components. Additionally, "structural batteries" and "carbon nanofibers" emerge as significant themes, highlighting innovations in battery design and materials. Finally, in the fifth time period (2022–2023), the themes include "carbon fibres," "anodes", and "lithium-ion batteries," demonstrating a mature focus on integrating carbon fibres into lithium-ion batteries to enhance their performance. The emergence of "energy efficiency" and "structural batteries" as consistent themes throughout the periods underscores the importance of these areas in the ongoing development of advanced energy storage solutions.

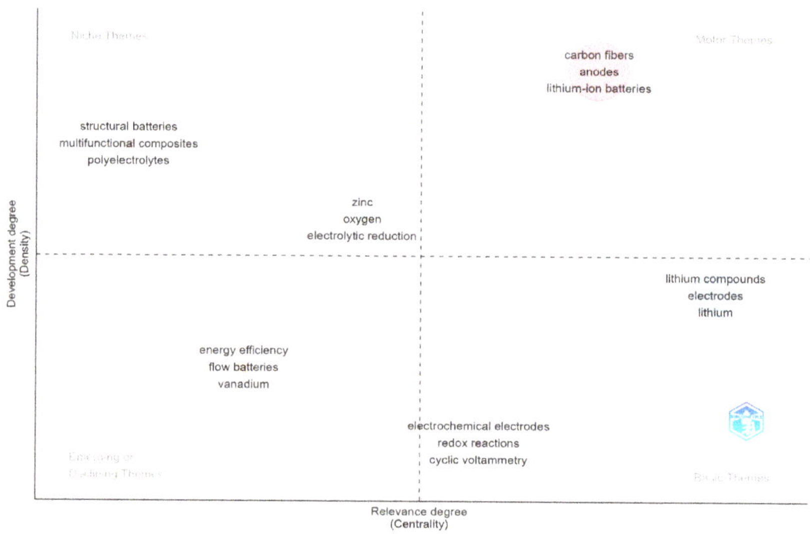

Figure 13. Thematic map of the topic.

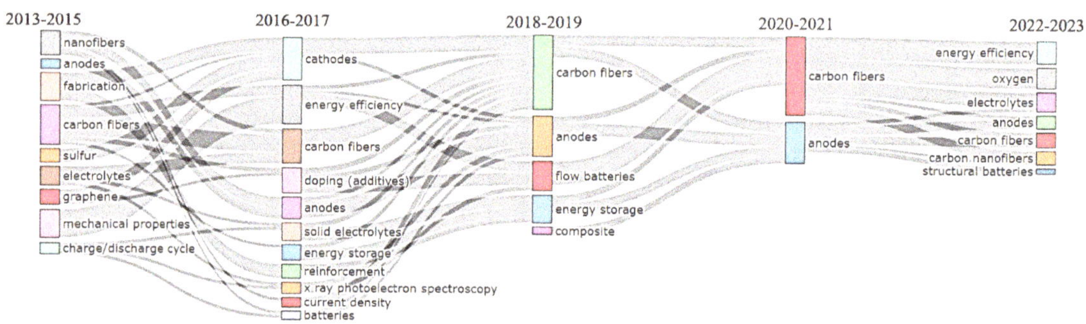

Figure 14. Thematic evolution of the topic.

Considering the bibliometric analyses of the paper, Figure 15 presents key trends and future perspectives in energy storage technologies based on carbon fiber-based batteries. The bibliometric analysis highlights three primary areas: the development of carbon fiber-based anodes for lithium-ion batteries, advancements in next-generation carbon fiber-based cathodes, and progress in solid electrolytes. Each domain faces significant challenges, such as dendrite formation and volume expansion in anodes, as well as the urgent need for cobalt-free alternatives in cathodes due to the high costs and extraction difficulties associated with cobalt. Efforts in solid electrolytes focus on balancing high ionic conductivity with strong mechanical properties to ensure the multifunctionality of these composite materials. Moreover, gel-based electrolytes, due to their semi-solid nature, help to reduce electrode polarization and maintain stable performance under high charge/discharge rates, making them highly suitable for demanding energy storage applications. Additionally, the geographical distribution of research contributions, with leading countries showing high citation impacts, although international collaboration remains limited. This suggests that while some nations dominate high-impact research, global collaboration needs to be strengthened to accelerate progress in this field.

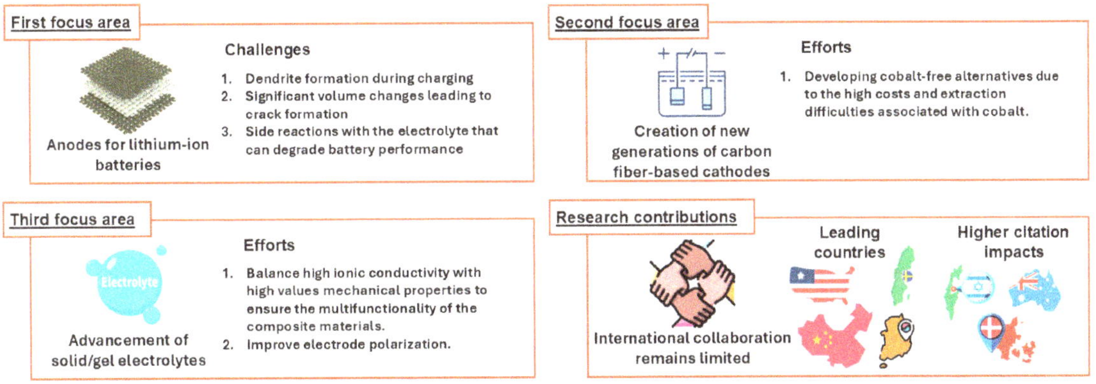

Figure 15. Key trends and future perspectives in energy storage technologies based on carbon fi-ber-based batteries.

4. Conclusions

Carbon fiber-based batteries are poised to play a transformative role in the pursuit of more sustainable energy solutions. Their unique ability to merge structural functionality with energy storage not only offers significant weight reduction and efficiency gains but also opens up new avenues for innovation across various industries. By integrating carbon fibers into battery systems, particularly as anodes, many of the critical challenges in battery performance can be addressed more effectively.

Recent research has primarily focused on three main areas within the development of carbon fiber-based batteries. The most investigated area is the development of anodes for lithium-ion batteries, driven by the high global demand and the significant challenges these anodes present, such as dendrite formation during charging, significant volume changes leading to crack formation, and side reactions with the electrolyte that can degrade battery performance.

Analysis of the research contributions reveals that China, the United States, Sweden, South Korea, and Australia are leading countries in this field, with China significantly ahead in both publication volume and intra-country collaborations. Despite this fact, international collaboration remains limited, underscoring the need for increased global cooperation. Notably, countries such as Israel, Australia, and Denmark exhibit higher citation impacts, reflecting the high quality and influence of their research.

Carbon fiber-based batteries are still in a developmental stage, with substantial research needed to realize their full potential. Future research directions should prioritize optimizing fiber-matrix interfaces to enhance electron and ion transport, scaling manufacturing processes for broader application, and developing standardized protocols to ensure consistency and reliability. Increased international collaboration will be vital in accelerating technological progress and addressing existing challenges. As the field matures, carbon fiber-based batteries hold significant promise for advancing sustainable energy systems and contributing to a decarbonized future.

Author Contributions: Conceptualization, A.d.B. and D.V.; methodology, G.L.; formal analysis, A.d.B., D.V., G.L. and P.F.-A.; data curation, G.L.; writing—original draft preparation, A.d.B., G.L. and P.F.-A.; writing—review and editing, A.d.B., D.V., G.L. and P.F.-A.; supervision, D.V. All authors have read and agreed to the published version of the manuscript.

Funding: This research received no external funding.

Data Availability Statement: The original contributions presented in the study are included in the article, further inquiries can be directed to the corresponding authors.

Conflicts of Interest: The authors declare no conflicts of interest.

References

1. Bashir, T.; Ismail, S.A.; Song, Y.; Irfan, R.M.; Yang, S.; Zhou, S.; Zhao, J.; Gao, L. A Review of the Energy Storage Aspects of Chemical Elements for Lithium-Ion Based Batteries. *Energy Mater.* **2021**, *1*, 100019. [CrossRef]
2. Pan, X.; Shao, T.; Zheng, X.; Zhang, Y.; Ma, X.; Zhang, Q. Energy and Sustainable Development Nexus: A Review. *Energy Strategy Rev.* **2023**, *47*, 101078. [CrossRef]
3. Shi, S.; Zhou, D.; Jiang, Y.; Cheng, F.; Sun, J.; Guo, Q.; Luo, Y.; Chen, Y.; Liu, W. Lightweight Zn-Philic 3D-Cu Scaffold for Customizable Zinc Ion Batteries. *Adv. Funct. Mater.* **2024**, *34*, 2312664. [CrossRef]
4. Kosmidis, I.; Müller-Eie, D.; Delbosc, A. Electric Cars as a Path to Sustainable Travel Behaviour: Insights from Nord-Jæren. *Transp. Res. D Transp. Environ.* **2023**, *125*, 103982. [CrossRef]
5. Jannesar Niri, A.; Poelzer, G.A.; Zhang, S.E.; Rosenkranz, J.; Pettersson, M.; Ghorbani, Y. Sustainability Challenges throughout the Electric Vehicle Battery Value Chain. *Renew. Sustain. Energy Rev.* **2024**, *191*, 114176. [CrossRef]
6. Qiao, Z.; Bian, K.; Ding, C.; Zhao, Y. Recent Progress of Carbon-Fiber-Based Electrode Materials for Energy Storage. *Diam. Relat. Mater.* **2023**, *138*, 110208. [CrossRef]
7. Bouton, K.; Schneider, L.; Zenkert, D.; Lindbergh, G. A Structural Battery with Carbon Fibre Electrodes Balancing Multifunctional Performance. *Compos. Sci. Technol.* **2024**, *256*, 110728. [CrossRef]
8. Sanchez, J.S.; Xu, J.; Xia, Z.; Sun, J.; Asp, L.E.; Palermo, V. Electrophoretic Coating of $LiFePO_4$/Graphene Oxide on Carbon Fibers as Cathode Electrodes for Structural Lithium Ion Batteries. *Compos. Sci. Technol.* **2021**, *208*, 108768. [CrossRef]
9. Biswas, P.K.; Liyanage, A.A.H.; Jadhav, M.; Agarwal, M.; Dalir, H. Higher Strength Carbon Fiber Lithium-Ion Polymer Battery Embedded Multifunctional Composites for Structural Applications. *Polym. Compos.* **2022**, *43*, 2952–2962. [CrossRef]
10. Zenkert, D.; Harnden, R.; Asp, L.E.; Lindbergh, G.; Johansson, M. Multifunctional Carbon Fibre Composites Using Electrochemistry. *Compos. B Eng.* **2024**, *273*, 111240. [CrossRef]
11. del Bosque, A.; Muñoz, B.K.; Sánchez, M.; Ureña, A.; del Bosque, A.; Muñoz, B.K.; Sánchez, M.; Ureña, A. Thermomechanically Robust Ceramic/Polymer Nanocomposites Modified with Ionic Liquid for Hybrid Polymer Electrolyte Applications. *ACS Appl. Energy Mater.* **2022**, *5*, 4247–4258. [CrossRef]
12. Asp, L.E.; Bouton, K.; Carlstedt, D.; Duan, S.; Harnden, R.; Johannisson, W.; Johansen, M.; Johansson, M.K.G.; Lindbergh, G.; Liu, F.; et al. A Structural Battery and Its Multifunctional Performance. *Adv. Energy Sustain. Res.* **2021**, *2*, 2000093. [CrossRef]
13. Dharmasiri, B.; Stojcevski, F.; Usman, K.A.S.; Alex Qin, S.; Razal, J.M.; Doeven, E.H.; Francis, P.S.; Connell, T.U.; Yin, Y.; Andersson, G.G.; et al. Flexible Carbon Fiber Based Structural Supercapacitor Composites with Solvate Ionic Liquid-Epoxy Solid Electrolyte. *Chem. Eng. J.* **2023**, *455*, 140778. [CrossRef]
14. Sánchez-Romate, X.F.; Del Bosque, A.; Artigas-Arnaudas, J.; Muñoz, B.K.; Sánchez, M.; Ureña, A. A Proof of Concept of a Structural Supercapacitor Made of Graphene Coated Woven Carbon Fibers: EIS Study and Mechanical Performance. *Electrochim. Acta* **2021**, *370*, 137746. [CrossRef]
15. Asp, L.E.; Johansson, M.; Lindbergh, G.; Xu, J.; Zenkert, D. Structural Battery Composites: A Review. *Funct. Compos. Struct.* **2019**, *1*, 042001. [CrossRef]
16. Jiang, Q.; Beutl, A.; Kühnelt, H.; Bismarck, A. Structural Composite Batteries Made from Carbon Fibre Reinforced Electrodes/Polymer Gel Electrolyte Prepregs. *Compos. Sci. Technol.* **2023**, *244*, 110312. [CrossRef]
17. Lin, W.; Wang, F.; Wang, H.; Li, H.; Fan, Y.; Chan, D.; Chen, S.; Tang, Y.; Zhang, Y. Thermal-Stable Separators: Design Principles and Strategies Towards Safe Lithium-Ion Battery Operations. *ChemSusChem* **2022**, *15*, e202201464. [CrossRef]
18. Muñoz, B.K.; del Bosque, A.; Sánchez, M.; Utrilla, V.; Prolongo, S.G.; Prolongo, M.G.; Ureña, A. Epoxy Resin Systems Modified with Ionic Liquids and Ceramic Nanoparticles as Structural Composites for Multifunctional Applications. *Polymer* **2021**, *214*, 123233. [CrossRef]
19. Tripathi, A.K. Ionic Liquid Based Solid Electrolytes (Ionogels) for Application in Rechargeable Lithium Battery. *Mater. Today Energy* **2021**, *20*, 100643. [CrossRef]
20. Hubert, O.; Todorovic, N.; Bismarck, A. Towards Separator-Free Structural Composite Supercapacitors. *Compos. Sci. Technol.* **2022**, *217*, 109126. [CrossRef]
21. Li, L.; Zhang, D.; Deng, J.; Gou, Y.; Fang, J.; Cui, H.; Zhao, Y.; Cao, M. Carbon-Based Materials for Fast Charging Lithium-Ion Batteries. *Carbon* **2021**, *183*, 721–734. [CrossRef]
22. Zhang, T.; Ran, F.; Zhang, T.; Ran, F. Design Strategies of 3D Carbon-Based Electrodes for Charge/Ion Transport in Lithium Ion Battery and Sodium Ion Battery. *Adv. Funct. Mater.* **2021**, *31*, 2010041. [CrossRef]
23. Lobato-Peralta, D.R.; Okoye, P.U.; Alegre, C. A Review on Carbon Materials for Electrochemical Energy Storage Applications: State of the Art, Implementation, and Synergy with Metallic Compounds for Supercapacitor and Battery Electrodes. *J. Power Sources* **2024**, *617*, 235140. [CrossRef]
24. Choi, J.; Zabihi, O.; Ahmadi, M.; Naebe, M. Advancing Structural Batteries: Cost-Efficient High-Performance Carbon Fiber-Coated LiFePO4 Cathodes. *RSC Adv.* **2023**, *13*, 30633–30642. [CrossRef] [PubMed]

25. Yücel, Y.D.; Zenkert, D.; Lindström, R.W.; Lindbergh, G. $LiFePO_4$-Coated Carbon Fibers as Positive Electrodes in Structural Batteries: Insights from Spray Coating Technique. *Electrochem. Commun.* **2024**, *160*, 107670. [CrossRef]
26. Yücel, Y.D.; Adolfsson, E.; Dykhoff, H.; Pettersson, J.; Trey, S.; Wysocki, M.; Widenkvist Zetterström, E.; Zenkert, D.; Wreland Lindström, R.; Lindbergh, G. Enhancing Structural Battery Performance: Investigating the Role of Conductive Carbon Additives in LiFePO4-Impregnated Carbon Fiber Electrodes. *Compos. Sci. Technol.* **2024**, *251*, 110571. [CrossRef]
27. Bhar, M.; Vishwakarma, V.; Yalamanchili, K.; Martha, S.K. Unravelling Li-Ion Storage Capability of Cobalt Oxide Anode Recovered from Spent LiCoO2 Cathode via Carbothermal Reduction. *J. Electrochem. Soc.* **2023**, *170*, 090507. [CrossRef]
28. Guoping, W.; Qingtang, Z.; Zuolong, Y.; MeiZheng, Q. The Effect of Different Kinds of Nano-Carbon Conductive Additives in Lithium Ion Batteries on the Resistance and Electrochemical Behavior of the $LiCoO_2$ Composite Cathodes. *Solid. State Ion.* **2008**, *179*, 263–268. [CrossRef]
29. Zhang, Q.; Zeng, Y.; Chen, X.; Ding, Y.; Wang, H.; Wang, L.; Wang, H.; Li, X.; Xiao, J.; Zhang, T. Architectural Engineering of MoS_2/Fe_2O_3@carbon Fiber Heterostructures to Obtain Ultra-Stable Anodes for Sodium-Ion Batteries. *Electrochim. Acta* **2023**, *470*, 143363. [CrossRef]
30. Chen, X.; Zhang, Q.; Wang, H.; Wang, L.; Wang, H.; Zeng, Y. MoS_2 Nanoflowers Implanted in Nitrogen-Doped Carbon Fibers with Fe_2O_3 Nanoparticles Decorated for Enhanced Lithium Storage Performance. *J. Alloys Compd.* **2023**, *963*, 171195. [CrossRef]
31. Yang, X.; Qin, T.; Zhang, X.; Liu, X.; Wang, Z.; Zhang, W.; Zheng, W. Self-Crystallized Interlayer Integrating Polysulfide-Adsorbed TiO_2/TiO and Highly-Electron-Conductive TiO for High-Stability Lithium-Sulfur Batteries. *Chem. Res. Chin. Univ.* **2021**, *37*, 259–264. [CrossRef]
32. Wang, L.; Yang, Q.; Li, K.; Wei, M.; Wang, Q.; Hu, Z.; Zhen, M. Defects-Rich TiO_2(B) in-Situ Grown on Carbon Fiber Cloth as Self-Standing Electrode for High-Loading Lithium–Sulfur Batteries. *Mater. Lett.* **2024**, *357*, 135668. [CrossRef]
33. Yu, J.; Cai, D.; Si, J.; Zhan, H.; Wang, Q. MOF-Derived NiCo2S4 and Carbon Hybrid Hollow Spheres Compactly Concatenated by Electrospun Carbon Nanofibers as Self-Standing Electrodes for Aqueous Alkaline Zn Batteries. *J. Mater. Chem. A Mater.* **2022**, *10*, 4100–4109. [CrossRef]
34. Chen, X.; Pu, J.; Hu, X.; An, L.; Jiang, J.; Li, Y. Confinement Synthesis of Bimetallic MOF-Derived Defect-Rich Nanofiber Electrocatalysts for Rechargeable Zn-Air Battery. *Nano Res.* **2022**, *15*, 9000–9009. [CrossRef]
35. Wang, M.; Li, D.; Xu, H.; Wang, L.; Li, Y.; Li, G.; Li, J.; Han, W. Flexible $Ti_3C_2T_x$ MXene Bonded Bio-Derived Carbon Fibers Support Tin Disulfide for Fast and Stable Sodium Storage. *Small* **2024**, *20*, 2305530. [CrossRef]
36. Li, D.; Li, J.; Wang, M.; Li, G.; Milinevsky, G.; Wang, L.; Han, W. $T_3C_2T_x$ Boost SnSbS Composited Carbon Fibers Achieved Fast and Stable Free-Standing Quasi-Solid Sodium Full-Battery. *Chem. Eng. J.* **2023**, *465*, 142934. [CrossRef]
37. Deng, W.; Zhu, W.; Zhou, X.; Liu, Z. Graphene Nested Porous Carbon Current Collector for Lithium Metal Anode with Ultrahigh Areal Capacity. *Energy Storage Mater.* **2018**, *15*, 266–273. [CrossRef]
38. Yan, X.; Zhang, H.; Huang, M.; Jia, W.; Jiang, Y.; Chen, T.; Qu, M. Micro/Nano-Structure Construct of Carbon Fibers Reinforced Graphene/CNT Matrix Composites for Li-S Batteries. *Diam. Relat. Mater.* **2022**, *123*, 108888. [CrossRef]
39. Hu, Z.; Fu, Y.; Hong, Z.; Huang, Y.; Guo, W.; Yang, R.; Xu, J.; Zhou, L.; Yin, S. Composite Structural Batteries with Co_3O_4/CNT Modified Carbon Fibers as Anode: Computational Insights on the Interfacial Behavior. *Compos. Sci. Technol.* **2021**, *201*, 108495. [CrossRef]
40. Kalnaus, S.; Asp, L.E.; Li, J.; Veith, G.M.; Nanda, J.; Daniel, C.; Chen, X.C.; Westover, A.; Dudney, N.J. Multifunctional Approaches for Safe Structural Batteries. *J. Energy Storage* **2021**, *40*, 102747. [CrossRef]
41. Ekstedt, S.; Wysocki, M.; Asp, L.E. Structural Batteries Made from Fibre Reinforced Composites. *Plast. Rubber Compos.* **2010**, *39*, 148–150. [CrossRef]
42. Danzi, F.; Salgado, R.M.; Oliveira, J.E.; Arteiro, A.; Camanho, P.P.; Braga, M.H. Structural Batteries: A Review. *Molecules* **2021**, *26*, 2203. [CrossRef] [PubMed]
43. Adam, T.J.; Liao, G.; Petersen, J.; Geier, S.; Finke, B.; Wierach, P.; Kwade, A.; Wiedemann, M. Multifunctional Composites for Future Energy Storage in Aerospace Structures. *Energies* **2018**, *11*, 335. [CrossRef]
44. Chan, K.Y.; Jia, B.; Lin, H.; Hameed, N.; Lee, J.H.; Lau, K.T. A Critical Review on Multifunctional Composites as Structural Capacitors for Energy Storage. *Compos. Struct.* **2018**, *188*, 126–142. [CrossRef]
45. Fredi, G.; Dorigato, A.; Fambri, L.; Pegoretti, A. Evaluating the Multifunctional Performance of Structural Composites for Thermal Energy Storage. *Polymer* **2021**, *13*, 3108. [CrossRef]
46. Shirshova, N.; Qian, H.; Houllé, M.; Steinke, J.H.G.; Kucernak, A.R.J.; Fontana, Q.P.V.; Greenhalgh, E.S.; Bismarck, A.; Shaffer, M.S.P. Multifunctional Structural Energy Storage Composite Supercapacitors. *Faraday Discuss.* **2014**, *172*, 81–103. [CrossRef]
47. Page, M.J.; McKenzie, J.E.; Bossuyt, P.M.; Boutron, I.; Hoffmann, T.C.; Mulrow, C.D.; Shamseer, L.; Tetzlaff, J.M.; Akl, E.A.; Brennan, S.E.; et al. The PRISMA 2020 Statement: An Updated Guideline for Reporting Systematic Reviews. *Int. J. Surg.* **2021**, *88*, 105906. [CrossRef]
48. Ellegaard, O.; Wallin, J.A. The Bibliometric Analysis of Scholarly Production: How Great Is the Impact? *Scientometrics* **2015**, *105*, 1809–1831. [CrossRef]
49. Aria, M.; Cuccurullo, C. Bibliometrix: An R-Tool for Comprehensive Science Mapping Analysis. *J. Inf.* **2017**, *11*, 959–975. [CrossRef]
50. Zhang, R.; Chen, X.; Shen, X.; Zhang, X.Q.; Chen, X.R.; Cheng, X.B.; Yan, C.; Zhao, C.Z.; Zhang, Q. Coralloid Carbon Fiber-Based Composite Lithium Anode for Robust Lithium Metal Batteries. *Joule* **2018**, *2*, 764–777. [CrossRef]

51. Zhu, Y.; Han, X.; Xu, Y.; Liu, Y.; Zheng, S.; Xu, K.; Hu, L.; Wang, C. Electrospun Sb/C Fibers for a Stable and Fast Sodium-Ion Battery Anode. *ACS Nano* **2013**, *7*, 6378–6386. [CrossRef] [PubMed]
52. Fu, L.; Tang, K.; Song, K.; Van Aken, P.A.; Yu, Y.; Maier, J. Nitrogen Doped Porous Carbon Fibres as Anode Materials for Sodium Ion Batteries with Excellent Rate Performance. *Nanoscale* **2014**, *6*, 1384–1389. [CrossRef] [PubMed]
53. Zhang, W.; Pang, W.K.; Sencadas, V.; Guo, Z. Understanding High-Energy-Density Sn_4P_3 Anodes for Potassium-Ion Batteries. *Joule* **2018**, *2*, 1534–1547. [CrossRef]
54. Zhong, Y.; Chao, D.; Deng, S.; Zhan, J.; Fang, R.Y.; Xia, Y.; Wang, Y.; Wang, X.; Xia, X.; Tu, J. Confining Sulfur in Integrated Composite Scaffold with Highly Porous Carbon Fibers/Vanadium Nitride Arrays for High-Performance Lithium–Sulfur Batteries. *Adv. Funct. Mater.* **2018**, *28*, 1706391. [CrossRef]
55. Liu, L.; Yin, Y.X.; Li, J.Y.; Li, N.W.; Zeng, X.X.; Ye, H.; Guo, Y.G.; Wan, L.J. Free-Standing Hollow Carbon Fibers as High-Capacity Containers for Stable Lithium Metal Anodes. *Joule* **2017**, *1*, 563–575. [CrossRef]
56. Pan, H.; Chen, J.; Cao, R.; Murugesan, V.; Rajput, N.N.; Han, K.S.; Persson, K.; Estevez, L.; Engelhard, M.H.; Zhang, J.G.; et al. Non-Encapsulation Approach for High-Performance Li–S Batteries through Controlled Nucleation and Growth. *Nature Energy* **2017**, *2*, 813–820. [CrossRef]
57. Shi, P.; Li, T.; Zhang, R.; Shen, X.; Cheng, X.B.; Xu, R.; Huang, J.Q.; Chen, X.R.; Liu, H.; Zhang, Q. Lithiophilic LiC6 Layers on Carbon Hosts Enabling Stable Li Metal Anode in Working Batteries. *Adv. Mater.* **2019**, *31*, 1807131. [CrossRef] [PubMed]
58. Wang, S.X.; Yang, L.; Stubbs, L.P.; Li, X.; He, C. Lignin-Derived Fused Electrospun Carbon Fibrous Mats as High Performance Anode Materials for Lithium Ion Batteries. *ACS Appl. Mater. Interfaces* **2013**, *5*, 12275–12282. [CrossRef]
59. Huang, Y.; Yang, H.; Xiong, T.; Adekoya, D.; Qiu, W.; Wang, Z.; Zhang, S.; Balogun, M.S. Adsorption Energy Engineering of Nickel Oxide Hybrid Nanosheets for High Areal Capacity Flexible Lithium-Ion Batteries. *Energy Storage Mater.* **2020**, *25*, 41–51. [CrossRef]
60. Man, J.; Liu, K.; Zhang, H.; Du, Y.; Yin, J.; Wang, X.; Sun, J. Dendrite-Free Lithium Metal Anode Enabled by Ion/Electron-Conductive N-Doped 3D Carbon Fiber Interlayer. *J. Power Sources* **2021**, *489*, 229524. [CrossRef]
61. Lin, N.; Jia, Z.; Wang, Z.; Zhao, H.; Ai, G.; Song, X.; Bai, Y.; Battaglia, V.; Sun, C.; Qiao, J.; et al. Understanding the Crack Formation of Graphite Particles in Cycled Commercial Lithium-Ion Batteries by Focused Ion Beam—Scanning Electron Microscopy. *J. Power Sources* **2017**, *365*, 235–239. [CrossRef]
62. Xu, J.; Deshpande, R.D.; Pan, J.; Cheng, Y.-T.; Battaglia, V.S. Electrode Side Reactions, Capacity Loss and Mechanical Degradation in Lithium-Ion Batteries. *J. Electrochem. Soc.* **2015**, *162*, A2026–A2035. [CrossRef]
63. Croy, J.R.; Long, B.R.; Balasubramanian, M. A Path toward Cobalt-Free Lithium-Ion Cathodes. *J. Power Sources* **2019**, *440*, 227113. [CrossRef]
64. Zhao, H.; Lam, W.Y.A.; Sheng, L.; Wang, L.; Bai, P.; Yang, Y.; Ren, D.; Xu, H.; He, X. Cobalt-Free Cathode Materials: Families and Their Prospects. *Adv. Energy Mater.* **2022**, *12*, 2103894. [CrossRef]
65. Kwon, S.J.; Kim, T.; Jung, B.M.; Lee, S.B.; Choi, U.H. Multifunctional Epoxy-Based Solid Polymer Electrolytes for Solid-State Supercapacitors. *ACS Appl. Mater. Interfaces* **2018**, *10*, 35108–35117. [CrossRef]

MDPI AG
Grosspeteranlage 5
4052 Basel
Switzerland
Tel.: +41 61 683 77 34

Applied Sciences Editorial Office
E-mail: applsci@mdpi.com
www.mdpi.com/journal/applsci

www.ingramcontent.com/pod-product-compliance
Lightning Source LLC
Chambersburg PA
CBHW040818250926
51850CB00007BA/1202